INTERNATIONAL UNION OF PURE AND APPLIED CHEMISTRY

ANALYTICAL CHEMISTRY DIVISION
COMMISSION ON SOLUBILITY DATA

SOLUBILITY DATA SERIES

Volume 19

CUMULATIVE INDEX
VOLUMES 1–18

SOLUBILITY DATA SERIES

Volume 1	H. L. Clever, *Helium and Neon*
Volume 2	H. L. Clever, *Krypton, Xenon and Radon*
Volume 3	M. Salomon, *Silver Azide, Cyanide, Cyanamides, Cyanate, Selenocyanate and Thiocyanate*
Volume 4	H. L. Clever, *Argon*
Volume 5/6	C. L. Young, *Hydrogen and Deuterium*
Volume 7	R. Battino, *Oxygen and Ozone*
Volume 8	C. L. Young, *Oxides of Nitrogen*
Volume 9	W. Hayduk, *Ethane*
Volume 10	R. Battino, *Nitrogen and Air*
Volume 11	B. Scrosati and C. A. Vincent, *Alkali Metal, Alkaline Earth Metal and Ammonium Halides. Amide Solvents*
Volume 12	C. L. Young, *Sulfur Dioxide, Chlorine, Fluorine and Chlorine Oxides*
Volume 13	S. Siekierski, T. Mioduski and M. Salomon, *Scandium, Yttrium, Lanthanum and Lanthanide Nitrates*
Volume 14	H. Miyamoto, M. Salomon and H. L. Clever, *Alkaline Earth Metal Halates*
Volume 15	A. F. M. Barton, *Alcohols with Water*
Volume 16/17	E. Tomlinson and A. Regosz, *Antibiotics: I. β-Lactam Antibiotics*
Volume 18	O. Popovych, *Tetraphenylborates*
Volume 19	C. L. Young, *Cumulative Index: Volumes 1–18*

Selected Volumes in Preparation

A. L. Horvath and F. W. Getzen, *Halogenated Benzenes, Toluenes and Phenols with Water*

C. L. Young and P. G. T. Fogg, *Ammonia, Amines, Phosphine, Arsine, Stibine, Silane, Germane and Stannane in Organic Solvents*

T. Mioduski and M. Salomon, *Scandium, Yttrium, Lanthanum and Lanthanide Halides in Nonaqueous Solvents*

T. P. Dirkse, *Copper, Silver, Gold, and Zinc, Cadmium and Mercury Oxides and Hydroxides*

W. Hayduk, *Propane, Butane and 2-Methylpropane*

H. L. Clever and W. Gerrard, *Hydrogen Halides in Nonaqueous Solvents*

H. L. Clever and C. L. Young, *Carbon Dioxide*

H. L. Clever and C. L. Young, *Methane*

NOTICE TO READERS

Dear Reader

If your library is not already a standing-order customer or subscriber to the Solubility Data Series, may we recommend that you place a standing order or subscription order to receive immediately upon publication all new volumes published in this valuable series. Should you find that these volumes no longer serve your needs, your order can be cancelled at any time without notice.

Robert Maxwell
Publisher at Pergamon Press

SOLUBILITY DATA SERIES

Editor-in-Chief
A. S. KERTES

Volume 19

CUMULATIVE INDEX VOLUMES 1–18

Compiled and Edited by

COLIN L. YOUNG

Department of Chemistry, University of Melbourne, Parkville, Victoria, Australia

PERGAMON PRESS

OXFORD · NEW YORK · TORONTO · SYDNEY · FRANKFURT

U.K.	Pergamon Press Ltd., Headington Hill Hall, Oxford OX3 0BW, England
U.S.A.	Pergamon Press Inc., Maxwell House, Fairview Park, Elmsford, New York 10523, U.S.A.
CANADA	Pergamon Press Canada Ltd., Suite 104, 150 Consumers Road, Willowdale, Ontario M2J 1P9, Canada
AUSTRALIA	Pergamon Press (Aust.) Pty. Ltd., P.O. Box 544, Potts Point, N.S.W. 2011, Australia
FEDERAL REPUBLIC OF GERMANY	Pergamon Press GmbH, Hammerweg 6, D-6242 Kronberg-Taunus, Federal Republic of Germany

First edition 1985

British Library Cataloguing in Publication Data

Cumulative index: volumes 1–18.—(Solubility data series; v. 19)
1. Solubility—Tables—Indexes
I. Young, Colin L. II. Series
016.5413'42 Z5524.S64
ISBN 0-08-032495-9

Printed in Great Britain by A. Wheaton & Co. Ltd., Exeter

CONTENTS

INTERNATIONAL UNION OF PURE AND APPLIED CHEMISTRY

IUPAC Secretariat: Bank Court Chambers, 2–3 Pound Way,
Cowley Centre, Oxford OX4 3YF, UK

FOREWORD

If the knowledge is undigested or simply wrong, more is not better

How to communicate and disseminate numerical data effectively in chemical science and technology has been a problem of serious and growing concern to IUPAC, the International Union of Pure and Applied Chemistry, for the last two decades. The steadily expanding volume of numerical information, the formulation of new interdisciplinary areas in which chemistry is a partner, and the links between these and existing traditional subdisciplines in chemistry, along with an increasing number of users, have been considered as urgent aspects of the information problem in general, and of the numerical data problem in particular.

Among the several numerical data projects initiated and operated by various IUPAC commissions, the *Solubility Data Project* is probably one of the most ambitious ones. It is concerned with preparing a comprehensive critical compilation of data on solubilities in all physical systems, of gases, liquids and solids. Both the basic and applied branches of almost all scientific disciplines require a knowledge of solubilities as a function of solvent, temperature and pressure. Solubility data are basic to the fundamental understanding of processes relevant to agronomy, biology, chemistry, geology and oceanography, medicine and pharmacology, and metallurgy and materials science. Knowledge of solubility is very frequently of great importance to such diverse practical applications as drug dosage and drug solubility in biological fluids, anesthesiology, corrosion by dissolution of metals, properties of glasses, ceramics, concretes and coatings, phase relations in the formation of minerals and alloys, the deposits of minerals and radioactive fission products from ocean waters, the composition of ground waters, and the requirements of oxygen and other gases in life support systems.

The widespread relevance of solubility data to many branches and disciplines of science, medicine, technology and engineering, and the difficulty of recovering solubility data from the literature, lead to the proliferation of published data in an ever increasing number of scientific and technical primary sources. The sheer volume of data has overcome the capacity of the classical secondary and tertiary services to respond effectively.

While the proportion of secondary services of the review article type is generally increasing due to the rapid growth of all forms of primary literature, the review articles become more limited in scope, more specialized. The disturbing phenomenon is that in some disciplines, certainly in chemistry, authors are reluctant to treat even those limited-in-scope reviews exhaustively. There is a trend to preselect the literature, sometimes under the pretext of reducing it to manageable size. The crucial problem with such preselection - as far as numerical data are concerned - is that there is no indication as to whether the material was excluded by design or by a less than thorough literature search. We are equally concerned that most current secondary sources, critical in character as they may be, give scant attention to numerical data.

On the other hand, tertiary sources - handbooks, reference books and other tabulated and graphical compilations - as they exist today are comprehensive but, as a rule, uncritical. They usually attempt to cover whole disciplines, and thus obviously are superficial in treatment. Since they command a wide market, we believe that their service to the advancement of science is at least questionable. Additionally, the change which is taking place in the generation of new and diversified numerical data, and the rate at which this is done, is not reflected in an increased third-level service. The emergence of new tertiary literature sources does not parallel the shift that has occurred in the primary literature.

With the status of current secondary and tertiary services being as briefly stated above, the innovative approach of the *Solubility Data Project* is that its compilation and critical evaluation work involve consolidation and reprocessing services when both activities are based on intellectual and scholarly reworking of information from primary sources. It comprises compact compilation, rationalization and simplification, and the fitting of isolated numerical data into a critically evaluated general framework.

The *Solubility Data Project* has developed a mechanism which involves a number of innovations in exploiting the literature fully, and which contains new elements of a more imaginative approach for transfer of reliable information from primary to secondary/tertiary sources. *The fundamental trend of the Solubility Data Project is toward integration of secondary and tertiary services with the objective of producing in-depth critical analysis and evaluation which are characteristic to secondary services, in a scope as broad as conventional tertiary services.*

Fundamental to the philosophy of the project is the recognition that the basic element of strength is the active participation of career scientists in it. Consolidating primary data, producing a truly critically-evaluated set of numerical data, and synthesizing data in a meaningful relationship are demands considered worthy of the efforts of top scientists. Career scientists, who themselves contribute to science by their involvement in active scientific research, are the backbone of the project. The scholarly work is commissioned to recognized authorities, involving a process of careful selection in the best tradition of IUPAC. This selection in turn is the key to the quality of the output. These top experts are expected to view their specific topics dispassionately, paying equal attention to their own contributions and to those of their peers. They digest literature data into a coherent story by weeding out what is wrong from what is believed to be right. To fulfill this task, the evaluator must cover *all* relevant open literature. No reference is excluded by design and every effort is made to detect every bit of relevant primary source. Poor quality or wrong data are mentioned and explicitly disqualified as such. In fact, it is only when the reliable data are presented alongside the unreliable data that proper justice can be done. The user is bound to have incomparably more confidence in a succinct evaluative commentary and a comprehensive review with a complete bibliography to both good and poor data.

It is the standard practice that the treatment of any given solute-solvent system consists of two essential parts: I. Critical Evaluation and Recommended Values, and II. Compiled Data Sheets.

The Critical Evaluation part gives the following information:

(i) a verbal text of evaluation which discusses the numerical solubility information appearing in the primary sources located in the literature. The evaluation text concerns primarily the quality of data after consideration of the purity of the materials and their characterization, the experimental method employed and the uncertainties in control of physical parameters, the reproducibility of the data, the agreement of the worker's results on accepted test systems with standard values, and finally, the fitting of data, with suitable statistical tests, to mathematical functions;

(ii) a set of recommended numerical data. Whenever possible, the set of recommended data includes weighted average and standard deviations, and a set of smoothing equations derived from the experimental data endorsed by the evaluator;

(iii) a graphical plot of recommended data.

The Compilation part consists of data sheets of the best experimental data in the primary literature. Generally speaking, such independent data sheets are given only to the best and endorsed data covering the known range of experimental parameters. Data sheets based on primary sources where the data are of a lower precision are given only when no better data are available. Experimental data with a precision poorer than considered acceptable are reproduced in the form of data sheets when they are the only known data for a particular system. Such data are considered to be still suitable for some applications, and their presence in the compilation should alert researchers to areas that need more work.

The typical data sheet carries the following information:

(i) components - definition of the system - their names, formulas and Chemical Abstracts registry numbers;
(ii) reference to the primary source where the numerical information is reported. In cases when the primary source is a less common periodical or a report document, published though of limited availability, abstract references are also given;
(iii) experimental variables;
(iv) identification of the compiler;
(v) experimental values as they appear in the primary source. Whenever available, the data may be given both in tabular and graphical form. If auxiliary information is available, the experimental data are converted also to SI units by the compiler.

Under the general heading of Auxiliary Information, the essential experimental details are summarized:

(vi) experimental method used for the generation of data;
(vii) type of apparatus and procedure employed;
(viii) source and purity of materials;
(ix) estimated error;
(x) references relevant to the generation of experimental data as cited in the primary source.

This new approach to numerical data presentation, formulated at the initiation of the project and perfected as experience has accumulated, has been strongly influenced by the diversity of background of those whom we are supposed to serve. We thus deemed it right to preface the evaluation/compilation sheets in each volume with a detailed discussion of the principles of the accurate determination of relevant solubility data and related thermodynamic information.

Finally, the role of education is more than corollary to the efforts we are seeking. The scientific standards advocated here are necessary to strengthen science and technology, and should be regarded as a major effort in the training and formation of the next generation of scientists and engineers. Specifically, we believe that there is going to be an impact of our project on scientific-communication practices. The quality of consolidation adopted by this program offers down-to-earth guidelines, concrete examples which are bound to make primary publication services more responsive than ever before to the needs of users. The self-regulatory message to scientists of the early 1970s to refrain from unnecessary publication has not achieved much. A good fraction of the literature is still cluttered with poor-quality articles. The Weinberg report (in 'Reader in Science Information', ed. J. Sherrod and A. Hodina, Microcard Editions Books, Indian Head, Inc., 1973, p. 292) states that 'admonition to authors to restrain themselves from premature, unnecessary publication can have little effect unless the climate of the entire technical and scholarly community encourages restraint...' We think that projects of this kind translate the climate into operational terms by exerting pressure on authors to avoid submitting low-grade material. The type of our output, we hope, will encourage attention to quality as authors will increasingly realize that their work will not be suited for permanent retrievability unless it meets the standards adopted in this project. It should help to dispel confusion in the minds of many authors of what represents a permanently useful bit of information of an archival value, and what does not.

If we succeed in that aim, even partially, we have then done our share in protecting the scientific community from unwanted and irrelevant, wrong numerical information.

A. S. Kertes

PREFACE

This CUMULATIVE INDEX for the first eighteen volumes of the Solubility Data Series has been produced in the hope that it will enable workers to locate data for a system of interest together with data on closely related systems without the necessity of consulting several indexes of individual volumes.

Further cumulative indexes will be issued after a reasonable number of additional volumes in the series have been published. Publication of a COMPREHENSIVE CUMULATIVE INDEX is planned at the end of the complete series of some 100 volumes. The editor would welcome comments on any aspects of the present cumulative index.

The index consists of three sections: a System Index, an Author Index and a *Chemical Abstracts* Registry Number Index. The system index is arranged in a similar manner to the system indexes in individual volumes. Page numbers preceded by E refer to evaluation text whereas those not preceded by E refer to compiled tables. In general, compounds are listed as in *Chemical Abstracts*: for example toluene is listed as Benzene, methyl-; and dimethyl sulfoxide is listed as Methane, sulfinylbis-. Gas-liquid systems are listed by solvent. Solid-liquid systems are listed under the solid but not usually under the solvent as well. Liquid-liquid systems appear only in Volume 15 and are listed under the alcohol, not under water.

I am indebted to members of the Commission on Solubility Data and editors of the individual volumes for suggestions and their help in checking parts of the indexes. The help of Jane Chipperfield and James Young in production of the index is gratefully acknowledged.

Melbourne
February 1985

COLIN YOUNG

SYSTEM INDEX

Page numbers preceded by E refer to evaluation text whereas those not preceded by e refer to compiled tables. In general compounds are indexed as listed in Chemical Abstracts; for example toluene is listed as Benzene, methyl- and dimethylsulfoxide is listed as Methane, sulfinylbis-. Gas-liquid systems are listed by solvent. Solid-liquid systems are listed under the solid but not usually under the solvent as well. Liquid-liquid systems appear only in volume 15 and are listed under the alcohol not under water.

Abdominal muscle		
see rat abdominal muscle		
Acetamide		
+ potassium bromide	vol 11	E249, 250
+ potassium chloride	vol 11	E247, 248
+ potassium iodide	vol 11	E251, E252,
+ sodium iodide	vol 11	E245, 246
+ water	vol 13	447
Acetamide (aqueous)		
+ hydrogen	vol 5/6	95
+ nitric acid, erbium salt	vol 13	447
+ nitrogen	vol 10	E84, 93 253, 254
Acetamide, N-(aminocarbonyl)-		
+ water	vol 13	220, 370, 446
Acetamide,N-(aminocarbonyl)-, (aqueous)		
+ nitric acid, cerium salt	vol 13	220
+ nitric acid, erbium salt	vol 13	446
+ nitric acid, samarium salt	vol 13	370
Acetamide,N,N-dimethyl-		
+ calcium bromide	vol 11	E339, 340
+ calcium chloride	vol 11	E337, 338
+ calcium iodide	vol 11	E341, 342
+ lithium bromide	vol 11	E313, 314
+ lithium chloride	vol 11	E309, E310, 311, 312
+ lithium iodide	vol 11	E315, 316
+ magnesium chloride	vol 11	E333, 334
+ magnesium iodide	vol 11	E335, 336
+ potassium bromide	vol 11	E327, 328
+ potassium chloride	vol 11	E325, 326
+ potassium iodide	vol 11	E329, 330
+ silver azide + sodium azide + nitric acid, sodium salt	vol 3	21
+ silver tetraphenylborate (1-) + nitric acid, sodium salt + sodium tetraphenylborate (1-)	vol 18	153
+ sodium bromide	vol 11	E319, 320
+ sodium chloride	vol 11	E317, 318
+ sodium iodide	vol 11	E321, 322
+ strontium chloride	vol 11	E343, 344
+ sulfur dioxide	vol 12	276, 297
+ tetraphenylarsonium tetraphenylborate (1-)	vol 18	E194, 195 - 197
+ thiocyanic acid,potassium salt	vol 11	E331, 332
+ thiocyanic acid, silver salt + thiocyanic acid, potassium salt + nitric acid, potassium salt	vol 3	211
+ thiocyanic acid,sodium salt	vol 11	E323, 324
Acetamide,N-methyl-		
+ ammonium bromide	vol 11	E288, E289, 290, 291
+ ammonium chloride	vol 11	E284, E285, 286, 287
+ ammonium iodide	vol 11	E292, 293
+ argon	vol 4	229
+ barium bromide	vol 11	E307, 308
+ barium chloride	vol 11	E305, 306
+ calcium bromide	vol 11	E298, 299

Acetamide,N-methyl-
+ calcium chloride vol 11 E296, 297
+ ethane vol 9 E195 -E199, 215
+ helium vol 1 105
+ lithium chloride vol 11 E255, 256, 257
+ nitrogen vol 10 261
+ potassium bromide vol 11 E274, E275, 276, 277
+ potassium chloride vol 11 E271, 272, 273
+ potassium iodide vol 11 E278, E279, 280, 281
+ sodium bromide vol 11 E262, 263 - 265
+ sodium chloride vol 11 E258, E259, 260, 261
+ sodium iodide vol 11 E266, 267, 268
+ strontium bromide vol 11 E302, 303, 304
+ strontium chloride vol 11 E300, 301
+ thiocyanic acid,ammonium salt vol 11 E294, 295
+ thiocyanic acid,potassium salt vol 11 E282, 283
+ thiocyanic acid,sodium salt vol 11 E269, 270
+ water vol 13 370, 446
Acetate,butyl
see acetic acid,butyl ester
Acetate,ethyl
see acetic acid,ethyl ester
Acetic acid
+ argon vol 4 E169 -E170, 204
+ chlorine vol 12 E354 -E366, 386, 387
+ ethane vol 9 E195 -E199, 208
+ hydrogen vol 5/6 E227, E228, 229
+ nitrogen vol 10 221
+ nitrous oxide vol 8 E161, 206 - 208
+ oxygen vol 7 453
+ ozone vol 7 492
+ radon-222 vol 2 E291, 292 - 293
+ sulfur dioxide vol 12 198, 200, 201
+ xenon vol 2 171
Acetic acid (aqueous)
+ ammonium tetraphenylborate (1-) + ethane, tris(hydroxymethyl)amino- vol 18 78
+ argon vol 4 E80 - 84, 103 - 104
+ cesium tetraphenylborate (1-) + ethane, tris(hydroxymethyl)amino- vol 18 60
+ nitrogen vol 10 E45 - E47, E54, 77
+ nitrous oxide vol 8 E116, 132
+ potassium tetraphenylborate (1-) + ethane, tris(hydroxymethyl)amino- vol 18 19
+ rubidium tetraphenylborate (1-) + ethane, tris(hydroxymethyl)amino- vol 18 47
+ silver tetraphenylborate (1-) + ethane, tris(hydroxymethyl)amino- vol 18 146
+ sulfur dioxide vol 12 199
+ thallium tetraphenylborate (1-) + ethane, tris(hydroxymethyl)amino- vol 18 179
Acetic acid (glacial)
+ krypton vol 2 84
Acetic acid (ternary)
+ nitric oxide vol 8 323
Acetic acid (ternary and multicomponent)
+ nitrogen vol 10 357 - 360
Acetic acid, anhydride
+ chlorine vol 12 388
+ ozone vol 7 492
+ sulfur dioxide vol 12 E189 -E191, 220 - 223

Acetic acid, bromo-, sodium salt (aqueous)		
+ iodic acid, barium salt	vol 14	263
Acetic acid, butyl ester		
+ krypton	vol 2	85
+ oxygen	vol 7	453
+ sulfur dioxide	vol 12	218
+ tris(o-phenanthroline)ruthenium (11)tetraphenylborate (1-)	vol 18	120
Acetic acid, chloro-(aqueous)		
+ hydrogen	vol 5/6	91
Acetic acid, cobalt salt (aqueous)		
+ nitrogen	vol 10	E45 - E47, E54, 77
Acetic acid, copper (2+) salt (aqueous)		
+ nitrogen	vol 10	E45 - E47, E54, 77
Acetic acid, cyano-, sodium salt		
+ iodic acid, calcium salt	vol 14	E75, E76, 114
Acetic acid, ethyl ester		
+ ethane	vol 9	E195 -E199, 201
+ hydrogen	vol 5/6	E230, 233, 234
+ nitric oxide	vol 8	E336, 345
+ nitrogen	vol 10	224
+ oxygen	vol 7	304, 453
+ pyridinium,4-(aminocarbonyl)-1-[[[2-carboxy-8-oxo-7-(phenylsulfoaceyl)amino]-5-thia-1-azabicyclo[4,2,0]oct-2-en-3-yl]-methyl]- hydroxide, inner salt, monosodium salt	vol 16/17	E733, 737
+ radon-222	vol 2	312
+ sulfur dioxide	vol 12	218, 230
+ tris(o-phenanthroline)ruthenium (11)tetraphenylborate (1-)	vol 18	127
Acetic acid, ethyl ester (aqueous)		
+ iodic acid, barium salt	vol 14	E251, E252, 309
+ iodic acid, calcium salt	vol 14	E74, 174
Acetic acid, ethyl ester (ternary)		
+ nitric oxide	vol 8	310
+ nitrogen	vol 10	E105, 107, 108
Acetic acid, hydroxy-, monosodium salt (aqueous)		
+ iodic acid, calcium salt	vol 14	E75, E76, 106 - 108
Acetic acid, lanthanum salt		
+ water	vol 13	66
Acetic acid, lanthanum salt (aqueous)		
+ nitric acid, lanthanum salt	vol 13	66
Acetic acid, manganese (2+) salt (aqueous)		
+ nitrogen	vol 10	E45 - E47, E54, 77
Acetic acid, mercury (2+) salt (aqueous)		
+ nitrogen	vol 10	E45 - E47, E54, 77
Acetic acid, methoxy-, sodium salt (aqueous)		
+ iodic acid, calcium salt	vol 14	E75, E76, 113
Acetic acid, methyl ester		
+ ethane	vol 9	E195 -E199, 204, E232, E233, 245
+ hydrogen	vol 5/6	E230, 232
+ nitrogen	vol 10	223
+ nitrous oxide	vol 8	209
+ oxygen	vol 7	302, 304, 453
+ ozone	vol 7	492
+ sulfur dioxide	vol 12	E189 -E191, 214, 215, 218
Acetic acid, methyl ester (ternary)		
+ sulfur dioxide	vol 12	179, 216

Acetic acid, 2-methylpropyl ester
+ hydrogen vol 5/6 E231, 235
+ nitrogen vol 10 225
+ tris(o-phenanthroline)ruthenium (II)tetraphenylborate (1-)
vol 18 133
Acetic acid, nickel (2+) salt (aqueous)
+ nitrogen vol 10 E45 - E47, E54, 77
Acetic acid, pentyl ester
+ ethane vol 9 E232, E233, 245
+ hydrogen vol 5/6 E231, 236
+ nitrogen vol 10 226
+ nitrous oxide vol 8 E161, 208, 210, 211
+ radon vol 2 313
Acetic acid, phenylmethyl ester
+ sulfur dioxide vol 12 253
Acetic acid, propyl ester
+ oxygen vol 7 453
+ tris(*o*-phenanthroline)ruthenium (II)tetraphenylborate (1-)
vol 18 137
Acetic acid, sodium salt (aqueous)
+ nitrogen vol 10 E45 - E47, E54, 77
+ oxygen vol 7 E71, 162, 451
Acetic acid, trifluoro-
+ ozone vol 7 492
Acetic acid, zinc (2+) salt (aqueous)
+ nitrogen vol 10 E45 - E47, E54, 77
Acetic anhydride
see acetic acid, anhydride
Acetone
see 2-propanone
Acetonitrile
+ cesium tetraphenylborate vol 18 64 - 66
+ nitric oxide vol 8 E336, 348
+ potassium tetraphenylborate vol 18 E31 - 32, 33 - 36
+ rubidium tetraphenylborate vol 18 49
+ silver azide + tetraethylammonium azide tetraethylammonium nitrate vol 3 20
+ silver tetraphenylborate (1-) + sodium tetraphenylborate
vol 18 152
+ silver tetraphenylborate (1-) + sodium tetraphenylborate + perchloric acid, sodium salt vol 18 151
+ sulfur dioxide vol 12 267, 268
+ tetraphenylarsonium tetraphenylborate (1-)
vol 18 E187, 188 - 190
+ tetraphenylarsonium tetraphenylborate (1-) + nitric acid, silver salt + tetraphenylarsonium nitrate
vol 18 191
+ tetraphenylphosphonium tetraphenylborate (1-)
vol 18 229
+ thiocyanic acid, silver salt + thiocyanic acid, potassium salt + nitric acid, potassium salt
vol 3 208
Acetonitrile (aqueous)
+ sulfur dioxide vol 12 110
Acetophenone
see ethanone, 1-phenyl-

Acetylcholine chloride
see ethanaminium, 2-(acetyloxy)-N,N,N-trimethyl-, chloride
Aconitane-7,8-diol,4-[[(2-aminobenzoyl)oxy]methyl]-20-ethyl-1,6,14,16-tetramethoxy-(1b,6b,14a,16b)-
+ oxygen vol 7 406
Acetylene tetrachloride
see ethane, 1,1,2,2-tetrachloro-
Acetylene
see ethyne
Acids
see under individual acids
Acrylic acid
see 2-propenoic acid
Active amyl alcohol
see 1-propanol, 2,2-dimethyl-
Adenosine, 5-(tetrahydrogen phosphate)
+ oxygen vol 7 404
Adenosine, 5-(trihydrogen phosphate)
+ oxygen vol 7 404
ADP
see adenosine, 5-(trihydrogen phosphate)
Adrenals
see guinea pig adrenals
D-Alanine
+ hydrogen vol 5/6 103
D-Alanine (aqueous)
+ nitrogen vol 10 308
L-Alanine
+ oxygen vol 7 400
Alanine (aqueous)
+ iodic acid, barium salt vol 14 E252, E253, 285
+ iodic acid, barium salt + iodic acid, potassium salt vol 14 E252, E253, 287
+ iodic acid, calcium salt vol 14 E75, 142, 178
+ iodic acid, calcium salt + iodic acid, potassium salt vol 14 E77, 142
+ sodium hydroxide + iodic acid, barium salt vol 14 E252, E253, 286
DL-Alanine, monosodium salt (aqueous)
+ iodic acid, calcium salt vol 14 E76, E77, 115
Alanyl glycine, monosodium salt (aqueous)
+ iodic acid, calcium salt vol 14 E76, E77, 123
Albumen
see egg albumen, human albumen, serum albumen
Alcohols (c12-c16 aliphatic) (quaternary)
+ hydrogen vol 5/6 473, 474
Alcohols (c12-c16 aliphatic) (ternary)
+ hydrogen vol 5/6 471, 472
Alcohols
see under individual alcohols
Alkali halides
see under individual halides
Allyl alcohol
see 2-propen-1-ol
Aluminium chloride(aqueous)
+ argon vol 4 E33, 42
+ hydrogen vol 5/6 E30, 54
+ oxygen vol 7 E63, 99
Aluminium nitrate
see nitric acid, aluminium salt
Aluminium oxide(aqueous)
+ argon vol 4 77
+ hydrogen vol 5/6 E30, 53
+ nitrogen vol 10 E45 - E48, 60
+ oxygen vol 7 E63, 101, 389

Aluminium sulfate
see sulfuric acid, aluminium salt
Aminocyclohexane
see cyclohexanamine
7-Aminoacetoxycephalosporanic acid
see 5-thia-1-azabicyclo[4,2,0]oct-2-ene-2-carboxylic acid, 7-amino-3-(hydroxymethyl)-8-oxo, acetate ester
7-Aminodeacetoxycephalosporanic acid
see 5-thia-1-azabicyclo[4,2,0]oct-2-ene-2-carboxylic acid, 7-amino-3-methyl-8-oxo-
Aminoethane
see ethanamine
2-Aminoethanol
see ethanol, 2-amino-
N-(Aminoiminomethyl)-N-methylglycine
see glycine, N-(aminoiminomethyl)-, N-methyl-
6-Aminopenicillanic acid
see 4-thia-1-azabicyclo[3,2,0]heptane-2-carboxylic acid, 6-amino-3,3-dimethyl-7-oxo-
1-Aminopropane
see 1-propanamine
2-Aminopropane
see 2-propanamine
N,N´-bis-(3-Aminopropyl)-ethylenediamine
see 1,3-propanediamine, N-(2-aminoethyl)-N-(3-aminopropyl)-
N-(3-Aminopropyl)-1,3-propanediamine
See 1,3-Propanediamine, N-(3-aminopropyl)-
Amino-N-(aminoiminomethyl)-N-methylglycine
see glycine, amino-N-(aminoiminomethyl)-N-methyl-

Ammonia		
+ argon	vol 4	
+ helium	vol 1	E255, 282
+ hydrogen	vol 5/6	E622, 623 - 629
+ hydrogen deuteride	vol 5/6	E299
+ krypton	vol 2	105
+ nitrogen	vol 10	E476, 477 - 483
+ thiocyanic acid, silver salt	vol 3	233
Ammonia (aqueous)		
+ cyanic acid, silver salt	vol 3	E47, E54 - 55, 59, 66
+ cyanic acid, silver salt + cyanic acid, silver potassium salt + nitric acid, silver salt	vol 3	66
+ iodic acid, barium salt	vol 14	E247, 293
+ iodic acid, calcium salt	vol 14	E69, 87, 88
+ silver azide	vol 3	E1, E7, E8, 17
+ thiocyanic acid, silver salt	vol 3	E105, E109 - 110 116, 132
Ammonia (ternary)		
+ deuterium	vol 5/6	296
+ hydrogen	vol 5/6	634

Ammonium bisulfite
see sulfurous acid, monoammonium salt

Ammonium bromide		
+ acetamide, N-methyl-	vol 11	E288, E289, 290, 291
+ formamide	vol 11	E64, 65, 66
+ formamide, N,N-dimethyl-	vol 11	E226, 227, 288
+ formamide, N-methyl-	vol 11	E134, 135 - 137

Ammonium bromide (aqueous)
+ argon vol 4 E33 - 34, 39 - 40
+ ethane vol 9 E29, E30, 36
+ nitrous oxide vol 8 E29, 48
+ sulfur dioxide vol 12 E37 - E39, 55
Ammonium chloride
+ acetamide, N-methyl- vol 11 E284, E285, 286, 287
+ formamide vol 11 E61, 62, 63
+ formamide, N,N-dimethyl- vol 11 E222, 223 - 225
+ formamide, N-methyl- vol 11 E130, 131 - 133
Ammonium chloride (aqueous)
+ argon vol 4 E33, 38
+ ethane vol 9 E29, E30, 35
+ helium vol 1 E20, 25
+ hydrogen vol 5/6 E28, 43, 44
+ iodic acid, calcium salt vol 14 E70, E73, 89, 90
+ krypton vol 2 E12, E13, 14
+ neon vol 1 E141, 145
+ nitrous oxide vol 8 E29, 46, 47
+ radon-222 vol 2 E238 -E241, 242
+ sulfur dioxide vol 12 E37 - E39, 54
Ammonium hydroxide (aqueous)
+ hydrogen vol 5/6 E28, 42
+ iodic acid, barium salt vol 14 E247, 294
+ oxygen vol 7 97
Ammonium iodate
see iodic acid, ammonium salt
Ammonium iodide
+ acetamide, N-methyl- vol 11 E292, 293
+ formamide vol 11 E67, 68, 69
+ formamide, N,N-dimethyl- vol 11 E229, 230
+ formamide, N-methyl- vol 11 E138, 139
Ammonium nitrate
see nitric acid, ammonium salt
Ammonium penicillin G
see 4-thia-1-azabicyclo[3,2,0]heptane-2-carboxylic acid, 3,3-dimethyl-7-oxo-6-[(phenyl acetyl)amino], monoammonium salt
Ammonium sulfate
see sulfuric acid, ammonium salt
Ammonium tetrabutyl-, bromide
see 1-butanaminium, N,N,N-tributyl-, bromide
Ammonium tetraethyl bromide
see ethanaminium, N,N,N-triethyl-, bromide
Ammonium tetramethyl bromide
see methanaminium, N,N,N-trimethyl-, bromide
Ammonium tetramethyl iodide
see methanaminium, N,N,N-trimethyl-, iodide
Ammonium tetraphenylborate
+ N-methyl-2-pyrrolidinone vol 18 80
+ water vol 18 E72 - 73, 74, 75
Ammonium tetraphenylborate (aqueous)
+ 2-propanone vol 18 79
+ Tris(hydroxymethyl)aminoethane +acetic acid vol 18 78
+ sulfuric acid, sodium salt vol 18 76, 77
Ammonium tetrapropyl bromide
see 1-propanaminium, N,N,N-tripropyl-, bromide
Ammonium thiocyanate
see thiocyanic acid, ammonium salt

Amoxillin trihydrate
see 4-thia-1-azabicyclo[3,2,0]heptane-2-carboxylic acid, 6-[(amino(4-hydroxyphenyl)acetyl)amino]-3,3-dimethyl-7-oxo, trihydrate
Ampicillin anhydrate
see 4-thia-1-azabicyclo[3,2,0]heptane-2-carboxylic acid, 6-[(aminophenylacetyl)amino]-3,3-dimethyl-7-oxo-
Ampicillin sodium
see 4-thia-1-azabicyclo[3,2,0]heptane-2-carboxylic acid, 6-[(aminophenylacetyl)amino]-3,3-dimethyl-7-oxo-, monosodium salt
Ampicillin trihydrate
see 4-thia-1-azabicyclo[3,2,0]heptane-2-carboxylic acid, 6-[(aminophenylacetyl)amino]-3,3-dimethyl-7-oxo- trihydrate
Ampicillin
see 4-thia-1-azabicyclo[3,2,0]heptane-2-carboxylic acid, 6-[(aminophenylacetyl)amino]-3,3-dimethyl-7-oxo-

AMSCO-123-15		
+ argon	vol 4	142
+ helium	vol 1	115
+ krypton	vol 2	70
+ nitrogen	vol 10	330
+ oxygen	vol 7	449
+ xenon	vol 2	169

Amyl acetate
see acetic acid, pentyl ester
iso-Amyl acetate
see 1-Butanol, 3-methyl-, acetate
Amyl alcohol
see 1-pentanol
iso-Amyl alcohol
see 1-butanol, 3-methyl-
tert-Amyl alcohol
see 2-butanol, 2-methyl-
Aniline
see benzenamine

Aniline nitrate (aqueous)		
+ nitric acid, cerium salt	vol 13	215
+ nitric acid, dysprosium salt	vol 13	E416, E417, 424
+ nitric acid, lanthanum salt	vol 13	104, 105
+ nitric acid, praseodymium salt	vol 13	271
Aniline nitrate		
+ water	vol 13	104, 105, 215, 424

Animal oil
see oil, animal
Anisole
see Benzene, methoxy-

Antifoam		
+ oxygen	vol 7	423
Apiezon GW oil		
+ helium	vol 1	116
L-Arabinose		
+ nitrogen	vol 10	310
Argentate (1-), bis(cyano-c)-potassium		
+ eicosahydrodibenzo(b.k)(1,4,7,10,13,16)-hexaoxacyclo-octadecin + 1,4-dichlorobenzene	vol 3	83
+ water-d_2	vol 3	89
+ water	vol 3	74

Argentate (1-), bis(cyano-C)-potassium (aqueous)
+ cyanic acid, silver salt + ammonia + nitric acid, silver salt vol 3 66 - 67
+ cyanic acid, silver salt + silver chloride + potassium chloride vol 3 E45, 63
+ cyanic acid, silver salt + thiocyanic acid, silver salt + thiocyanic acid, potassium salt vol 3 E46, 65, E102, 122 - 123
+ cyanic acid, silver salt + sodium hydroxide vol 3 E47, 75 - 76, 79 - 80
Argon
+ helium vol 1 110, E283, 284
+ hydrogen vol 5/6 E571, 572 - 580
+ hydrogen deuteride vol 5/6 E299
+ neon vol 1 249, E359, 360
Argon (multicomponent)
+ nitrogen vol 10 482
Arsenious sulfide (aqueous colloidal)
+ nitrous oxide vol 8 145, 148
Arsonium, tetraphenyl-, nitrate
+ tetraphenylarsonium tetraphenylborate + nitric acid silver salt + acetonitrile vol 18 191
Artificial seawater
see seawater
ATP
see adenosine, 5-(tetrahydrogen phosphate)
Azidocillin
see 4-thia-1-azabicyclo[3,2,0]heptane-2-carboxylic acid, 6-[2-azido-2-phenyl-acetamido]-3,3-dimethyl-7-oxo-

Barium bromate			
see Bromic acid, barium salt			
Barium bromide			
+ acetamide, N-methyl-	vol 11		E307, 308
+ formamide	vol 11		E86, 87
Barium bromide (aqueous)			
+ bromic acid, barium salt	vol 14		240
+ chloric acid, barium salt + nitric acid, barium salt	vol 14	E210,	221, 222
+ iodic acid, barium salt	vol 14		291
+ neon	vol 1		E141, 158
Barium chlorate			
see Chloric acid, barium salt			
Barium chloride			
+ acetamide, N-methyl-	vol 11		E305, 306
+ formamide	vol 11		E84, 85
+ water	vol 14		236, 237, 289, 290
Barium chloride (aqueous)			
+ argon	vol 4		E33, 43
+ bromic acid, barium salt	vol 14	E227,	236, 237
+ chloric acid, barium salt	vol 14	E210,	217, 218
+ chlorine	vol 12		E349, 351
+ hydrogen	vol 5/6		60
+ helium	vol 1		E20, 27
+ iodic acid, barium salt	vol 14		289, 290
+ krypton	vol 2	E12,	E13, 19
+ neon	vol 1		E141, 157
+ nitrogen	vol 10	E49,	E45 - E47, 62, 63
+ nitrous oxide	vol 8		E32, 75
+ oxygen	vol 7	E66,	127, 128
Barium hydroxide			
+ water	vol 14		223
Barium hydroxide (aqueous)			
+ chloric acid, barium salt	vol 14		E210, 223
Barium iodate			
see Iodic acid, barium salt			
Barium iodide (aqueous)			
+ neon	vol 1		E141, 159
Barium nitrate			
see Nitric acid, barium salt			
Barium thiocyanate			
see Thiocyanic acid, barium salt			
Barracudina oil			
+ oxygen	vol 7		443
Beef tallow			
+ radon-222	vol 2		331
Benethamine penicillin G			
see 4-thia-1-azabicyclo[3,2,0]heptane-2-carboxylic acid, 3,3-dimethyl-7-oxo-6-[(phenylacetyl)amino]- complexed with N-(phenylmethyl)benzene ethanamine (1:1)			
Benethamine penicillin V			
see 4-thia-1-azabicyclo[3,2,0]heptane-2-carboxylic acid, 3,3-dimethyl-7-oxo-6-[(phenoxyacetyl)amino]- complexed with N-(phenylmethyl)benzene ethanamine (1:1)			
Benzaldehyde			
+ argon	vol 4	E253 -E255,	296
+ nitrous oxide	vol 8		212, 213
+ radon-219	vol 2		E339, 342
Benzamide			
+ water	vol 13		108
Benzamide (aqueous)			
+ nitric acid, lanthanum salt	vol 13		108

Benzathine penicillin V
see 4-thia-1-azabicyclo[3,2,0]heptane-2-carboxylic acid, 3,3-dimethyl-7-oxo-6-[(phenoxyacetyl)amino]- complexed with N,N´-bis(phenylmethyl)-1,2-ethanediamine (2:1)

Benzathine penicillin G
see 4-thia-1-azabicyclo[3,2,0]heptane-2-carboxylic acid, 3,3-dimethyl-7-oxo-6-[(phenylacetyl)amino]- complexed with N,N´-bis(phenylmethyl)-1,2-ethanediamine (2:1)

Benzathine penicillin P
see 4-thia-1-azabicyclo[3,2,0]heptane-2-carboxylic acid, 3,3-dimethyl-7-oxo-6-[(4-methylphenoxyacetyl)amino]- complexed with N,N´-bis(phenylmethyl)-1,2-ethanediamine (2:1)

Benzenamine		
+ ethane	vol 9	E195 -E199, 212
+ hydrogen	vol 5/6	264
+ nitrogen	vol 10	265, 266
+ nitrous oxide	vol 8	214, 215
+ oxygen	vol 7	305, 354, 454
+ radon -222	vol 2	320
+ sulfur dioxide	vol 12	E1, E2, 287
+ xenon	vol 2	188, 189
Benzenamine, N,N-diethyl-		
+ oxygen	vol 7	354
+ sulfur dioxide	vol 12	295, 296
Benzenamine, N,N-diethyl- (ternary)		
+ sulfur dioxide	vol 12	297
Benzenamine, N,N-dimethyl-		
+ oxygen	vol 7	354
+ sulfur dioxide	vol 12	E261, E262, 291 - 293, 297
Benzenamine, ar,ar-dimethyl-		
+ sulfur dioxide	vol 12	287, 297
Benzenamine, ar,ar-dimethyl- (ternary)		
+ sulfur dioxide	vol 12	297
Benzenamine, N-ethyl-		
+ oxygen	vol 7	354
+ sulfur dioxide	vol 12	294
Benzenamine, ar-methyl-		
+ sulfur dioxide	vol 12	287
Benzenamine, N-methyl-		
+ oxygen	vol 7	354
+ sulfur dioxide	vol 12	290
Benzenamine, 2-methyl-		
+ oxygen	vol 7	354
Benzenamine, 3-methyl-		
+ oxygen	vol 7	354
Benzene		
+ argon	vol 4	E158, 159 - 161, 211, 248 - 250, E253 -E255, 279
+ chlorine	vol 12	E354 -E366, 375 - 379
+ deuterium	vol 5/6	E281, 287
+ ethane	vol 9	E138 -E141, 155 - 157 E232, E233, 238 - 243
+ helium	vol 1	E68, 70, E255
+ hydrogen	vol 5/6	E159, E160 161 - 167, E405, E406, 407 - 414
+ krypton	vol 2	62

Benzene				
	+ neon	vol 1		E214, 215
	+ nitric oxide	vol 8		E336, 339
	+ nitrogen	vol 10	E162,	163 - 168,
			252,	443, 444
	+ nitrous oxide	vol 8	E160,	180 - 182
	+ oxygen	vol 7		E250, E251,
				252 - 256,
				305, 453
	+ radon-219	vol 2		E339, 341
	+ radon-222	vol 2		264
	+ sulfur dioxide	VOL 12		E145 -E147,
				148 - 153
	+ xenon	vol 2		166
Benzene (quaternary)				
	+ hydrogen	vol 5/6		554, 555
Benzene (ternary)				
	+ ethane	vol 9		163 - 165
	+ hydrogen	vol 5/6		168, 253
	+ nitrogen	vol 10		E104, 111,
			112,	300 - 302
Benzene, bis(1-methylethyl)-				
	+ sulfur dioxide	vol 12		164
Benzene, bis(trifluoromethyl)-				
	+ oxygen	vol 7		322
Benzene, bromo-				
	+ argon	vol 4		219
	+ helium	vol 1		96
	+ krypton	vol 2		93
	+ neon	vol 1		241
	+ oxygen	vol 7		311, 453
	+ sulfur dioxide	vol 12		311
	+ xenon	vol 2		179
Benzene, butyl-				
	+ oxygen	vol 7		260
Benzene, chloro-				
	+ argon	vol 4		217
	+ chlorine	vol 12		E354 -E366,
				419 - 422
	+ helium	vol 1		95
	+ hydrogen	vol 5/6		E240, 251
	+ krypton	vol 2		91
	+ neon	vol 1		240
	+ nitrogen	vol 10		240
	+ nitrous oxide	vol 8		225
	+ oxygen	vol 7	311,	312, 453
	+ sulfur dioxide	vol 12	E298,	308 - 310
	+ tris(o-phenanthroline)ruthenium (11)tetraphenylborate			
		vol 18		121
	+ xenon	vol 2		178
Benzene, (chloromethyl-)				
	+ chlorine	vol 12	E354 -E366,	426
Benzene, 1-chloro-2-methyl-				
	+ chlorine	vol 12		427
Benzene, 1-chloro-3-methyl-				
	+ chlorine	vol 12		428
Benzene, 1-chloro-4-methyl-				
	+ chlorine	vol 12		429
Benzene, 1-chloro-2-nitro-				
	+ chlorine	vol 12	E354 -E366,	439
Benzene, 1-chloro-3-nitro-				
	+ chlorine	vol 12	E354 -E366,	440
Benzene, 1-chloro-4-nitro-				
	+ chlorine	vol 12	E354 -E366,	441
Benzene, chloro(trifluoromethyl)- (ternary+quaternary)				
	+ chlorine	vol 12		430, 431

Benzene, 1,4-dibromo-			
+ krypton	vol 2		94
Benzene, 1,4-dibromo- (ternary)			
+ argon	vol 4		220
+ krypton	vol 2		94
Benzene, 1,2-dichloro-			
+ chlorine	vol 12	E354 -E366,	417
+ sulfur dioxide	vol 12		309
Benzene, 1,4-dichloro-			
+ argentate (1-), bis(cyano-C)-potassium + eicosahydrodibenzo (b,k)(1,4,7,10,13,16)-hexaoxacyclooctadecin	vol 3		83
+ krypton	vol 2		92
Benzene, 1,4-dichloro- (ternary)			
+ argon	vol 4		218
+ krypton	vol 2		92
Benzene (dichloromethyl-)			
+ chlorine	vol 12	E354 -E366,	425
Benzene, 1,2-dichloro-4-nitro-			
+ chlorine	vol 12	E354 -E366,	435
Benzene, 1,4-dichloro-2-nitro-			
+ chlorine	vol 12	E354 -E366,	436
Benzene, dichloro(trifluoromethyl)- (quaternary)			
+ chlorine	vol 12		430
Benzene, 1,4-diiodo-			
+ krypton	vol 2		96
Benzene, 1,4-diiodo- (ternary)			
+ argon	vol 4		222
+ krypton	vol 2		96
Benzene, 1,3-dimethoxy-			
+ sulfur dioxide	vol 12		252
Benzene, 1,4-dimethoxy-			
+ sulfur dioxide	vol 12		252
Benzene, dimethyl-			
+ argon	vol 4	E253 - 255,	281
+ hydrogen	vol 5/6		177
+ nitrogen	vol 10		171
+ radon-222	vol 2	270,	271
+ sulfur dioxide	vol 12		151
Benzene, 1,2-dimethyl-			
+ argon	vol 4		165
+ chlorine	vol 12		382
+ helium	vol 1		76
+ krypton	vol 2		64
+ neon	vol 1		221
+ oxygen	vol 7		258
Benzene, 1,3-dimethyl-			
+ argon	vol 4		166
+ chlorine	vol 12		383
+ helium	vol 1	E77,	78
+ hydrogen	vol 5/6	176, E419, 420 -	423
+ krypton	vol 2		65
+ neon	vol 1	E222,	223
+ nitric oxide	vol 8	E336,	341
+ oxygen	vol 7		258
+ sulfur dioxide	vol 12	E146, E147, 158 -	160
Benzene 1,3-dimethyl (ternary)			
+ hydrogen	vol 5/6	558,	559
Benzene 1,4-dimethyl-			
+ chlorine	vol 12		384
+ helium	vol 1		80
+ krypton	vol 2	66,	67

Benzene 1,4-dimethyl-				
+ neon	vol 1			225
+ oxygen	vol 7	258,	259,	305
				453
+ xenon	vol 2			168
Benzene, 1,4-dimethyl- (ternary)				
+ argon	vol 4	167,	E168,	218,
			220,	222
+ krypton	vol 2	92,	94,	96
Benzene, diphenyl-				
see terphenyl				
Benzene, ethoxy-				
+ sulfur dioxide	vol 12			249
Benzene, ethyl-				
+ chlorine	vol 12			385
+ oxygen	vol 7		258,	453
Benzene, fluoro-				
+ argon	vol 4			215
+ helium	vol 1			93
+ krypton	vol 2			89
+ neon	vol 1			237
+ oxygen	vol 7			311
+ xenon	vol 2			177
Benzene, hexafluoro-				
+ argon	vol 4			216
+ helium	vol 1			92
+ krypton	vol 2			90
+ neon	vol 1			236
+ nitrogen	vol 10			242
+ oxygen	vol 7		314,	315
Benzene, iodo-				
+ argon	vol 4			221
+ helium	vol 1			97
+ krypton	vol 2			95
+ neon	vol 1			242
+ oxygen	vol 7		311,	453
+ sulfur dioxide	vol 12			312
+ xenon	vol 2			180
Benzene, methoxy-				
+ sulfur dioxide	vol 12			248
Benzene, 1,1-(methoxymethylene)bis-				
+ sulfur dioxide	vol 12		E146,	E147,
		167,	168,	248
Benzene, methyl-				
+ argon	vol 4	E162,	163 -	164,
		E253 -	255,	280
+ chlorine	vol 12	E354 -	E366,	380,
			381,	391
+ deuterium	vol 5/6		E281,	288
+ ethane	vol 9	E138 -	E141,	154
+ helium	vol 1		E73,	74
+ hydrogen	vol 5/6	E169,	170 -	175,
		E415,	416 -	418
+ krypton	vol 2			63
+ neon	vol 1		E218,	219
+ nitric oxide	vol 8		E336,	340
+ nitrogen	vol 10		E162,	169,
			170,	445
+ oxygen	vol 7	256,	257,	453
+ radon-219	vol 2		E339,	341
+ radon-222	vol 2	E265,	266 -	269
+ sulfur dioxide	vol 12	E145 -	E147,	151,
			154 -	157
+ xenon	vol 2	167,	168,	191

Benzene, methyl- (aqueous)			
+ butyltrisopentylammonium tetraphenylborate + 2-propanol	vol 18		86
+ silver tetraphenylborate + 2-propanol	vol 18		148
Benzene, methyl- (ternary)			
+ hydrogen	vol 5/6	569,	570
+ xenon	vol 2		191
Benzene, 1,1´-methylenebis-			
+ ethane	vol 9	E195 -E199,	203
+ hydrogen	vol 5/6	183,	431
+ nitrogen	vol 10		173
Benzene 1,1 methylene bis- (ternary)			
+ hydrogen	vol 5/6	560,	561
Benzene, 1,1´-methylenebis-(methyl)-			
+ sulfur dioxide	vol 12	E146, E147, 166,	167
Benzene, 1,1-methylenebis(1-methylethyl)-			
+ sulfur dioxide	vol 12	E146, E147, 167,	168
Benzene (1-methylethenyl)-			
+ hydrogen	vol 5/6	E178, 179 -	181
Benzene (1-methylethenyl) (ternary)			
+ hydrogen	vol 5/6		183
Benzene, (1-methylethyl)-			
+ hydrogen	vol 5/6	182,	425
+ oxygen	vol 7		260
Benzene (1-methylethyl) (ternary)			
+ hydrogen	vol 5/6		183
Benzene, 1-methyl-2-nitro-			
+ sulfur dioxide	vol 12		289
Benzene, 1-methyl-4-propyl-			
+ oxygen	vol 7		260
Benzene, (1-methylpropyl)-			
+ oxygen	vol 7		260
Benzene nitro-			
+ argon	vol 4		228
+ chlorine	vol 12	E354 -E366,	391
+ ethane	vol 9	E195 -E199, 210,	211
+ helium	vol 1		106
+ hydrogen	vol 5/6		264
+ krypton	vol 2		104
+ neon	vol 1		246
+ nitric oxide	vol 8	E336,	349
+ nitrogen	vol 10	264,	265
+ oxygen	vol 7	353,	454
+ radon-222	vol 2		190
+ sulfur dioxide	vol 12	E261, E262, 264, 282 - 286,	297
+ xenon	vol 2		190
Benzene 1,1´-[oxybis (methylene)] bis-			
+ argon	vol 4	E253 - 255,	295
+ oxygen	vol 7		453
Benzene perfluoro-			
see Benzene, hexafluoro-			
Benzene, propyl-			
+ oxygen	vol 7		260
Benzene (trichloromethyl-)			
+ chlorine	vol 12	E354 -E366,	441
Benzene, 1,2,4-trichloro-			
+ chlorine	vol 12	E354 -E366,	418

Benzene, (trifluoromethyl)-		
+ chlorine	vol 12	E354 -E366, 423
Benzene, (trifluoromethyl)- (ternary + quaternary)		
+ chlorine	vol 12	430, 431
Benzene, 1,3,5-trimethyl-		
+ hydrogen	vol 5/6	424
+ krypton	vol 2	69
+ sulfur dioxide	vol 12	E146, E147, 161 - 163
+ xenon	vol 2	168
Benzene 1,3,5-trimethyl (ternary)		
+ hydrogen	vol 5/6	563
Benzene alcohol		
see benzenemethanol		
Benzeneacetonitrile		
+sulfur dioxide	vol 12	267
1,2-Benzenedicarboxylic acid, dibutyl ester		
+ chlorine	vol 12	E354 -E366, 389
+ krypton	vol 2	85
1,2-Benzenedicarboxylic acid, diethyl ester		
+ chlorine	vol 12	E354 -E366, 389
1,2-Benzenedicarboxylic acid, dimethyl ester		
+ chlorine	vol 12	E354 -E366, 389
+ sulfur dioxide	vol 12	237
1,2-Benzenedicarboxylic acid, dimonyl ester		
+ chlorine	vol 12	E354 -E366, 389
1,2-Benzenedicarboxylic acid, diotyl ester		
+ chlorine	vol 12	E354 -E366, 389
1,2-Benzenedicarboxylic acid, diundecyl ester		
+ chlorine	vol 12	E354 -E366, 389
1,2-Benzenediol		
+ hydrogen	vol 5/6	475
+ sulfur dioxide	vol 12	236
Benzenemethanamine		
+ sulfur dioxide	vol 12	293
Benzenemethanol		
+ argon	vol 4	E253 - E255, 293
+ ethane	vol 9	E166, E167, 194
+ sulfur dioxide	vol 12	196, 244
+ tris(o-phenanthroline)ruthenium (11)tetraphenylborate	vol 18	117
Benzoic acid, methyl ester		
+ sulfur dioxide	vol 12	245
Benzoic acid, 2-hydroxy-, methyl ester		
+ sulfur dioxide	vol 12	246, 247
Benzoic acid, ethyl ester (ternary)		
+ nitric acid	vol 8	310
Benzoic acid, phenyl methyl ester		
+ sulfur dioxide	vol 12	253
Benzoic acid, 2-hydroxy-, monosodium salt (aqueous)		
+ iodic acid, calcium salt	vol 14	E75, E76, 124
Benzonitrile		
+ sulfur dioxide	vol 12	267, 288, 297
Benzotrifluoride		
see Benzene, (trifluoromethyl)-		
Benzo (B) thiophene		
+ hydrogen	vol 5/6	517, 518
Benzyl acetate		
see Acetic acid, phenylmethyl ester		
Benzyl alcohol		
see Benzenemethanol		
Benzylamine		
see Benzenemethanamine		
Benzyl benzoate		
see Benzoic acid, phenylmethyl ester		

Benzyl cyanide
see Benzeneacetonitrile
Benzyl ether
see Benzene, 1,1-[oxybis(methylene)]bis-
Beryllium nitrate
see Nitric acid, beryllium salt
Betamethasone sodium phosphate
see Pregna-1,4-diene-3,20-dione,9-fluoro-11,17-dihydroxy-60oxy)-,disodium salt,(11beta,16beta)-
Bicyclo[2,2,1]heptan-2-one,1,7,7-trimethyl-
+ sulfur dioxide vol 12 E190, E191, 254 - 257
1,1-Bicyclohexyl
+ ethane vol 9 E138 -E141, 150
+ hydrogen vol 5/6 158, 394
+ nitrogen vol 10 161
Bis-2-chloroethyl ether
see Ethane, 1,1´-oxybis(2-chloro)-
Bis(1-methylethyl)benzene
see Benzene, bis(1-methylethyl)-
Blood
see Bovine blood, Dog blood, Human blood, Human blood-hypr subjects, Human blood plasma, Human blood-thyrotoxic subjects, Rabbit blood.
Blood, bovine
+ nitrous oxide vol 8 E226, E227, 228
Blood cells
see Human red blood cells
Blood, cod
+ nitrogen vol 10 292
Blood, dog
+ ethane vol 9 E195 -E199, 231
+ nitrogen vol 10 464
+ nitrous oxide vol 8 E226, E227, 232 235, 237
+ xenon vol 2 207
Blood, eel
+ argon vol 4 251
+ nitrogen vol 10 292
Blood, human
Blood, human
+ krypton vol 2 123
+ nitrogen vol 10 286, 292
+ nitrous oxide vol 8 E226, E227, 230 - 234, 236, 238, 239
+ oxygen vol 7 372 - 375
+ radon-222 vol 2 337
+ xenon-133 vol 2 212 - 214
+ xenon vol 2 211
Blood, human, plasma
+ nitrogen vol 10 284, 285
+ oxygen vol 7 376
Blood, human, red cells
see Human red blood cells
Blood, ox
+ nitrogen vol 10 289, 465
Blood, ox, plasma
+ nitrogen vol 10 289, 295
Blood, rabbit
+ nitrogen vol 10 293
+ nitrous oxide vol 8 E226, E227, 240
Blood, rat (venous)
+ radon-222 vol 2 336
Blood, trout
+ nitrogen vol 10 292

Bone marrow
see Guinea pig bone marrow
Borane, trimethyl-
+ nitrogen vol 10 497
Bovine albumen
+ oxygen vol 7 392, 397
Bovine brain (saline homogenate)
+ krypton vol 2 132
+ xenon vol 2 224
Bovine globulin
+ nitrous oxide vol 8 E226, 227
Bovine α-globulin
+ oxygen vol 7 387, 397
Bovine β-globulin
+ oxygen vol 7 387
Bovine β-globulin (in phosphate buffer)
+ nitrous oxide vol 8 E226, E227
Bovine γ-globulin (aqueous)
+ krypton vol 2 122
+ nitrous oxide vol 8 E226, E227
+ oxygen vol 7 387, 397
+ xenon vol 2 210
Bovine hemoglobin (aqueous)
+ krypton vol 2 122
+ xenon vol 2 210
Bovine hemoglobin (in phosphate buffer)
+ nitrous oxide vol 8 E226, E227
Bovine serum
+ nitrous oxide vol 8 E226, E227, 229
Bovine serum albumen (in phosphate buffer)
+ nitrous oxide vol 8 E226, E227
Bovine serum albumen (aqueous)
+ krypton vol 2 122
+ xenon vol 2 210
Brain
see Guinea pig brain, Homogenized brain, Human fetal brain, Rabbit brain
Brain, rabbit
+ nitrogen vol 10 293
+ nitrous oxide vol 8 E226, E227, 240
Brain matter
see Dog brain grey matter, Dog brain white matter, Human brain white matter
Brine
+ oxygen vol 7 144 - 146
Bromic acid, barium salt
+ water vol 14 E225 -E227, 228 - 232, 238, 239
Bromic acid, barium salt (aqueous)
+ barium bromide vol 14 240
+ barium chloride vol 14 E227, 236, 237
+ bromic acid, potassium salt vol 14 231
+ chloric acid, barium salt vol 14 E227, 238, 239
+ glycine vol 14 235
+ nitric acid, barium salt vol 14 234
+ nitric acid, barium salt vol 14 241
+ nitric acid, magnesium salt vol 14 233
+ nitric acid, potassium salt vol 14 232
+ potassium chloride vol 14 230
Bromic acid, magnesium salt
+ water vol 14 E6, E7, 8 - 10
Bromic acid, magnesium salt (aqueous)
+ nitric acid, magnesium salt vol 14 10

Bromic acid, potassium salt (aqueous)
+ bromic acid, barim salt vol 14 231
+ iodic acid, barium salt vol 14 269
Bromic acid, strontium salt
+ water vol 14 E186 -E188, 189, 190
Bromide, ammonium
see Ammonium bromide
Bromide, calcium
see Calcium bromide
Bromide, barium
see Barium bromide
Bromide, cesium
see Cesium bromide
Bromide, cetyl trimethyl ammonium
see 1-Hexadecanaminium, N,N,N-trimethyl-, bromide
Bromide, lithium
see Lithium bromide
Bromide, rubidium
see Rubidium bromide
Bromide, potassium
see Potassium bromide
Bromide, sodium
see Sodium bromide
Bromide, strontium
see Strontium bromide
Bromide, tetrabutyl ammonium
see 1-Butanaminium, N,N,N-tributyl-, bromide
Bromide, tetraethyl ammonium
see Ethanaminium, N,N,N-triethyl-, bromide
Bromide, tetramethyl ammonium
see Methanaminium, N,N,N-trimethyl-, bromide
Bromide, tetrapropyl ammonium
see 1-Propanaminium, N,N,N-tripropyl-, bromide
Bromobenzene
see Benzene, bromo-
Bromoform
see Methane, tribromo-
1-Bromoheptane
see Heptane, 1-bromo-
1-Bromo-octane
see Octane, 1-bromo-
1-Bromo-1,1,2,2,3,3,4,4,5,6,6,6-dodecafluoro-5-(trifluoromethyl)hexane
see Hexane, 1-bromo-1,1,2,2,3,3,4,4,5,6,6,6-dodecafluoro -5-(trifluoromethyl)
1-Bromo-1,1,2,2,3,3,4,4,5,5,6,6,7,7,8,8,8-heptadecafluoro-octane
see Octane, 1-bromo-1,1,2,2,3,3,4,4,5,5,6,6,7,7,8,8,8- hepadecafluoro-
1,3-Butadiene
+ nitrogen vol 10 418
1,3-Butadiene,1,1,2,3,4,4-hexachloro-
+ chlorine vol 12 E354 -E366, 415
Butanal
+ argon vol 4 E169 -E170
1-Butanamine
+ oxygen vol 7 453
+ sulfur dioxide vol 12 277
1-Butanamine, N-butyl-
+ hydrogen vol 5/6 E483, 507
1-Butanamine, N,N-dibutyl-
+ hydrogen vol 5/6 E483, 509
+ sulfur dioxide vol 12 277
2-Butanamine nitrate
+ water vol 13 326

2-Butanamine nitrate (aqueous)		
+ nitric acid, neodymium salt	vol 13	326
1-Butanamine 1,1,2,2,3,3,4,4,4-nonafluoro-N,N-bis(nonafluorobutyl)-		
+ air	vol 10	532
+ argon	vol 4	E230 - 231
+ deuterium	vol 5/6	E281, 294
+ ethane	vol 9	E138 -E141, 142, 143
+ helium	vol 1	107
+ helium-3	vol 1	108
+ hydrogen	vol 5/6	267
+ krypton	vol 2	97
+ nitrogen	vol 10	267 - 269
+ oxygen	vol 7	356 - 359, 386
+ xenon	vol 2	185
1-Butanaminium, N,N,N-tributyl-, bromide (aqueous)		
+ argon	vol 4	E33 - 34, 39 - 40
+ ethane	vol 9	E29, E30, 40
+ helium	vol 1	E20, 26
Butane		
+ ethane	vol 9	E110 -E112, 116
+ hydrogen	vol 5/6	E442, 462
+ hydrogen deuteride	vol 5/6	E299
+ nitrogen	vol 10	E409, 410 - 413, 418
Butane (ternary)		
+ ethane	vol 9	137
+ nitrogen	vol 10	385, 386, 414, 415
isoButane		
see Propane, 2-methyl-		
Butane, 1-chloro-		
+ sulfur dioxide	vol 12	E1, E2
Butane, 2-chloro-		
+ sulfur dioxide	vol 12	E1, E2
Butane, 1,4-dichloro-		
+ sulfur dioxide	vol 12	E1, E2
Butane, 2,2´-dimethyl-		
+ ethane	vol 9	E77, E78, 105
Butane, epoxy-		
+ tris(O-phenanthroline)ruthenium (11) tetraphenylborate (1-)	vol 18	138
Butane, 2-methyl-		
+ nitrogen	vol 10	421
Butane, 1,1,2,2,3,3,4,4-octafluoro-1,4-bis[1,2,2,2-tetrafluoro-1-(tri-fluoromethyl)-ethoxy]-		
+ nitrogen	vol 10	250
+ oxygen	vol 7	337
Butane, 1,1´-oxybis-		
+ sulfur dioxide	vol 12	251
Butane, 1,1´-oxybis(2,1-ethanediyloxy)bis-		
+ oxygen	vol 7	453
1,4-Butanediamine bis-tetraphenylborate		
+ water	vol 18	82
Butanedioic acid, hydroxy-, monosodium salt (aqueous)		
+ iodic acid, calcium salt	vol 14	E75, E76, 117, 118
1,4-Butanediol		
+ hydrogen	vol 5/6	E442, 462
(R,R)-1,2,3,4-Butanetetrol(aqueous)		
+ nitrogen	vol 10	309
Butanoic acid		
+ argon	vol 4	E169 - 170, 205
+ radon-222	vol 2	299, 300

Butanoic acid (aqueous)		
+ argon	vol 4	E80 - 84, 105
Butanoic acid, 3-hydroxy-, monosodium salt (aqueous)		
+ iodic acid, calcium salt	vol 14	E75, E76, 119
Butanoic acid, 1,2,3-propanetriyl ester		
+ radon-222	vol 2	315
Butanoic acid, sodium salt (aqueous)		
+ oxygen	vol 7	451
1-Butanol		
+ argon	vol 4	E169 - 170, E185 - 186, 187 - 190, E253 - 255, 285
+ ethane	vol 9	E166, E167, 179 - 181
+ hydrogen	vol 5/6	E187, 201, 202, E442, 452 - 455
+ krypton	vol 2	76
+ nitrogen	vol 10	E175, E177, E178, 197 - 200
+ nitrous oxide	vol 8	197
+ oxygen	vol 7	E264, 278, 280, 304, 453
+ radon-222	vol 2	280
+ water	vol 15	E32 - E38, 39 - 91
1-Butanol (aqueous)		
+ argon	vol 4	E81
2-Butanol		
+ argon	vol 4	E169 - 170, 190, E253 - 255, 286
+ hydrogen	vol 5/6	E442, 459
+ radon-222	vol 2	281
+ water	vol 15	E94 - E99, 100 - 117
2-Butanol (aqueous)		
+ argon	vol 4	E81
+ tris(O-phenanthroline)ruthenium (11) tetraphenylborate (1-)	vol 18	119
iso-Butanol		
see 1-Propanol, 2-methyl-		
sec-Butanol		
see 2-Butanol		
tert-Butanol		
see 2-Propanol, 2-methyl-		
1-Butanol-d		
+ water-d_2	vol 15	92, 93
2-Butanol-d		
+ water-d_2	vol 15	118, 119
1-Butanol, 2,2-dimethyl-		
+ water	vol 15	E228, 229, 230
2-Butanol, 2,3-dimethyl-		
+ water	vol 15	E231, 232, 233
2-Butanol, 3,3-dimethyl-		
+ water	vol 15	E234, 235, 236
1-Butanol, 2-ethyl-		
+ water	vol 15	E237, 238, 239
2-Butanol, 2-methyl-		
+ water	vol 15	E131, E132, 133 - 139
1-Butanol, 2-methyl-		
+ argon	vol 4	E253 -E255, 288
+ water	vol 15	E126, E127, 128-130

1-Butanol, 3-methyl-
+ oxygen vol 7
+ nitrous oxide vol 8 185, 198, 199
+ radon-222 vol 2 283
+ sulfur dioxide vol 12 193
+ tris(O-phenanthroline)ruthenium (11) tetraphenylborate (1-) vol 18 311
+ water vol 15 E140 - E142, 143 - 158
2-Butanol,3-methyl-
+ water vol 15 E159, 160, 161
1-Butanol,3-methyl-,acetate
+ tris(o-phenanthroline)ruthenium (11)tetraphenylborate (1- vol 18 132
+ sulfur dioxide vol 12 218
2-Butanol, 2,3,3-trimethyl-
+ water vol 15 299
2-Butanone
+ argon vol 4 E169 -E170, E253 - E255, 299
+ oxygen vol 7 294, 453
+ sulfur dioxide vol 12 210
+ tris(o-phenanthroline)ruthenium (11)tetraphenylborate (1- vol 18 118
2-Butanone, 3,3-dimethyl-
+ tris(o-phenanthroline)ruthenium (11)tetraphenylborate (1- vol 18 125
1-Butene,3-methyl-
+ sulfur dioxide vol 12 E1, E2
(Z)-2-Butenedioic acid dimethyl ester
+ sulfur dioxide vol 12 237
2-Butoxyethanol
see Ethanol, 2-butoxy-
Butrolactone
see 2(3H)-furanone, dihydro-
Butter fat
+ radon-222 vol 2 335
Butter oil
+ oxygen vol 7
Butter oil
see Oil, butter
Butter,cocoa
see Cocoabutter
Butylacetate
see Acetic acid, butyl ester
iso-Butyl acetate
see Acetic acid, 2-methylpropyl ester
Butyl alcohol
see 2-Propanol,2-methyl-
sec-Butyl alcohol
see 2-Butanol
Butylaminc
see 1-Butanamine
Butylammonium tetraphenylborate(1-)
+ water vol 18 83
Butylbenzene
see Benzene,butyl-
Butyl carbinol
see 1-Pentanol
iso-Butyl carbinol
see 1-Butanol, 3-methyl-
sec-Butyl carbinol
see 1-Butanol, 2-methyl-
Butyl cellosolve
see Ethanol,2-butoxy-

Butyl ether
see Butane, 1,1´-oxybis-
Butylmethyl carbinol
see 2-Hexanol
iso-Butylmethyl carbinol
see 2-Pentanol, 4-methyl-
sec-Butylmethyl carbinol
see 2-Pentanol, 3-methyl-
Butyl propionate
see Propanoic acid, butyl ester
Butyltriisopentylammonium tetraphenylborate
+ lithium chloride + ethanol vol 18 87
+ methanol vol 18 88
Butyltriisopentylammonium tetraphenylborate (aqueous)
+ lithium chloride + ethanol vol 18 85
+ methylbenzene + 2-propanone vol 18 86
+ sodium hydroxide vol 18 84

Entry	Volume	Pages
Cadmium bromide (aqueous)		
+ sulfur dioxide	vol 12	E37 - E39, 79
Cadmium chloride (aqueous)		
+ sulfur dioxide	vol 12	E37 - E39, 78
Cadmium iodide (aqueous)		
+ sulfur dioxide	vol 12	E37 - E39, 80
Cadmium nitrate		
see Nitric acid, cadmium salt		
Cadmium sulfate		
see Sulfuric acid, cadmium salt		
Calcium bisulfite		
see Sulfurous acid, calcium salt		
Calcium bromide		
+ acetamide, N,N-dimethyl-	vol 11	E339, 340
+ acetamide, N-methyl-	vol 11	E298, 299
+ formamide	vol 11	E78, 79
+ formamide, N,N-dimethyl-	vol 11	E243, 244
+ formamide, N-methyl-	vol 11	E146, 147
Calcium bromide (aqueous)		
+ neon	vol 1	E141, 152
Calcium chlorate		
see Chloric acid, calcium salt		
Calcium chloride		
+ acetamide, N,N-dimethyl-	vol 11	E337, 338
+ acetamide, N-methyl-	vol 11	E296, 297
+ formamide	vol 11	E74, 75 - 77
+ formamide, N,N-dimethyl-	vol 11	E240, 241, 242
+ formamide, N-methyl-	vol 11	E144, 145
+ water	vol 14	53 - 56
Calcium chloride (aqueous)		
+ argon	vol 4	E33, 43
+ chloric acid, calcium salt	vol 14	E45, 53 - 58
+ chloric acid, calcium salt + chloric acid, potassium salt	vol 14	E47, 57, 58
+ chloric acid, calcium salt + potassium chloride	vol 14	57, 58
+ ethane	vol 9	E30, 43
+ glycine + iodic acid, calcium salt	vol 14	E77, 158
+ hydrogen	vol 5/6	E31, 59
+ iodic acid, barium salt	vol 14	277
+ iodic acid, calcium salt	vol 14	E70, E73, 155 - 157
+ neon	vol 1	E141, 151
+ nitrogen	vol 10	346
+ nitrous oxide	vol 8	E31, 70 - 72
+ oxygen	vol 7	E66, 125, 126
Calcium hydroxide (aqueous)		
+ iodic acid, calcium salt	vol 14	E69, 102, 154
+ iodic acid, calcium salt + sucrose	vol 14	159
+ sodium chloride + iodic acid, calcium salt	vol 14	102
Calcium iodate		
see Iodic acid, calcium salt		
Calcium iodide		
+ acetamide, N,N-dimethyl-	vol 11	E341, 342
Calcium iodide (aqueous)		
+ neon	vol 1	E141, 153
Calcium lactate		
+ iodic acid, magnesium salt	vol 14	E17, 33, 34
Calcium nitrate		
see Nitric acid, calcium salt		

Calcium pencilin V
see 4-thia-1-azabicyclo[3,2,0]heptane-2-carboxylic acid, 3,3-dimethyl-7-oxo-6-[(phenoxyacetyl)amino], calcium salt
Calcium sulfite
see Sulfurous acid, calcium salt
Calcium thiocyanate
see Thiocyanic acid, calcium salt
Camphor
see bicyclo(2,2,1)heptan-2-one,1,7,7-trimethyl-

Capelin oil		
+ oxygen	vol 7	362, 363
Caproic acid see hexanoic acid		
Caprylic alcohol see 2-octanol		
Carbon dioxide		
+ argon	vol 4	E253 - 255, 320 - 321
+ helium	vol 1	E298, 299
+ hydrogen	vol 5/6	E610, 611 - 618
+ hydrogen deuteride	vol 5/6	E299
+ nitrogen	vol 10	E488, 489 - 494
+ oxygen	vol 7	468 - 472
Carbon dioxide (ternary)		
+ krypton	vol 2	98, 99
+ krypton-85	vol 2	98, 99
+ oxygen	vol 7	434, 435
+ xenon	vol 2	181, 182
+ xenon-133	vol 2	181, 182
Carbon dioxide (multicomponent)		
+ air	vol 10	516
+ nitrogen	vol 10	76, 466, 495, 498 - 501
Carbon disulfide		
+ argon	vol 4	E223, 224 - 225
+ chlorine	vol 12	E354 -E366, 392, 411
+ deuterium	vol 5/6	E281, 291
+ ethane	vol 9	E138 -E141, 161, 162
+ helium	vol 1	98
+ hydrogen	vol 5/6	E255, 256 - 258
+ krypton	vol 2	100
+ neon	vol 1	243
+ nitrogen	vol 10	E253, 254 - 257
+ nitrous oxide	vol 8	259
+ oxygen	vol 7	345, 434
+ radon-219	vol 2	E339, 342
+ radon-222	vol 2	319
+ sulfur dioxide	vol 12	328
+ xenon	vol 2	183
Carbonic acid, dipotassium salt (aqueous)		
+ hydrogen	vol 5/6	E33, 76
+ nitrogen	vol 10	E45 - E47, E55, 81, 82
+ nitrous oxide	vol 8	E38, 112
+ oxygen	vol 7	E74, 139
Carbonic acid, disodium salt (aqueous)		
+ oxygen	vol 7	E71, 139, 163
+ nitrogen	vol 10	71
+ nitrous oxide	vol 8	E35, 93
Carbonic acid, monopotassium salt (aqueous)		
+ nitrous oxide	vol 8	E38, 112

Carbonic acid, monosodium salt (aqueous)
+ nitrogen vol 10 71
+ nitrous oxide vol 8 E35, 93
+ oxygen vol 7 E71, 160
Carbonic acid, lithium salt (aqueous)
+ oxygen vol 7 E66, E67, 139
Carbon monoxide
+ argon vol 4 E253 - 255, 319
+ helium vol 1 E254, 295
+ hydrogen vol 5/6 E601, 602 - 608
+ hydrogen deuteride vol 5/6 E299
Carbon monoxide (multicomponent)
+ hydrogen vol 5/6 E434, 471 - 474, 528, 529, 631 - 633
Carbon tetrachloride
see Methane, tetrachloro-
Caroxin-D
see Butane,1,1,2,2,3,3,4,4-octafluoro-1,4-bis-(1,2,2,2-tetraflouro-1-(trifluoroethyl)ethoxy)-
Caroxin-F
see hexane,1,1,1,2,2,3,3,4,4,5,5,6,6-tridecafluoro-6-(1,2,2,2,-tetrafluoromethyl)ethoxy)-
Castor oil
+ oxygen vol 7 305
Cat blood
+ krypton-85 vol 12 125
Cefsulodin sodium
see Pyridinium,4-(aminocarbonyl)-1[[2-carboxy-8-oxo-7-(phenyl-sulfoacetyl)-amino]-5-thia-1-azabicyclo[4,2,0]oct-2-en-3-yl]-methyl-hydroxide, inner salt, monosodium salt
Celestone
see Pregna-1,4-diene-3,20-dione,9-fluoro-11,17-dihydroxy-16-methyl-21-(phosphonooxy)-disodium salt, (11beta,16beta)-
Cellosolve
see Ethanol,2-ethoxy-
Cephalexin monohydrate
see 5-thia-1-azabicyclo[4,2,0]oct-2-ene-2-caboxylic acid-7[(aminophenylacetyl)amino]-3-methyl-8-oxo, monohydrate
Cephalin (in benzene soln.)
+ argon vol 4 248
+ nitrogen vol 10 302
+ oxygen vol 7 394
Cephaloglycin dihydrate
see 5-thia-1-azabicyclo[4,2,0]oct-2-ene-2-carboxylic acid, 3[(acetyloxy)methyl]-7-[(aminophenylacetyl)amino]-8-oxo, dihydrate
Cephalomothin sodium
see 5-thia-1-azabicyclo[4,2,0]oct-2-ene-2-carboxylic acid, 3[(acetyloxy)methyl]-8-oxo-7-[(2-thienylacetyl)amino], monosodium salt
Cephaloridine
see Pyridium,1-[[2-carboxy-8-oxo-7-(2-thienylacetyl) amino-5-thia-1-azabicyclo[4,2,0]oct-2-en-3-yl]-methyl]-hydroxide,inner salt (6R-trans)
Cephradine monohydrate
see 5-thia-1-azabicyclo(4,2,0)oct-2-ene-2-caboxylic acid, 7[(aminophenylacetyl)amino]-3-methyl-8-oxo, monohydrate
Cerelose
see D-glucose
Cerium ammonium nitrate
see Nitric acid, cerium ammonium salt

Cerium chloride
+ water vol 13 192
Cerium chloride (aqueous)
+ nitric acid, cerium salt vol 13 192
Cerium cobalt nitrate
see Nitric acid, cerium cobalt salt
Cerium manganese nitrate
see Nitric acid, cerium manganese salt
Cerium nickel nitrate
see Nitric acid, cerium nickel
Cerium nitrate monohydrate
see Nitric acid, cerium salt, monohydrate
Cerium nitrate
see Nitric acid, cerium salt
Cesium bromide
+ formamide vol 11 E56, 57
+ formamide, N,N-dimethyl- vol 11 E216, 217 - 219
Cesium bromide (aqueous)
+ argon vol 4 E33 - 63, 67, 75
+ oxygen vol 7 E75, 134
Cesium chloride
+ formamide vol 11 E54, 55
+ formamide, N,N-dimethyl- vol 11 E212, 213 - 215
Cesium chloride (aqueous)
+ argon vol 4 E33 - 36, 66, 74
+ ethane vol 9 E33, 59
+ neon vol 1 E141, 171
+ nitrous oxide vol 8 E38, E39, 114, 115
+ oxygen vol 7 E75, 133, 152
Cesium fluoride (aqueous)
+ oxygen vol 7 E75, 132
Cesium hydroxide (aqueous)
+ oxygen vol 7 E75, 131
Cesium iodide
+ formamide vol 11 E58, 59 - 60
+ formamide, N,N-dimethyl- vol 11 E220, 221
Cesium iodide (aqueous)
+ oxygen vol 7 E75, 135
Cesium nitrate
see Nitric acid, cesium salt
Cesium sulfate
see Sulfuric acid, cesium salt
Cesium tetraphenylborate
+ acetonitrile vol 18 64 - 66
+ ethane,1,1-dichloro- vol 18 67
+ ethane,1,2-dichloro- vol 18 68
+ formamide vol 18 70
+ methanol vol 18 71
Cesium tetraphenylborate (aqueous)
+ lithium chloride + ethanol vol 18 69
+ lithium chloride + sodium hydroxide + methanol vol 18 61, 62
+ 2-propanone vol 18 63
+ sulfuric acid, sodium salt vol 18 55, 56
+ tris(hydroxymethyl)aminomethane + acetic acid vol 18 60
+ water vol 18 E52 - 53, 54, 57 - 59
Cetyl trimethyl ammonium bromide
see 1-hexadecanaminium,N,N,N-trimethyl-, bromide

Charcoal suspension
+ nitrous oxide vol 8 158, 159
Chinese hampster blood
+ krypton-85 vol 2 125
Chloroform
see Methane, trichloro-
Chloral hydrate
see 1,1-ethanediol,2,2,2-trichloro-
Chloric acid, barium salt
+ water vol 14 E208 -E212, 213 - 214, 219, 220
Chloric acid, barium salt (aqueous)
+ barium chloride vol 14 E210, 217, 218
+ barium hydroxide vol 14 E210, 223
+ bromic acid, barium salt vol 14 E227, 238, 239
+ chloric acid, sodium salt vol 14 E210, 216
+ ethanol vol 14 E210, 224
+ nitric acid, barium salt + barium bromide vol 14 E210, 221, 222
+ sodium chloride vol 14 E210, 215
Chloric acid, calcium salt
+ water vol 14 E44 - E47, 48 - 50
Chloric acid, calcium salt (aqueous)
+ calcium chloride vol 14 E45, 53 - 58
+ calcium chloride + chloric acid, potassium salt vol 14 E47, 57, 58
+ calcium chloride + potassium chloride vol 14 57, 58
+ chloric acid, potassium salt vol 14 E45, 51, 52
Chloric acid, lithium salt (aqueous)
+ iodic acid, calcium salt + nitric acid, lithium salt vol 14 93
Chloric acid, magnesium salt
+ water vol 14 E1 - E3, 4, 5
Chloric acid, potassium salt (aqueous)
+ calcium chloride + chloric acid, calcium salt vol 14 E47, 57, 58
+ chloric acid, calcium salt vol 14 E45, 51, 52
+ iodic acid, barium salt vol 14 E253, 267
Chloric acid, sodium salt
+ water vol 14 216
Chloric acid, sodium salt (aqueous)
+ chloric acid, barium salt vol 14 E210, 216
Chloric acid, strontium salt
+ water vol 14 E179 -E181, 182 - 184
Chloric acid, strontium salt (aqueous)
+ strontium bromide vol 14 185
Chloride
see Aluminum chloride, Ammonium chloride, Barium chloride, Calcium chloride, Cesium chloride, Iron chloride, Lithium chloride, Lanthanum chloride, Magnesium chloride, Potassium chloride, Rubidium chloride , Sodium chloride and Strontium chloride
Chlorine hexafluoride
+ nitrogen vol 10 474
Chlorine pentafluoride
+ nitrogen vol 10 472, 473
Chlorine trifluoride
+ nitrogen vol 10 471

Chlorine
+ oxygen vol 7 465
Chloroacetic acid
see Acetic acid, chloro-
Chlorobenzene
see Benzene, chloro-
1-Chlorobutane
see Butane, 1-chloro-
2-Chlorobutane
see Butane, 2-chloro-
Chlorodifluoromethane
see Methane, chlorodifluoro-
Chloroethane
see Ethane, chloro-
Chloroform
see Methane, trichloro-
1-Chloroheptane
see Heptane, 1-chloro-
1-Chlorohexane
see Hexane, 1-chloro-
Chloromethane
see Methane, chloro-
(Chloromethyl)benzene
see Benzene, (chloromethyl)-
1-Chloropentane
see Pentane, 1-chloro-
2-Chlorophenol
see Phenol,.2-chloro-
Chloroprocaine penicillin O
see 4-thia-1-azabicyclo[3,2,0]heptane-2-carboxylic acid, 3,3-dimethyl-7-oxo-6-[2-allylthioacetamido]- complexed with 2-(diethylamino)ethyl-4-amino-2-chlorobenzoate (1:1)
1-Chloropropane
see Propane, 1-chloro-
2-Chloropropane
see Propane, 2-chloro-
Chlorothene
+ nitrous oxide vol 8 220
2-Chlorotoluene
see Benzene, 1-chloro-2-methyl-
3-Chlorotoluene
see Benzene, 1-chloro-3-methyl-
4-Chlorotoluene
see Benzene, 1-chloro-4-methyl-
Chlorotrifluoromethane
see Methane, chlorotrifluoro-
1-Chloro-1,1,2,2,3,3,4,4,5,6,6,6-dodecafluoro-5-(trifluoromethyl)-hexane,
see hexane, 1-chloro-1,1,2,2,3,3,4,4,5,6,6,6-dodecafluoro 5-(trifluoromethyl)-
1-Chloro-2-methylbenzene
see Benzene, 1-chloro-2-methyl-
1-Chloro-3-methylbenzene
see Benzene,1-chloro-3-methyl-
1-Chloro-2-methylpropane
see Propane, 1-chloro-2-methyl-
2-Chloro-2-methylpropane
see Propane, 2-chloro-2-methyl-
1-Chloro-2-nitrobenzene
see Benzene,1-chloro-2-nitro-
1-Chloro-3-nitrobenzene
see Benzene,1-chloro-3-nitro

1-Chloro-4-nitrobenzene
see Benzene,1-chloro-4-nitro-
2-Chloro-6-(trichloromethyl)pyridine
see Pyridine,2-chloro-6-(trichloromethyl)-
5-Chloro-2-(trichloromethyl)pyridine
see Pyridine,5-chloro-2-(trichloromethyl)-
Chloro(trifluoromethyl)benzene
see Benzene,chloro(trifluoromethyl)-
Cholest-5-en-3β-ol
+ nitrogen vol 10 300
Cholest-5-en-3β-ol (in benzene soln.)
+ argon vol 4 249
+ oxygen vol 7 396
Cholest-5-en-3β-ol (in 2-methyl-1-propanol soln.)
+ oxygen vol 7 393
Cholesterol
see Cholest-5-en-3β-ol
Cholic acid
see Cholan-24-oic acid, 3,7,12-trihydroxy-
Choline chloride
see Ethanaminium, 2-hydroxy-N,N,N-trimethyl-, chloride
Chondroitin sulfuric acid
+ oxygen vol 7 397, 405
Chorex
see Ethane, 1,1´-oxybis (2-chloro-
Choroid layer
see Rabbit choroid layer
Chromic acid, potassium salt (aqueous)
+ nitric acid, yttrium salt vol 13 24
Chromic sulfate
see Sulfuric acid, chromium salt
Chromium, dichlorodioxo-
+ chlorine vol 12 E354 -E366, 443, 444
Chromium sulfate
see Sulfuric acid, chromium salt
Ciclacillin anhydrate
see 4-thia-1-azabicyclo[3,2,0]heptane-2-carboxylic acid, 6-[[(1-aminocyclohexyl)carbonyl]amino]-3,3-dimethyl-7-oxo-
Ciclacillin dihydrate
see 4-thia-1-azabicyclo[3,2,0]heptane-2-carboxylic acid, [[(1-aminocyclohexyl)carbonyl]amino]-3,3-dimethyl-7-oxo, dihydrate
Citric acid
see 1,2,3-propanetricarboxylic acid, 2-hydroxy-
Clemizole penicillin G
see 4-thia-1-azabicyclo[3,2,0]heptane-2-carboxylic acid, 3,3-dimethyl-7-oxo-6-(2-phenylacetamido)-complexed with 1-(p-chlorobenzyl)-2-(1-pyrrolidinyl-methyl)benzimidazole (1:1)
Cloxacillin sodium monohydrate
see 4-thia-1-azabicyclo[3,2,0]heptane-2-carboxylic acid, 6-[[[3-(2-chlorophenyl)-5-methyl-4-isoxazoyl]carbonyl]amino]-3,3-dimethyl-7-oxo, monosodium salt, monodihydrate
Cobalt (1+), chlorobis(1,2-ethanediamine-N,N)(thiocyanato-N)-, bromide, (OC-6-23)- (aqueous)
+ argon vol 4 E33 - 34, 41
Cobalt acetate
see Acetic acid, cobalt (2+) salt
Cobalt nitrate
see Nitric acid, cobalt salt
Cobalt sulfate
see Sulfuric acid, cobalt salt

Cobaltous sulfate
see Sulfuric acid, cobalt salt
Cocoa butter
+ radon-222 vol 2 331
Cod blood
see Blood, cod
Cod liver oil
+ oxygen vol 7 362, 363
Colza oil
+ radon-222 vol 2 332
Copper acetate
see Acetic acid, copper (2+) salt
Copper bromide (aqueous)
+ nitric oxide vol 8 E266, 316
Copper bromide (in ethanol)
+ nitric oxide vol 8 E266, 321, 325
Copper chloride (aqueous)
+ nitric oxide vol 8 E266, 315
Copper chloride (aqueous, ternary)
+ nitric oxide vol 8 E266, 314
Copper chloride (in acetic acid)
+ nitric oxide vol 8 E266, 323
Copper chloride (in ethanol)
+ nitric oxide vol 8 E266, 318, 320, 326, 327
Copper chloride (in formic acid)
+ nitric oxide vol 8 E266, 322
Copper chloride (in methanol)
+ nitric oxide vol 8 E266, 319
Copper chloride (in 2-propanone)
+ nitric oxide vol 8 E266, 324
Copper nitrate
see Nitric acid, copper salt
Copper sulfate
see Sulfuric acid, copper (2+) salt (1,1)
Cordycepic acid
see D-mannitol
Corn oil
+ oxygen vol 7 366
Corn steep liquor
+ oxygen vol 7 420
Corticosteroids
+ oxygen vol 7 380, 381
Cortisone acetate
see Pregn-4-ene-3,11,20-trione,21-(acetyloxy)-17-hydroxy-
Cotton seed oil
see oil, cotton seed
CP-38,371
see 4-thia-1-azabicyclo[3,2,0]hept-2-yl-5-tetrazole, 6-[D-2-amino-2-(4-aminophenyl)-acetamido]-3,3-dimethyl-7-oxo, trihydrate
Creatine
see Gylcine, amino-N-(aminoiminomethyl)-N-methyl-
Creosote oil
+ hydrogen vol 5/6 524, 525
1,2-Cresol
see 1,2-Benzenediol
1,3-Cresol
see 1,3-Benzenediol
o-Cresol
see 1,2-Benzenediol
m-Cresol
see 1,3-Benzenediol

CTAB		
see 1-Hexadecanaminium, N,N,N-trimethyl-, bromide		
Cumene		
see Benzene,(1-methylethyl)-		
Cupric bromide		
see Copper bromide		
Cupric chloride		
see Copper chloride		
Cuprous nitrate		
see Nitric acid, copper salt		
Cuprous sulfate		
see Sulfuric acid, copper (2+) salt (1,1)		
Cyanate, sodium		
see Hydrocyanic acid, sodium salt		
Cyanate, potassium		
see Hydrocyanic acid, potassium salt		
Cyanide, potassium		
see Potassium cyanide		
Cyanide, silver		
see Silver cyanide		
Cyanide, sodium		
see Sodium cyanide		
Cyclohexanamine		
+ ethane	vol 9	E195 -E199, 202
+ hydrogen	vol 5/6	264
Cyclohexanamine, N,N-dimethyl-		
+ hydrogen	vol 5/6	E483, 506
Cyclohexanamine nitrate		
+ water	vol 13	328
Cyclohexanamine nitrate (aqueous)		
+ nitric acid, neodymium salt	vol 13	328
+ nitric acid, praseodymium salt	vol 13	266, 267
Cyclohexane		
+ argon	vol 4	E145, 146 - 148, E253 - 255, 277
+ chlorine	vol 12	374
+ deuterium	vol 5/6	
+ ethane	vol 9	E138 -E141, 148, 149
+ helium	vol 1	E59, 60
+ hydrogen	vol 5/6	E148, 149 - 152, E380, E381, 382 - 390
+ krypton	vol 2	E52, 53 - 54
+ neon	vol 1	E204, 205
+ nitric oxide	vol 8	E336, 337
+ nitrogen	vol 10	E148, 149 - 154, 439, 440
+ nitrous oxide	vol 8	179
+ oxygen	vol 7	241 - 243
+ radon-222	vol 2	263
+ sulfur dioxide	vol 12	140 - 142
+ xenon	vol 2	E162, 163 - 164
Cyclohexane (quaternary)		
+ hydrogen	vol 5/6	554, 555
Cyclohexane (ternary)		
+ hydrogen	vol 5/6	153, 154
+ nitrogen	vol 10	E104, 117, 118, 353, 354
Cyclohexane, decafluorobis (trifluoromethyl)-		
+ nitrogen	vol 10	246
Cyclohexane, 1,2-dimethyl-, cis-		
+ argon	vol 4	153
+ helium	vol 1	64
+ krypton	vol 2	57
+ neon	vol 1	210
+ nitrogen	vol 10	157
+ oxygen	vol 7	245

Cyclohexane, 1,2-dimethyl-, trans-		
+ argon	vol 4	154
+ helium	vol 1	65
+ krypton	vol 2	58
+ neon	vol 1	211
+ nitrogen	vol 10	158
+ oxygen	vol 7	247
Cyclohexane, 1,3-dimethyl-		
+ helium	vol 1	66
+ krypton	vol 2	59
+ neon	vol 1	212
+ oxygen	vol 7	246
Cyclohexane, 1,3-dimethyl-, cis-		
+ nitrogen	vol 10	160
Cyclohexane, 1,3-dimethyl-, trans-		
+ nitrogen	vol 10	160
Cyclohexane, 1,3-dimethyl, (cis- and trans- mixture)		
+ argon	vol 4	155
Cyclohexane, 1,4-dimethyl-		
+ helium	vol 1	67
+ krypton	vol 2	60
+ neon	vol 1	213
+ oxygen	vol 7	248
Cyclohexane, 1,4-dimethyl-, cis-		
+ nitrogen	vol 10	159
Cyclohexane, 1,4-dimethyl-, trans-		
+ nitrogen	vol 10	159
Cyclohexane, 1,4-dimethyl, (cis- and trans- mixture)		
+ argon	vol 4	156
Cyclohexane, perfluoromethyl-		
see Cyclohexane, undecafluoro (trifluoromethyl)-		
Cyclohexane, methyl-		
+ argon	vol 4	E149, 150 - 152, E253 - 255, 277
+ helium	vol 1	62
+ hydrogen	vol 5/6	391, 392
+ krypton	vol 2	55
+ neon	vol 1	208
+ nitrogen	vol 10	155, 441, 442
+ oxygen	vol 7	244
+ xenon	vol 2	165
Cyclohexane, methyl- (ternary)		
+ nitrogen	vol 10	355, 356
Cyclohexane, 1-methylethyl- (ternary)		
+ hydrogen	vol 5/6	564
Cyclohexane, 1-methylethyl- (quaternary)		
+ hydrogen	vol 5/6	565, 566
Cyclohexane, undecafluoro (trifluoromethyl)-		
+ argon	vol 4	E207, 208 - 209, 253 - 255, 307
+ helium	vol 1	91
+ krypton	vol 2	86
+ neon	vol 1	235
+ nitrogen	vol 10	244
+ oxygen	vol 7	320
+ xenon	vol 2	172
Cyclohexanol		
+ argon	vol 4	E169 - 170, E201, E253 - 255, 292
+ ethane	vol 9	E166, E167, 189
+ helium	vol 1	88
+ hydrogen	vol 5/6	E189, 210, 211
+ krypton	vol 2	82
+ neon	vol 1	233

Cyclohexanol		
+ nitrogen	vol 10	207
+ nitrous oxide	vol 8	185, 202
+ oxygen	vol 7	E267, 289
+ sulfur dioxide	vol 12	196, 197
+ water	vol 15	E210 -E213, 214 - 224
Cyclohexanol (multicomponent)		
+ nitrogen	vol 10	357
Cyclohexanol, 3-methyl-		
+ sulfur dioxide	vol 12	238
Cyclohexanone		
+ argon	vol 4	E253 - 255, 304
+ hydrogen	vol 5/6	225
+ nitrogen	vol 10	220, 449, 450
+ nitrous oxide	vol 8	201
+ oxygen	vol 7	297
+ sulfur dioxide	vol 12	E190, E191, 238 - 240
Cyclohexanone (ternary and multicomponent)		
+ nitrogen	vol 10	357 - 360
Cyclohexanone, 2-methyl-		
+ sulfur dioxide	vol 12	240
Cyclohexanone, 3-methyl-		
+ sulfur dioxide	vol 12	238
Cyclohexene		
+ hydrogen	vol 5/6	149
+ nitrogen	vol 10	154
+ oxygen	vol 7	241
Cyclohexylamine		
see Cyclohexanamine		
Cyclooctane		
+ helium	vol 1	63
+ krypton	vol 2	56
+ neon	vol 1	209
+ nitrogen	vol 10	156
+ oxygen	vol 7	249
Cyclooctanone		
+ argon	vol 4	157
Cyclopentanamine		
+ deuterium	vol 5/6	E281, 295
+ hydrogen	vol 5/6	262
Cycloserine		
see 3-Isoxazolidinone, 4-amino-		
Cyclotetrasiloxane, octamethyl-		
+ argon	vol 4	239
+ ethane	vol 9	E195 -E199, 227
+ helium	vol 1	109
+ hydrogen	vol 5/6	268, 520
+ krypton	vol 2	107
+ neon	vol 1	248
+ nitrogen	vol 10	275
Cysteine		
+ oxygen	vol 7	400

Decahydronaphthalene		
see Naphthalene decahydro-		
Decalin		
see Naphthalene decahydro-		
Decane		
+ air	vol 10	517
+ argon	vol 4	E128, 129 - 131
+ chlorine	vol 12	E354 -E366, 372
+ ethane	vol 9	E77, E78, 93, 94, E110 -E112, 129 - 131
+ helium	vol 1	E51, 52
+ hydrogen	vol 5/6	E122, 130, 139, E355, 368 - 370
+ krypton	vol 2	E35, 40 - 42
+ neon	vol 1	E196, 197
+ nitrogen	vol 10	E119 -E121, 131, 139, 140, E419, 435, 436
+ nitrous oxide	vol 8	E160, 172
+ oxygen	vol 7	E214, E215, 230 231, 238, 239
+ sulfur dioxide	vol 12	E116, E117, 128, 129
+ xenon	vol 2	159
Decanoic acid		
+ radon-222	vol 2	306
1-Decanol		
+ argon	vol 4	E169 -E170, 198
+ hydrogen	vol 5/6	208
+ helium	vol 1	87
+ krypton	vol 2	80
+ neon	vol 1	232
+ nitrogen	vol 10	E176 -E178, 213, 214, 324
+ nitrous oxide	vol 8	204
+ oxygen	vol 7	E266, E267, 286, 287, 439, 453
+ water	vol 15	E402, E403, 404 - 411
Detergent (aqueous)		
+ nitrogen	vol 10	352
Deuterium		
+ helium-3	vol 1	E254, 302
+ helium	vol 1	E254, 304
Dextrin (aqueous)		
+ nitrous oxide	vol 8	253, 255
Dextrin (colloidal)		
+ hydrogen	vol 5/6	114
Dextrin		
+ oxygen	vol 7	406
Dextrose		
see D-glucose		
1,2-Diaminoethane		
see 1,2-ethanediamine		
1,6-Diaminohexane		
see 1,6-hexanediamine		
1,2-Diaminopropane		
see 1,2-propanediamine		
1,3-Diaminopropane		
see 1,3-propanediamine		
1,3-Propanediamine		
+ hydrogen	vol 5/6	E482, 497

Dibenzo(b,k)(1,4,7,10,13,16)hexaoxacyclooctadecin,eicosahydro-
+ argentate (1-), bis(cyano-C)-potassium + 1,4-dichlorobenzene vol 3 83
Dibenzyl ether
see Benzene, 1,1´-[oxybis(methylene)]bis-
Diborane
+ hydrogen vol 5/6 630
+ nitrogen vol 10 496
1,4-Dibromobenzene
see Benzene, 1,4-dibromo-
1,2-Dibromoethane
see Ethane, 1,2-dibromo-
1,2-Dibromo-1,1,2,3,3,3-hexafluoropropane
see Propane, 1,2-dibromo-1,1,2,3,3,3-hexafluoro-
Dibutylamine
see 1-butanamine, N-butyl-
Dibutylether
see Butane, 1,1´-oxybis-
Diisobutyl ketone
see 3-Pentanone, 2,2,4,4-tetramethyl-
Dibutylphthalate
see 1,2-benzene dicarboxylic acid, dibutyl ester
1,2-Dichlorobenzene
see Benzene, 1,2-dichloro-
1,4-Dichlorobenzene
see Benzene,1,4-dichloro-
o-Dichlorobenzene
see Benzene, 1,2-dichloro-
p-Dichlorobenzene
see Benzene, 1,4-dichloro-
1,4-Dichlorobutane
see Butane, 1,4-dichloro-
3,5-Dichloro-2-(dichloromethyl)pyridine
see Pyridine, 3,5-dichloro-2-(dichloromenthyl)-
Dichlorodifluoromethane
see Methane, dichlorodifluoro-
Dichlorodioxochromium
see Chromium, dichlorodioxo
1,1-Dichloroethane
see Ethane, 1,1-dichloro-
1,2-Dichloroethane
see Ethane, 1,2-dichloro-
1,1-Dichloroethene
see Ethene, 1,1-dichloro-
cis-1,2-Dichloroethene
see Ethene, 1,2-dichloro-,cis-
trans-1,2-Dichloroethene
see Ethene, 1,2-dichloro-,trans-
1,1-Dichloroethylene
see Ethene, 1,1-dichloro-
cis-1,2-Dichloroethylene
see Ethene, 1,2-dichloro-,cis-
trans-1,2-Dichloroethylene
see Ethene, 1,2-dichloro-,trans-
1,2-Dichloro-4-nitrobenzene
see Benzene, 1,2-dichloro-4-nitro-
1,4-Dichloro-2-nitrobenzene
see Benzene, 1,4-dichloro-2-nitro-
Dichloromethane
see Methane, dichloro-
(Dichloromethyl)benzene
see Benzene,(dichloromethyl)-
1,2-Dichloropropane
see Propane, 1,2-dichloro-

1,2-Dichloro-1,1,2,2-tetrafluoroethane
see Ethane, 1,2-dichloro-1,1,2,2-tetrafluoro-
Dichloro(trifluoromethyl)benzene
see Benzene, dichloro(trifloromethyl)-
3,5-Dichloro-2-(trichloromenthyl)pyridine
see Pyridine, 3,5-dichloro-2-(trichloromethyl)-
Dicloxacillin sodium monohydrate
see 4-thia-1-azabicyclo[3,2,0]heptane-2-carboxylic acid, 6-[[[3-(2,6-dichlorophenyl)-5-methyl-4-isoxazoyl] carbonyl]amino]-3,3-dimethyl-7-oxo, monosodium salt monohydrate
Dicumyl methane
see Benzene, 1,1´-methylenebis (1-methyl ethyl)-
Dicyclohexyl-18-crown-6
see Dibenzo(b,k)-(1,4,7,10,13,16)hexaoxacyclooctadecin, eicosahydro-
Diethanolamine
see Ethanol, 2,2´-iminobis-
Diether ether
see Ethane, 1,1-oxybis-
Diethylamine nitrate
see Ethanamine, N-ethyl-, nitrate
Diethylaniline
see Benzenamine, N,N-diethyl-
Diethyl carbinol
see 3-Pentanol
Diethylene glycol
see Ethanol, 2,2´-oxybis-
Diethylene glycol dibutyl ether
see Butane, 1,1´-oxybis(2,1-ethanediyloxy)bis-
Diethylenetriamine trinitrate
+ water vol 13 395
Diethylenetriamine trinitrate (aqueous)
+ nitric acid, gadolinium salt vol 13 395
Diethylether
see Ethane, 1,1´-oxybis-
Diethylketone
see 3-Pentanone
Diethylmethyl carbinol
see 3-Pentanol, 3-methyl-
2,2-Diethyl-1-pentanol
see 1-Pentanol, 2,2-diethyl-
Diethyl phenylamine
see Benzenamine, N,N-diethyl-
Diethyl propanedioate (ternary)
+ nitric oxide vol 8 310
Difluorodioxouranium
see Uranium, difluorodioxo-
Diglycolamine
see Ethanol, 2-(2-aminoethoxy)-
Diglyme
see Ethane, 1,1-oxybis(2-methoxy-
9,10-Dihydrophenanthrene
see Phenanthrene, 9,10-dihydro-
2,3-Dihydropyran
see Pyran, 2,3-dihydro-
Dihydroxyethane
see 1,2-Ethanediol
1,4-Diiodobenzene
see Benzene, 1,4-diiodo-
Diisopropanolamine
see 2-Propanol, 1,1´-iminobis-
Diisopropyl benzene
see Benzene, bis(1-methylethyl)-

Diisopropyl carbinol
see 3-pentanol, 2,4-dimethyl-
Diisopropyl ether
see Propane, 1,1´-oxybis-
Diisopropyl phenylmethane
see Benzene, 1,1-methylenebis(1-methylethyl)-
Dilithium carbonate
see Carbonic acid, lithium salt
1,3-Dimethoxybenzene
see Benzene, 1,3-dimethoxy-
Dimethylacetamide
see Acetamide, N,N-dimethyl-
Dimethylamine
see Methanamine, N-methyl-
Dimethylamine nitrate
see Methanamine, N-methyl-, nitrate
Dimethylaminocyclohexamine
see Cyclohexamine, N,N-dimethyl-
1-Dimethylamino-3-propylamine
see 1,3-Propanediamine, N,N-dimethyl-
Dimethylammonium tetraphenylborate (1-)
+ water vol 18 89
N,N-Dimethylaniline
see Benzenamine, N,N-dimethyl-
Dimethylbenzene
see Benzene, dimethyl-
1,2-Dimethylbenzene
see Benzene, 1,2-dimethyl-
1,3-Dimethylbenzene
see Benzene, 1,3-dimethyl-
1,4-Dimethylbenzene
see Benzene, 1,4-dimethyl-
2,2-Dimethylbutane
see Butane, 2,2-dimethyl-
2,2-Dimethyl-1-butanol
see 1-Butanol, 2,2-dimethyl-
2,3-Dimethyl-2-butanol
see 2-Butanol, 2,3-dimethyl-
3,3-Dimethyl-2-butanol
see 2-Butanol, 3,3-dimethyl-
3,3-Dimethyl-2-butanone
see 2-Butanone, 3,3-dimethyl-
Dimethylcyclohexane
see Cyclohexane, dimethyl-
1,2-Dimethylcyclohexane
see Cyclohexane, 1,2-dimethyl-
1,3-Dimethylcyclohexane
see Cylcohexane, 1,3-dimethyl-
1,4-Dimethylcyclohexane
see Cyclohexane, 1,4-dimethyl-
Dimethylformamide
see Formamide, N,N-dimethyl-
Dimethyl glutarate
see Pentanedioic acid, dimethyl ester
2,6-Dimethyl-4-heptanol
see 4-Heptanol, 2,6-dimethyl-
3,5-Dimethyl-4-heptanol
see 4-Heptanol, 3,5-dimethyl-
2,3-Dimethylhexane
see Hexane, 2,3-dimethyl-
2,4-Dimethylhexane
see Hexane, 2,4-dimethyl-
1,1-Dimethylhydrazine
see Hydrazine, 1,1-dimethyl-

1,2-Dimethylhydrazine
see Hydrazine, 1,2-dimethyl-
Dimethyl ketone
see 2-Propanone
Dimethyl maleate
see (z)-2-Butenedioic acid, dimethyl ester
2,2-Dimethyl-1-pentanol
see 1-Pentanol, 2,2-dimethyl-
2,4-Dimethyl-1-pentanol
see 1-Pentanol, 2,4-dimethyl-
4,4-Dimethyl-1-pentanol
see 1-Pentanol, 4,4-dimethyl-
2,3-Dimethyl-2-pentanol
see 2-Pentanol, 2,3-dimethyl-
2,4-Dimethyl-2-pentanol
see 2-Pentanol, 2,4-dimethyl-
2,2-Dimethyl-3-pentanol
see 3-Pentanol, 2,2-dimethyl
2,3-Dimethyl-3-pentanol
see 3-Pentanol, 2,3-dimethyl-
2,4-Dimethyl-3-pentanol
see 3-Pentanol, 2,4-dimethyl-
Dimethyl phthalate
see 1,2-Benzenecarboxylic acid, dimethyl ester
Dimethylpropyl carbinol
see 2-Pentanol, 2-methyl-
Dimethylisopropyl carbinol
see 2-Butanol, 2,3-dimethyl-
N,N-Dimethyl-1,3-propanediamine
see 1,3-propanediamine, N,N-dimethyl-
2,2-Dimethyl-1-propanol
see 1-Propanol, 2,2-dimethyl-
Dimethylphenylamine
see benzenamine, N,N-dimethyl-
1,5-Dimethyl-2-pyrrolidinone
see 2-Pyrrolidinone, 1,5-dimethyl-
Dimethylsulfoxide
see Methane, 1,1-sulfinylbis-
Dimethyl sulfate
see Sulfuric acid, dimethyl ester
Dinitrogen tetroxide
see Nitrogen oxide (N2O4)
1,4-Dioxane
+ ethane vol 9 E195 -E199, 207
+ hydrogen vol 5/6 E214, 219, 220
+ nitric acid, cerium salt vol 13 E221, 224
+ nitrogen vol 10 229
+ oxygen vol 7 301, 305, 453
+ sulfur dioxide vol 12 226, 228, 229
1,4-Dioxane (aqueous)
+ argon vol 4 E80 - 84, 102
+ ethane vol 9 E64, 66
+ iodic acid, barium salt vol 14 E251, E252, 307, 308
+ iodic acid, calcium salt vol 14 E74, 170, 171
+ thiocyanic acid, silver salt + thiocyanic acid, potassium salt + nitric acid, potassium salt vol 3 E188, 189
+ thiocyanic acid, silver salt + thiocyanic acid, potassium salt + perchloric acid, potassium salt vol 3 E188, 190 - 191, 192
p-Dioxane
see 1,4-Dioxane

1,3-Dioxolan-2-one, 4-methyl-
+ argon vol 4 E253 - 255, 305
+ ethane vol 9 E195 -E199, 223, 224
+ hydrogen vol 5/6 477, 478
+ nitrogen vol 10 448
+ oxygen vol 7 183, 184, 186, 187
+ sulfur dioxide vol 12 E190, E191, 224, 225
+ silver tetraphenylborate + sodium tetraphenylborate + perchloric acid, sodium salt vol 18 172
+ tetraphenylarsonium tetraphenylborate (1-) vol 18 223
Dioxosulfatouranium
see Uranium, dioxosulfato-
Diphenylbenzene
see Terphenyl
Diphenylmethane
see Benzene, 1,1´-methylenebis-
Dipotassium carbonate
see Carbonic acid, potassium salt
Dipropylene glycol
see Propanol, oxybis-
Dipropyl ether
see Propane, 1,1´-oxybis-
Disodium carbonate
see Carbonic acid, sodium salt
Disodium glutamate
see DL-glutamic acid, disodium salt
Disodium hydrogen phosphate
see Phosphoric acid, disodium salt
Dithioous acid, disodium salt (aqueous)
+ oxygen vol 7 E70, 161
DMSO
see Methane, sulfinylbis-
Docosane
+ ethane vol 9 E77, E78, 104
Dodecane
+ air vol 10 519
+ argon vol 4 E133, 134 - 135
+ ethane vol 9 E77, E78, 95, 96, E110 -E112, 132 - 135
+ helium vol 1 55
+ hydrogen vol 5/6 142, E356, 371
+ krypton vol 2 44 - 46
+ neon vol 1 200
+ nitrogen vol 10 E119, 142, 437
+ nitrous oxide vol 8 E160, 174
+ oxygen vol 7 E214, E215, 233, 239
+ sulfur dioxide vol 12 E116, E117, 131, 132
+ xenon vol 2 160
Dodecanoic acid
+ radon-222 vol 2 308
Dodecanoic acid, ethyl ester
+ sulfur dioxide vol 12 239
1-Dodecanol
+ hydrogen vol 5/6 E189, 209
+ nitrous oxide vol 8 205
+ nitrogen vol 10 E177, E178, 215, 324

1-Dodecanol			
+ oxygen	vol 7	E267, 288,	439
+ water	vol 15	E413, 414 -	417
Dog blood			
see Blood, dog			
Dog brain grey matter			
+ xenon	vol 2		221
Dog brain white matter			
+ xenon	vol 2		221
Dog erthrocytes			
+ xenon	vol 2		207
Dog fat			
+ krypton	vol 2	E114, E115,	119
+ nitrogen	vol 10		281
+ xenon	vol 2	E199, 204,	221
Dog hemoglobin (aqueous)			
+ xenon	vol 2	E199, 204,	221
Dog homogenized brain			
+ nitrous oxide	vol 8	E266, E227,	242
Dog methemoglobin (aqueous)			
+ xenon	vol 2		208
Dog myocardium			
+ nitrous oxide	vol 8	E226, E227,	246
Dog tissues			
+ xenon	vol 2		221
Dow Corning Antifoam A			
+ xenon	vol 2		192
Dow Corning silicone oil			
see Silicone oil			
Dowtherm A			
+ xenon	vol 2		192
Dysprosium nitrate			
see Nitric acid, dysprosium salt			

EDTA		
see Glycine, N,N´-1,2-ethanediylbis(N-carboxymethyl)-		
Eel blood		
see Blood, eel		
Egg albumen		
+ nitrous oxide	vol 8	E226, E227, 251, 252
+ oxygen	vol 7	389
Eicosane		
+ ethane	vol 9	E77, E78, 102, 103, E110 -E112, 136
Electrolytes		
see Under individual electrolyte		
Electrolyte (aqueous)		
+ krypton	vol 2	E12 - 13, 14 - 24, 124 - 125
+ radon	vol 2	E238 - 241, 242 - 255, 340
+ xenon	vol 2	E149, 150 - 154
1-Ephenamine penicillin G		
see 4-thia-1-azabicyclo[3,2,0]heptane-2-carboxylic acid, 3,3-dimethyl-7-oxo-6-(2-phenylacetamido)- complexed with (-)-2-(methylamino)-1,2-diphenylethanol (1:1)		
Epicillin anhydrate		
see 4-thia-1-azabicyclo[3,2,0]heptane-2-carboxylic acid, 6-[(amino-1,4-cyclohexadien-1-ylacetyl)amino]- 3,3-dimethyl-7-oxo-		
1,2-Epoxyethylene		
see Oxirane		
Erbium nitrate		
see Nitric acid, erbium salt		
Erythritol		
see (R,R)-1,2,3,4-Butanetetrol		
Erythrocytes		
see Rabbit erythrocytes and Ox erythrocytes		
Ethanamine		
+ hydrogen	vol 5/6	E481, E482, 486 - 488
Ethanamine, N,N-diethyl-		
+ hydrogen	vol 5/6	E482, 503
+ water	vol 13	325
Ethanamine, N,N-diethyl- (aqueous)		
+ nitric acid, lanthanum salt	vol 13	95
+ nitric acid, praseodymium salt	vol 13	261
Ethanamine, N,N-diethyl-, nitrate		
+ water	vol 13	96, 97
Ethanamine, N,N-diethyl-, nitrate (aqueous)		
+ nitric acid, neodymium salt	vol 13	325
Ethanamine, N-ethyl-		
+ water	vol 13	208
Ethanamine, N-ethyl- (aqueous)		
+ nitric acid, cerium salt	vol 13	208
Ethanamine, N-ethyl-, nitrate		
+ water	vol 13	93, 94
Ethanamine, N-ethyl-, nitrate (aqueous)		
+ nitric acid, lanthanum salt	vol 13	93, 94 96, 97
+ nitric acid, neodymium salt	vol 13	324
+ nitric acid, praseodymium salt	vol 13	260
Ethanaminium, 2-acetoxy-N,N,N-trimethyl-, chloride		
+ oxygen	vol 7	397, 403
Ethanaminium, 2-hydroxy-, N,N,N-tris(2-hydroxyethyl)-, bromide (aqueous)		
+ ethane	vol 9	E29, E30, 42

Ethanaminium, 2-hydroxy-, N,N,N-trimethyl-, chloride
+ oxygen vol 7 403
Ethanaminium, N,N,N-triethyl-, bromide (aqueous)
+ argon vol 4 E33 - 34, 39 - 40
+ ethane vol 9 E29, E30, 38
+ helium vol 4 E20, 25
+ krypton vol 2 E12, E13, 15
+ oxygen vol 7 E63, 355
Ethane
+ argon vol 4 E253 - 255, 269 - 270
+ helium vol 1 E254, 276
+ hydrogen vol 5/6 E332, 333 - 338
Ethane (multicomponent)
+ hydrogen vol 5/6 552, 553
Ethane (ternary)
+ hydrogen vol 5/6 532 - 536, 543, 544
Ethane, chloro-
+ sulfur dioxide vol 12 E1, E2
1,2-Ethanediamine
+ hydrogen vol 5/6 E482, 488, 489
+ nitrous oxide vol 8 139
+ nitric acid, lanthanum salt vol 13 E109, 143
1,2-Ethanediamine, N,N-bis(2-aminoethyl)-
+ sulfur dioxide vol 12 297
1,2-Ethanediamine dinitrate
+ water vol 13 98, 99, 209, 210, 394, 421
1,2-Ethanediamine dinitrate (aqueous)
+ nitric acid, cerium salt vol 13 209, 210
+ nitric acid, dysprosium salt vol 13 421
+ nitric acid, gadolinium salt vol 13 394
+ nitric acid, lanthanum salt vol 13 98, 99
+ nitric acid, praseodymium salt vol 13 262, 263
Ethane, 1,2-dibromo-
+ chlorine vol 12 E354 -E366, 412
+ nitrous oxide vol 8 220, 223, 224
Ethane, 1,1-dichloro-
+ cesium tetraphenylboraate vol 18 67
+ sulfur dioxide vol 12 E1, E2
+ tetra-N-butylammonium tetraphenylborate vol 18 99
+ tetraethylammonium tetraphenylborate vol 18 103
+ tetramethylammonium tetraphenylborate (1-) vol 18 109
+ tetraphenylarsonium tetraphenylborate (1-) vol 18 192
+ tetraphenylphosphonium tetraphenylborate (1-) vol 18 230
+ tetrapropylammonium tetraphenylborate vol 18 112
Ethane, 1,2-dichloro-
+ cesium tetraphenylborate vol 18 68
+ hydrogen vol 5/6 E239, 248
+ nitrous oxide vol 8 220
+ oxygen vol 7 452
+ ozone vol 7 492
+ rubidium tetraphenylborate vol 18 50
+ sulfur dioxide vol 12 E1, E2, 306, 307

Ethane, 1,2-dichloro-
+ tetrabutylammonium tetraphenylborate vol 18 100
+ tetraethylammonium tetraphenylborate vol 18 104
+ tetramethylammonium tetraphenylborate (1-) vol 18 110
+ tetraphenylarsonium tetraphenylborate (1-) vol 18 193
+ tetraphenylphosphonium tetraphenylborate (1-) vol 18 231
+ tetrapropylammonium tetraphenylborate vol 18 113
+ tris(O-phenanthroline)ruthenium (11) tetraphenylborate (1-) vol 18 124

Ethane, 1,2-dichloro-1,1,2,2-tetrafluoro-
+ ozone vol 7 489

Ethane, 1,2-dimethoxy-
+ hydrogen vol 5/6 E442, 460
+ sulfur dioxide vol 12 234

Ethane, 1,1´-oxybis-
+ air vol 10 521
+ argon vol 4 E253 - 255, 294
+ ethane vol 9 E232, E233, 234
+ hydrogen vol 5/6 E213, 216, 217
+ nitric acid, cerium salt vol 13 E221, 222, 223
+ nitric oxide vol 8 E336, 344
+ nitrous oxide vol 8 E161, 196
+ oxygen vol 7 298, 299, 304, 453
+ radon-222 vol 2 288
+ sulfur dioxide vol 12 E190, E191, 231, 232, 234

Ethane, 1,1-oxybis-(2-chloro)-
+ ethane vol 9 E195 -E199, 209
+ tris(O-phenanthroline)ruthenium (11) tetraphenylborate (1- vol 18 122

Ethane, 1,1´-oxybis (2-ethoxy)-
+ oxygen vol 7 453

Ethane, 1,1´-oxybis (2-methoxy)-
+ sulfur dioxide vol 252
+ thiocyanic acid, silver salt + thiocyanic acid, tetrabutylammonium salt + perchloric acid, tetrabutylammonium salt vol 3 209 - 210

Ethane, 1,1,2,2-tetrachloro-
+ chlorine vol 12 E354 -E366, 414
+ helium vol 1 94
+ hydrogen vol 5/6 E240, 249
+ neon vol 1 239
+ oxygen vol 7 452
+ ozone vol 7 489, 492
+ sulfur dioxide vol 12 E1, E2

Ethane, 1,1,1-trichloro-
+ sulfur dioxide vol 12 E1, E2

Ethane, 1,1,2-trichloro-
+ sulfur dioxide vol 12 E1, E2

Ethane, 1,1,2-trichloro-1,2,2-trifluoro-
+ argon vol 4 210 - 211
+ ethane vol 9 E138 -E141, 146, 147
+ hydrogen vol 5/6 E241, 252
+ neon vol 1 238
+ oxygen vol 7 452
+ ozone vol 7 489
+ xenon vol 2 174

Ethane, 1,1,2-trichloro,-1,2,2-trifluoro (ternary)		
+ ethane	vol 9	163 - 165
+ hydrogen	vol 5/6	253
1,2-Ethanedioic acid (aqueous)		
+ nitrous oxide	vol 8	E116, 129 - 131
1,2-Ethanedioic acid, lanthanium salt (aqueous)		
+ nitric acid, lanthanum salt	vol 13	67, 68
1,2-Ethanedioic acid, neodynium salt (aqueous)		
+ nitric acid, neodynium salt	vol 13	316
1,2-Ethanediol		
+ argon	vol 4	E169 - 170, 199 - 200
+ ethane	vol 9	E166, E167, 190, 192
+ hydrogen	vol 5/6	E441, 445
+ oxygen	vol 7	305, 453
+ sulfur dioxide	vol 12	E189, E191, 203 - 206
+ thiocyanic acid, silver salt + thiocyanic acid, potassium salt + perchloric acid, potassium salt	vol 3	220
1,2-Ethanediol (aqueous)		
+ argon	vol 4	E80 - 84, 96
+ iodic acid, barium salt	vol 14	E251, E252, 303
+ iodic acid, calcium salt	vol 14	E74, 168
+ nitrous oxide	vol 8	122
1,2-Ethanediol (aqueous methanol)		
+ argon	vol 4	E80 - 84, 97
1,2-Ethanediol (aqueous 1,2,3-propanetriol)		
+ argon	vol 4	E80 - E84, 100
1,2-Ethanediol, diacetate		
+ sulfur dioxide	vol 12	241
1,1-Ethanediol, 2,2,2-trichloro-		
+ hydrogen	vol 5/6	92, 93
+ oxygen	vol 7	211
1,1-Ethanediol, 2,2,2-trichloro- (aqueous)		
+ nitrous oxide	vol 8	144
Ethanoic acid		
see Acetic acid		
Ethanolamine		
see Ethanol, 2-amino-		
Ethanol		
+ argon	vol 4	E169 - 170, E176, 177 - 179, E253 - 255, 283
+ butyltriisopentylammonium tetraphenylborate + lithium chloride	vol 18	87
+ cesium tetraphenylborate + lithium chloride	vol 18	69
+ ethane	vol 9	E166, E167, 172 - 175
+ helium	vol 1	E82, 83
+ hydrogen	vol 5/6	E187, 194 - 199
+ krypton	vol 2	E73, 74
+ neon	vol 1	E227, 228
+ nitric oxide	vol 8	E336, 343
+ nitrous oxide	vol 8	E161, 185, 186, 188 - 191
+ oxygen	vol 7	E262, E263, 270 - 276, 304, 452
+ radon-219	vol 2	E339, 342
+ radon-222	vol 2	E273, 274 - 277

Ethanol
+ rubidium tetraphenylborate + lithium chloride
vol 18 51
+ silver tetraphenylborate + perchloric acid, sodium salt + sodium tetraphenylborate vol 18 163
+ sulfur dioxide vol 12 176, E181, 182 - 188
+ tetraphenylarsonium tetraphenylborate (1-)
vol 18 208
+ tris(O-phenanthroline)ruthenium (11) tetraphenylborate (1-)
vol 18 126
+ thiocyanic acid, silver salt + thiocyanic acid, lithium salt vol 3 217 - 218
+ thiocyanic acid, silver salt + thiocyanic acid, potassium salt vol 3 219

Ethanol (aqueous)
+ argentate bis(cyano-C-), sodium vol 3 82
+ argon vol 4 E80 - 84, 86 - 91
+ butyltriisopentylammonium tetraphenylborate + lithium chloride vol 18 85
+ ethane vol 9 E64, 69, 70
+ helium vol 1 35
+ hydrogen vol 5/6 97 - 99
+ iodic acid, barium salt vol 14 E251, E252, 299, 300
+ iodic acid, barium salt + nitric acid, sodium salt + sodu
vol 14 301
+ iodic acid, calcium salt vol 14 E74, 166
+ neon vol 1 181
+ nitrous oxide vol 8 118
+ oxygen vol 7 E190, E191, 195 - 202
+ radon-222 vol 2 256
+ sulfur dioxide vol 12 111, 112
+ potassium tetraphenylborate + lithium chloride
vol 18 25, 26
+ thiocyanic acid, silver salt + thiocyanic acid, potassium salt + nitric acid, potassium salt
vol 3 E193, 194 - 195
+ thiocyanic acid, silver salt + thiocyanic acid, potassium salt + perchloric acid, potassium salt
vol 3 E193, 196 - 197, 198

Ethanol (aqueous methanol)
+ argon vol 4 E81

Ethanol (aqueous potassium chloride)
+ argon vol 4 E34

Ethanol (ternary)
+ nitric oxide vol 8 294, 306, 312, 313, 318, 320, 321, 325, 326, 327
+ oxygen vol 7 213

Ethanol (multicomponent)
+ oxygen vol 7 213, 303, 426 - 428

Ethanol, 2-amino- (aqueous)
+ ethane vol 9 E74, 75

Ethanol, 2-amino-
+ ethane vol 9 E64, 69, 70
+ nitrous oxide vol 8 138
+ sulfur dioxide vol 12 277

Ethanol, 2-amino-, nitrate
+ water vol 13 391
Ethanol, 2-amino-, nitrate (aqueous)
+ nitric acid, gadolinium salt vol 13 391
Ethanol, 2-(2-aminoethoxy)-
+ ethane vol 9 E195 -E199, 225
Ethanol, 2-butoxy-
+ hydrogen vol 5/6 E442, 467
+ sulfur dioxide vol 12 217
Ethanol, 2,2-(1,2-ethanediybis(oxy))bis-
+ hydrogen vol 5/6 E443, 468
Ethanol, 2-ethoxy-
+ ethane vol 9 E195 -E199, 205
+ hydrogen vol 5/6 E443, 461
+ nitric acid, neodymium salt + nitric acid, praseodymium salt
vol 13 280
Ethanol, 2-ethoxy-, acetate
+ sulfur dioxide vol 12 241
Ethanol, 2,2´-iminobis- (aqueous)
+ ethane vol 9 E74, 76
+ nitrous oxide vol 8 141
Ethanol, 2,2´-iminobis-, nitrate (aqueous)
+ nitric acid, gadolinium salt vol 13 392
Ethanol, 2-methoxy-
+ hydrogen vol 5/6 E442, 450
+ nitric acid, neodymium salt + nitric acid, praseodymium salt
vol 13 279
+ sulfur dioxide vol 12 217
Ethanol, 2-(2-methoxyethoxy-)
+ hydrogen vol 5/6 E443, 463
Ethanol, 2-[2-(methoxyethoxy)-ethoxy]-
+ hydrogen vol 5/6 E443, 469
Ethanol, 2,2´,2"-nitrilotris- (aqueous)
+ nitrous oxide vol 8 143
Ethanol, 2,2´,2"-nitrilotris-, nitrate (aqueous)
+ nitric acid, gadolinium salt vol 13 393
Ethanol, 2,2´-oxybis-
+ hydrogen vol 5/6 E443, 468
+ nitrous oxide vol 8 126
+ sulfur dioxide vol 12 204
Ethanol, 2,2´-oxybis- (aqueous)
+ argon vol 4 E80 - 84, 101
Ethanol, 2-phenoxy-
+ sulfur dioxide vol 12 217
Ethanol, 2,2,2-trichloro-
+ sulfur dioxide vol 12 305
Ethanone, 1-phenyl-
+ sulfur dioxide vol 12 239
Ethene
+ hydrogen vol 5/6 E395, 396 - 399
+ nitrogen vol 10 E104, E105, 106-114
Ethene (multicomponent)
+ hydrogen vol 5/6 553
Ethene (ternary)
+ ethane vol 9 E23, 24 - 26
+ hydrogen vol 5/6 537 - 539, 543 - 546
Ethene, 1,1-dichloro-
+ sulfur dioxide vol 12 E1, E2
Ethene, 1,2-dichloro-, cis-
+ sulfur dioxide vol 12 E1, E2
Ethene, 1,2-dichloro-, trans-
+ sulfur dioxide vol 12 E1, E2

Ethene, 1,1,2,2-tetrachloro-
+ chlorine vol 12 E354 -E366, 413
+ sulfur dioxide vol 12 E1, E2
Ethene, tetrachloro-
+ oxygen vol 7 304, 452
Ethene, 1,1,2-trichloro-
+ oxygen vol 7 452
+ sulfur dioxide vol 12 E1, E2
2-Ethoxy-1,1,1,3,3,4,4,5,5,5-decafluoro-2-(trifluoromethyl)pentane
see Pentane, 2-ethoxy-1,1,1,3,3,4,4,5,5,5-decafluoro-2-(trifluoromethyl)-
2-Ethoxyethanol
see Ethanol, 2-ethoxy-
Ethyl acetate
see Acetic acid, ethyl ester
Ethyl acetoacetate
see 3-Oxobutanoic acid, ethyl ester
Ethylammonium tetraphenylborate
+ water vol 18 90
N-Ethylaniline
see Benzenamine, N-ethyl-
Ethylbenzene
see Benzene, ethyl-
Ethyl cellosolve
see Ethanol, 2-ethoxy-
Ethyl ethanoate
see Acetic acid, ethyl ester
Ethyl ether
see Ethane, 1,1′-oxybis-
Ethylene
see Ethene
Ethylenediamine
see 1,2-ethanediamine
Ethylenediamine dinitrate
see 1,2-ethanediamine, dinitrate
Ethylene glycol
see 1,2-Ethanediol
Ethylene glycol, dimethyl ether
see Ethane, dimethoxy-
Ethylene glycol, monoethyl monoacetate
see Ethanol, 2-ethoxy-, acetate
Ethylene oxide
see Ethane, 1,2-epoxy-
Ethyl formate
see Formic acid, ethyl ester
2-Ethyl hexanol
see 1-Hexanol, 2-ethyl-
Ethyl laurate
see Dodecanoic acid, ethyl ester
Ethyl malonate
see Diethyl propanedioate
Ethyl methyl carbinol
see 2-butanol, 2-methyl-
Ethyl palmitate
see Hexadecanoic acid, ethyl ester
3-Ethyl-3-pentanol
see 3-pentanol, 3-ethyl-
Ethylpropylketone
see 3-hexanone
Ethyl phenyl ether
see Benzene, ethoxy-
Ethyne (quaternary)
+ hydrogen vol 5/6 473, 474

Eucerin anhydricum
+ radon-222 vol 2 331
Eucerin cum aqua
+ radon-222 vol 2 331
Europium nitrate
see Nitric acid, europium salt
Eyes
see Guinea pig eyes

Fat
see Butter fat, Dog fat, Human fat, Rat fat, Guinea pig omental fat, Guinea pig subcutaneous fat
FC-47
see 1-Butanamine,1,1,2,2,3,3,4,4,4-nonafluoro-N,N-bis(nonfluoro-)
FC-75
+ air vol 10 290
FC-80
see furan, hexafluorotetrahydro(nonafluorobutyl)-
Fermentation medium
+ oxygen vol 7 418, 419
Ferrate (4-), hexakis (cyano-C)-, tetrapotassium (aqueous)
+ iodic acid, calcium salt vol 14 E70, E74, 147
+ radon-222 vol 2 290
Ferric chloride
see Iron chloride
Ferric hydroxide
see Iron hydroxide
Ferric oxide
see Iron (3+) oxide
Ferric sulfate
see Sulfuric acid, iron (3+) salt
Ferrocyanide, potassium
see Ferrate (4-), hexakis (cyano-C-)-, tetrapotassium salt
Ferrous ammonium sulfate
see Sulfuric acid, iron (2+) ammonium salt
Ferrous bromide
see Iron bromide
Ferrous chloride
see Iron chloride
Ferrous iodide
see Iron iodide
Ferrous nitrate
see Nitric acid, iron (2+) salt
Ferrous selenate
see Selenic acid, iron (2+) salt
Ferrous sulfate
see Sulfuric acid, iron (2+) salt
Flounder oil
+ oxygen vol 7 362, 363
Fluorocarbon, L-1822
+ oxygen vol 7 450
Fluoride
see Lithium fluoride, Potassium fluoride, Sodium fluoride
Fluorine
+ helium vol 1 E254, 306
Fluorobenzene
see benzene, fluoro-
Fluorocarbon fluid FC-47
see 1-Butanamine,1,1,2,2,3,3,4,4,4-nonafluoro-N,N-bis(nonfluoro-)
Fluorocarbon fluid FC-80
see furan, hexafluorotetrahydro(nonafluorobutyl)-
Fluorocarbon fluid, L-1822
+ oxygen vol 7 450
Formamide
+ ammonium bromide vol 11 E64, 65, 66
+ ammonium chloride vol 11 E61, 62, 63
+ ammonium iodide vol 11 E67, 68, 69
+ barium bromide vol 11 E86, 87
+ barium chloride vol 11 E84, 85
+ cesium bromide vol 11 E56, 57
+ cesium chloride vol 11 e54, 55
+ cesium iodide vol 11 E58, 59 - 60
+ cesium tetraphenylborate vol 18 70

Formamide		
+ calcium bromide	vol 11	E78, 79
+ calcium chloride	vol 11	E74, 75 - 77
+ lithium bromide	vol 11	E4, 5, 6
+ lithium chloride	vol 11	E1, 2, 3
+ magnesium chloride	vol 11	E72, 73
+ potassium bromide	vol 11	E35, 36 - 38
+ potassium chloride	vol 11	E28, E29, 30 - 34
+ potassium cyanide	vol 11	E46, 47
+ potassium iodide	vol 11	E39, E40, 41 - 45
+ potassium tetraphenylborate	vol 18	37
+ rubidium bromide	vol 11	E52, 53
+ silver azide + sodium azide + nitric acid, sodium salt	vol 3	26
+ silver azide + tetraethylammonium azide + nitric acid, tetraethylammonium salt	vol 3	26
+ silver tetraphenylborate + nitric acid, sodium salt + sodium tetraphenylborate	vol 18	169
+ sodium bromide	vol 11	E14, E15, 16 - 20
+ sodium chloride	vol 11	E9, 10 - 13
+ sodium fluoride	vol 11	E7, 8
+ sodium iodide	vol 11	E21, 22 - 25
+ strontium bromide	vol 11	E82, 83
+ strontium chloride	vol 11	E80, 81 - 83
+ sulfur dioxide	vol 12	263
+ tetraphenylarsonium tetraphenylborate (1-)	vol 18	E209, 210 - 212
+ thiocyanic acid, ammonium salt	vol 11	E70, 71
+ thiocyanic acid, potassium salt	vol 11	E48, 49 - 51
+ thiocyanic acid, silver salt + thiocyanic acid, potassium salt + nitric acid, potassium salt	vol 3	221
+ thiocyanic acid, silver salt + thiocyanic acid, tetraethylammonium salt + nitric acid, tetraethylammonium salt	vol 3	221
+ thiocyanic acid, sodium salt	vol 11	E26, 27
Formamide (aqueous alkali chlorides)		
+ argon	vol 4	E34 - 35
Formamide, N,N-dimethyl-		
+ ammonium bromide	vol 11	E226, 227, 228
+ ammonium chloride	vol 11	E222, 223 - 225
+ ammonium iodide	vol 11	E229, 230
+ ammonium thiocyanate	vol 11	E231, 232 - 234
+ cesium bromide	vol 11	E216, 217 - 219
+ cesium chloride	vol 11	E212, 213 - 215
+ cesium iodide	vol 11	E220, 221
+ calcium bromide	vol 11	E243, 244
+ calcium chloride	vol 11	E240, 241, 242
+ chlorine	vol 12	E354 -E366, 390
+ ethane	vol 9	E195 -E199, 220
+ hydrocyanic acid, potassium salt	vol 11	E206, 207
+ hydrocyanic acid, sodium salt	vol 11	E179, 180

Formamide, N,N-dimethyl-
+ lithium bromide vol 11 E156, 157, 158
+ lithium chloride vol 11 E151, 152 - 155
+ lithium fluoride vol 11 E148, 149, 150
+ magnesium chloride vol 11 E235, 236, 237
+ magnesium iodide vol 11 E238, 239
+ potassium bromide vol 11 E194, 197
+ potassium chloride vol 11 E186, 187 - 193
+ potassium cyanide vol 11 E204, 205
+ potassium iodide vol 11 E198, 199 - 203
+ silver azide + sodium azide + nitric acid, sodium salt vol 3 22
+ silver tetraphenylborate (1-) vol 18 E154
+ silver tetraphenylborate + perchloric acid, sodium salt + sodium tetraphenylborate vol 18 156
+ silver tetraphenylborate + nitric acid, sodium salt + sodium tetraphenylborate vol 18 155
+ silver tetraphenylborate + sodium tetraphenylborate vol 18 157
+ sodium bromide vol 11 E168, 169, 170
+ sodium chloride vol 11 E162, 163 - 167
+ sodium cyanide vol 11 E176 - 177, 178
+ sodium fluoride vol 11 E159, 160, 161
+ sodium iodide vol 11 E171, 172 - 175
+ sulfur dioxide vol 11 E261, E262, 269 - 275, 297
+ tetraphenylarsonium tetraphenylborate (1-) vol 18 E198, 199 - 202
+ thiocyanic acid, potassium salt vol 11 E208, 209 - 211
+ thiocyanic acid, silver salt + thiocyanic acid, potassium salt + nitric acid, potassium salt vol 3 212
+ thiocyanic acid, sodium salt vol 11 E181, 182 - 185
Formamide, N,N-dimethyl- (aqueous)
+ argon vol 4 E83
+ iodic acid, barium salt vol 14 E251, E252, 310
+ iodic acid, calcium salt vol 14 E74, 176
+ iodic acid, strontium salt vol 14 207
+ thiocyanic acid, silver salt + thiocyanic acid, potassium salt + perchloric acid, potassium salt vol 3 184
Formamide N,N-dimethyl-(aqueous)
+ argon vol 4 E83
Formamide, N-methyl-
+ ammonium bromide vol 11 E134, 135 - 137
+ ammonium chloride vol 11 E130, 131 - 133
+ ammonium iodide vol 11 E138, 139
+ calcium bromide vol 11 E146, 147
+ calcium chloride vol 11 E144, 145
+ lithium bromide vol 11 E93, 94, 95
+ lithium chloride vol 11 E88, 89 - 92
+ magnesium chloride vol 11 E142, 143
+ potassium bromide vol 11 E118, 119 - 122
+ potassium chloride vol 11 E113, 114 - 117
+ potassium iodide vol 11 E123, 124 - 127
+ sodium bromide vol 11 E101, 102 - 105
+ sodium chloride vol 11 E96, 97 - 100
+ sodium iodide vol 11 E106, 107 - 110
+ thiocyanic acid, ammonium salt vol 11 E140, 141
+ thiocyanic acid, potassium salt vol 11 E128, 129
+ thiocyanic acid, sodium salt vol 11 E111, 112

Formic acid		
+ nitric oxide	vol 8	E266, 322
+ nitrous oxide	vol 8	E116, 128
+ oxygen	vol 7	453
+ radon-222	vol 2	290
+ sulfur dioxide	vol 12	198, 200
Formic acid (aqueous)		
+ argon	vol 4	E83
Formic acid, ethyl ester		
+ oxygen	vol 7	453
+ sulfur dioxide	vol 12	202
Formic acid, methyl ester		
+ oxygen	vol 7	453
+ sulfur dioxide	vol 12	202
Formic acid, octyl ester		
+ oxygen	vol 7	453
Freon-11		
see Methane, trichlorofluoro-		
Freon-12		
see Methane, dichlorodifluoro-		
Freon-13		
see Methane, chlorotrifluoro-		
Freon-22		
see Methane, chlorodifluoro-		
Freon-113		
see Ethane, 1,1,2-trichloro-1,2,2-trifluoro-		
Freon-114		
see Ethane, 1,2-dichloro-1,1,2,2-tetrafluoro-		
D-Fructose		
+ nitrogen	vol 10	311
Fuel		
+ air	vol 10	540 - 543
Fuel hydrogenated		
see Hydrogenated fuel		
Fuel (hydrocarbon)		
+ air	vol 10	544
+ nitrogen	vol 10	327, 328
+ oxygen	vol 7	442, 445
Fuel (military)		
+ ethane	vol 9	E232, E233, 249 - 251
Fuel, jet		
+ air	vol 10	548, 549
+ nitrogen	vol 10	507, 508
Fuel, RJ-1		
+ air	vol 10	546, 547
+ nitrogen	vol 10	505, 506
2-Furancarboxaldehyde		
+ ethane	vol 9	E195 -E199, 219
Furan, heptafluorotetrahydro(nonafluorobutyl)-		
+ nitrogen	vol 10	247, 248
+ oxygen	vol 7	324, 327, 328, 385
2(3H)-Furanone, dihydro-		
+ oxygen	vol 7	185
2-Furanmethanol		
+ sulfur dioxide	vol 12	233
2-Furanmethanol, tetrahydro-		
+ sulfur dioxide	vol 12	233
Furan, tetrahydro-		
+ hydrogen	vol 5/6	E215, 219
+ nitrogen	vol 10	229
+ oxygen	vol 7	301
+ sulfur dioxide	vol 12	226, 227

Furan, tetrahydro- (aqueous)
+ iodic acid, barium salt vol 14 E251, E252, 306
+ iodic acid, calcium salt vol 14 E74, 173
+ iodic acid, strontium salt vol 14 206
Furfural alcohol
see 2-Furanmethanol
Furfuryl
see 2-Furancarboxaldehyde

Gadolinium cobalt nitrate
 see Nitric acid, cobalt gadolinium salt
Gadolinium nickel nitrate
 see Nitric acid, gadolinium nickel salt
Gadolinium nitrate
 see Nitric acid, gadolinium salt
Gadolinium zinc nitrate
 see Nitric acid, gadolinium zinc salt
Gas oil
 see oil gas
Gasoline
 + krypton vol 2 68
Gasoline, cracked
 + oxygen vol 7 304
Gasoline, olefin poor
 + oxygen vol 7 304
Gelatin (aqueous)
 + argon vol 4 245
 + hydrogen vol 5/6 115, 116
 + nitrogen vol 10 313
 + nitrous oxide vol 8 152 - 155
 + oxygen vol 7 389, 408
Geothermal brine
 see Brine
Germanium ,tetrachloro-
 + nitrogen vol 10 252
α-Globulin, bovine
 see bovine α-globulin
β-Globulin, bovine
 see bovine β-globulin
D-Glucitol (aqueous)
 + hydrogen vol 5/6 319, 320
Gluconic acid (aqueous)
 + oxygen vol 7 411
Glucono lactone
 see D-gluconic acid
α-D-glucopyranose, 1-(dihydrogenphosphate)
 + oxygen vol 7 407
α-D-glucopyranoside, β -D-fructofuranosyl
 + ethane vol 9 E64, 68
 + iodic acid, calcium salt + sodium chloride
 vol 14 103
 + oxygen vol 7 414 - 417, 426 - 428
 + nitrogen vol 10 314
4-O-α-D-glucopyranosyl-D-glucose
 + oxygen vol 7 417
4-O-β-D-glucopyranosyl-D-glucose
 + oxygen vol 7 417
D-glucose (aqueous)
 + hydrogen vol 5/6 109, 110
 + nitrogen vol 10 312
 + oxygen vol 7 397, 412, 413
α-glucose-1-phosphate
 see α-D-glucopyranose, 1-(dihydrogen phosphate)-
DL-Glutamic acid, disodium salt (aqueous)
 + iodic acid, calcium salt vol 4 E76, E77, 121
DL-Glutamic acid, monosodium salt (aqueous)
 + iodic acid, calcium salt vol 4 E76, E77, 122
Glycerin
 see 1,2,3-Propanetriol
Glycerol
 see 1,2,3-Propanetriol
Glycerol hexanoate
 see Hexanoic acid, 1,2,3-propanetriyl ester
Glycerol octanoate
 see Octanoic acid, 1,2,3-propanetriyl ester

Glycerol triacetate		
see 1,2,3-Propanetriol, triacetate		
Glycerol tributurate		
see Butanoic acid, 1,2,3-propanetriyl ester		
Glycine		
+ hydrogen	vol 5/6	96
+ oxygen	vol 7	400
Glycine (aqueous)		
+ bromic acid, barium salt	vol 14	235
+ calcium chloride + iodic acid, calcium salt	vol 14	E77, 158
+ iodic acid, barium salt	vol 14	284
+ iodic acid, barium salt + iodic acid, potassium salt	vol 14	E252, 284
+ iodic acid, barium salt	vol 14	E252, 281
+ iodic acid, calcium salt + iodic acid, potassium salt	vol 14	E77, 141
+ iodic acid, calcium salt	vol 14	E75, 177
+ nitrogen	vol 10	307
+ potassium chloride + iodic acid, calcium salt	vol 14	137
+ sodium hydroxide + iodic acid, barium salt	vol 14	E252, 282
+ potassium chloride + iodic acid, barium salt	vol 14	E252, 283
Glycine, monosodium salt (aqueous)		
+ iodic acid, calcium salt	vol 14	E77, 109, 110
Glycine, N-(aminoiminomethyl)-N-methyl- (aqueous)		
+ oxygen	vol 7	402
Glycine, N,N-dimethyl-, potassium salt (aqueous)		
+ ethane	vol 9	60
Glycine, N,N´-1,2-ethanediylbis(N-carboxymethyl)- (aqueous)		
+ krypton-85	vol 2	125
Glycine, N-glycyl- (aqueous)		
+ iodic acid, calcium salt	vol 14	E75
+ iodic acid, calcium salt + iodic acid, potassium salt	vol 14	E77, 143
+ iodic acid, barium salt	vol 14	288
Glycine, N-glycyl-, sodium salt (aqueous)		
+ iodic acid, calcium salt	vol 14	E76, E77, 120
Glycogen		
+ nitrous oxide	vol 8	254
+ oxygen	vol 7	405
Glycol		
see 1,2-ethanediol		
Glycosamine		
+ oxygen	vol 7	405
Guaiacol		
see phenol,2-methoxy-		
Guanidine monohydrochloride (aqueous)		
+ ethane	vol 9	E29, E30, 41
Guanidine mononitrate		
+ water	vol 13	92
Guanidine mononitrate (aqueous)		
+ nitric acid, lanthanum salt	vol 13	92
+ nitric acid, praseodymium salt	vol 13	259
Guanidine tetraphenylborate		
+ water	vol 18	91
Guanidinium chloride		
see Guanidine monohydrochloride		
Guinea pig adrenals		
+ krypton-85	vol 2	129

Guinea pig blood + krypton-85	vol 12	125, 130
Guinea pig bone marrow + krypton-85	vol 2	129
Guinea pig brain + krypton-85	vol 2	130
Guinea pig brain (in sodium chloride solution) + krypton	vol 2	133
Guinea pig brain (in sodium chloride solution) + krypton-85	vol 2	130
Guinea pig brain (whole) + krypton-85	vol 2	130
Guinea pig heart + krypton-85	vol 2	130
Guinea pig kidneys + krypton-85	vol 2	130
Guinea pig large intestine + krypton-85	vol 2	129
Guinea pig liver + krypton-85	vol 2	129
Guinea pig lungs + krypton-85	vol 2	129
Guinea pig lymph modes + krypton-85	vol 2	129
Guinea pig muscle + krypton-85	vol 2	130
Guinea pig omental fat + krypton-85	vol 2	130
Guinea pig ovaries + krypton-85	vol 2	130
Guinea pig seminal vesicles + krypton-85	vol 2	130
Guinea pig small intestine + krypton-85	vol 2	129
Guinea pig spleen + krypton-85	vol 2	130
Guinea pig stomach + krypton-85	vol 2	130
Guinea pig subcutaneous fat + krypton-85	vol 2	129
Guinea pig testes + krypton-85	vol 2	130
Guinea pig thymus + krypton-85	vol 2	129
Guinea pig thyroids + krypton-85	vol 2	129
Guinea pig uterus + krypton-85	vol 2	130

Harp seal oil		
+ oxygen	vol 7	362, 363
Hamster blood		
see Chinese hamster blood		
Heart		
see Guinea pig heart, Human fetal heart, Rabbit heart		
Heavy naphtha		
+ hydrogen	vol 5/6	522
Heavy water		
see water-D2		
Hemoglobin		
see also Bovine hemoglobin, Dog hemoglobin, Human hemoglobin, Whale hemoglobin		
Hemoglobin		
+ nitrogen	vol 10	287, 288, 290 291, 294, 296
+ oxygen	vol 7	377 - 379
Heptadecane		
+ ethane	vol 9	E77, E78, 98
+ oxygen	vol 7	E214, E215, 239
+ sulfur dioxide	vol 12	E116, E117, 138, 139
Heptadecane (ternary)		
+ sulfur dioxide	vol 12	316, 317
1-Heptadecanol		
+ water	vol 15	426
2,2,4,4,6,8,8-Heptamethylnonane		
see Nonane, 2,2,4,4,6,8,8-Heptamethyl-		
Heptane		
+ air	vol 10	517
+ argon	vol 4	E111, 112 - 113
+ chlorine	vol 12	E354 -E366, 367 - 369
+ deuterium	vol 5/6	E281, 283, 284, E300, E354
+ ethane	vol 9	E77, E78, 85 - 89, E110 -E112, 125, 126
+ helium	vol 1	42
+ hydrogen	vol 5/6	E120, 126, 130 - 133, E354, 364
+ krypton	vol 2	30
+ neon	vol 1	187
+ nitrogen	vol 10	E119, E120, 128 - 131, E425, 426 - 430
+ nitrous oxide	vol 8	E160, 167, 168
+ oxygen	vol 7	E214, E215, 219 - 221, 238, 239, 452
+ sulfur dioxide	vol 12	E116, E117, 122
+ xenon	vol 2	157
Heptane (ternary)		
+ nitrogen	vol 10	355, 356
Heptane, 1-bromo-		
+ air	vol 10	522
+ hydrogen	vol 5/6	E242, 254
+ nitrogen	vol 10	243
+ oxygen	vol 7	317
Heptane, 1-chloro-		
+ sulfur dioxide	vol 12	E1, E2

Heptane, 1,1,1,2,2,3,3-heptafluoro-4,4-bis(trifluoromethyl)-
+ oxygen vol 7 329
Heptane, 1,1,1,2,2,3,3-heptafluoro-6-methyl-4,4-bis(trifluoromethyl)-
+ oxygen vol 7 335
Heptane, hexadecafluoro-
+ argon vol 4 206
+ chlorine vol 12 E354 -E366, 416
+ deuterium vol 5/6 E281, 289
+ ethane vol 9 E138 -E141, 144, 145
+ helium vol 1 90
+ hydrogen vol 5/6 E237, 243, 244
+ nitrogen vol 10 245
+ oxygen vol 7 321
Heptane, 1,1,1,7,7,7-hexafluoro-2,2,6,6-tetrakis(trifluoromethyl)-
+ oxygen vol 7 339
Heptane, 3-methyl-
+ argon vol 4 120
+ helium vol 1 46
+ krypton vol 2 31
+ neon vol 1 191
Heptanoic acid
+ radon-222 vol 2 303
1-Heptanol
+ air vol 10 520
+ argon vol 4 E169 - 170, 195
+ ethane vol 9 144, 145
+ hydrogen vol 5/6 E189, 206, 207
+ nitrogen vol 10 E177, E178 208 - 210
+ nitrous oxide vol 8 203
+ oxygen vol 7 E265, 283, 284, 453
+ sulfur dioxide vol 12 194
+ water vol 15 E329 -E333, 334 - 351
2-Heptanol
+ water vol 15 352, 353
3-Heptanol
+ water vol 15 E354, 355, 356
4-Heptanol
+ water vol 15 357
4-Heptanol, 2,6-dimethyl-
+ water vol 15 E387, 388, 389
4-Heptanol, 3,5-dimethyl-
+ water vol 15 390
2-Heptanol, 2-methyl-
+ water vol 15 362
3-Heptanol, 3-methyl-
+ water vol 15 363
4-Heptanone
+ argon vol 4 E169 - E170
1-Heptene
+ hydrogen vol 5/6 E400, E404
Heptyl alcohol
see 1-heptanol
Herring oil
+ oxygen vol 7 361 - 363
Hexacarbonylbis (tributyl phosphine) dicobalt
+ hydrogen vol 5/6 569, 570
1,1,2,3,4,4-Hexachloro-1,3-butadiene
see 1,3-butadiene,1,1,2,3,4,4-hexachloro-
Hexadecafluoroheptane
see Heptane, hexadecafluoro-

Hexane, 1-chloro-1,1,2,2,3,3,4,4,5,6,6,6-dodecafluoro-5-(trifluoromethyl)-
+ air vol 10 524
+ oxygen vol 7 319
Hexane, 2,3-dimethyl-
+ argon vol 4 121
+ helium vol 1 47
+ krypton vol 2 32
+ neon vol 1 192
Hexane, 2,4-dimethyl-
+ argon vol 4 122
+ helium vol 1 48
+ krypton vol 2 33
+ neon vol 1 193
Hexane, 1,1,1,2,2,3,3,4,4,5,5,6,6-tridecafluoro-6-[1,2,2,2-tetrafluoro-1-(trifluoromethyl)ethoxy]-
+ air vol 10 249
+ oxygen vol 7 333
1,6-Hexanediamine
+ hydrogen vol 5/6 E483, 504
1,6-Hexanediamine dinitrate
+ water vol 13 327, 396, 422
1,6-Hexanediamine dinitrate (aqueous)
+ nitric acid, dysprosium salt vol 13 422
+ nitric acid, gadolinium salt vol 13 396
+ nitric acid, neodymium salt vol 13 327
Hexanoic acid
+ radon-222 vol 2 302
+ sulfur dioxide vol 12 213, 243
Hexanoic acid, sodium salt (aqueous)
+ oxygen vol 7 451
Hexanoic acid, 1,2,3-propanetriyl ester
+ radon-222 vol 2 316
1-Hexanol
+ argon vol 4 E169 - 170, 194, E253 - 255, 289
+ ethane vol 9 E166, E167, 185, 186
+ hydrogen vol 5/6 205, E442, 469
+ nitrogen vol 10 E177, E178, 205, 207
+ nitrous oxide vol 8 200
+ oxygen vol 7 E265, 282
+ sulfur dioxide vol 12 193
+ water vol 15 E263 -E268, 269 - 290
2-Hexanol
+ water vol 15 E291, 292 - 294
3-Hexanol
+ water vol 15 E295, 296 - 298
1-Hexanol, 2-ethyl-
+ hydrogen vol 5/6 E442, 470
+ water vol 15 E359, 360, 361
2-Hexanol, 2-methyl-
+ water vol 15 E321, 322, 323
3-Hexanol, 2-methyl-
+ water vol 15 324
3-Hexanol, 3-methyl-
+ water vol 15 E325, 326, 327
2-Hexanol, 5-methyl-
+ water vol 15 328
2-Hexanone
+ oxygen vol 7 296
3-Hexanone
+ argon vol 4 E169 - 170, E253 - 255, 302

1-Hexene			
+ hydrogen	vol 5/6	E400,	403
1-Hexen-3-ol			
+ water	vol 15		226
4-Hexen-3-ol			
+ water	vol 15		227
Hippuric acid sodium salt (aqueous)			
+ iodic acid, calcium salt	vol 14	E76, E77,	130
Histamine bis-tetraphenylborate			
+ water	vol 18		92
Holmium nitrate			
see Nitric acid, holmium salt			
Horse heart cyanometmyoglobin, metmyglobin and myoglobin (aqueous)			
+ xenon	vol 2		220
Human albumen (in sodium chloride aqueous solution)			
+ krypton	vol 2		126
+ xenon	vol 2	208, 216,	217
Human blood			
see Blood, human			
Human blood plasma			
see Blood, human, plasma			
Human brain (homogenized)			
+ nitrous oxide	vol 8	E226, E227,	242
Human brain grey matter			
+ xenon-133	vol 2	225,	226
Human brain white matter			
+ xenon-133	vol 2	225,	226
Human erythrocytes			
+ xenon-133	vol 2		215
Human fat			
+ krypton	vol 2	E114, E115, 116,	117
+ neon	vol 1		252
+ nitrogen	vol 10		281
+ radon-222	vol 2		335
+ xenon	vol 2	E199,	202
Human fetal brain			
+ nitrous oxide	vol 8	E226, E227,	244
Human fetal heart			
+ nitrous oxide	vol 8	E226, E227,	244
Human fetal liver			
+ nitrous oxide	vol 8	E226, E227,	244
Human fetal lung			
+ nitrous oxide	vol 8	E226, E227,	244
Human fetal scalp			
+ nitrous oxide	vol 8	E226, E227,	244
Human fetal skeletal muscle			
+ nitrous oxide	vol 8	E226, E227,	244
Human fetal skin			
+ nitrous oxide	vol 8	E226, E227,	244
Human fetal spleen			
+ nitrous oxide	vol 8	E226, E227,	244
Human hemaglobin (aqueous)			
+ krypton-85	vol 2		124
+ krypton	vol 2		123
+ xenon	vol 2	E199 - 200, 208, 209,	211
Human hyperlipidemic plasma			
+ nitrous oxide	vol 8	E226, E227,	241
Human liver tissue			
+ krypton-85	vol 2		131
+ xenon-133	vol 2		223

Human lung homogenate (blood free tissue)			
+ nitrous oxide	vol 8	E226, E227,	243
Human lung homogenate			
+ helium	vol 1		120
Human methemoglobin (aqueous)			
+ xenon	vol 2	E199 - 200,	208
Human methemoglobin (saline)			
+ xenon-133	vol 2	E199 - 200,	219
Human myocardium			
+ nitrous oxide	vol 8	E226, E227,	245
Human placenta			
+ nitrous oxide	vol 8	E226, E227,	244
Human plasma			
+ nitrous oxide	vol 8	E226, E227,	241
Human pulmonary tissue			
+ oxygen	vol 7		383
Human red blood cells			
+ krypton-85	vol 2		121
+ nitrous oxide	vol 8	E226,	E227,
		228,	241
+ xenon-133	vol 2	213, 214,	218
Human red blood ghosts			
+ hydrogen	vol 5/6		118
Human red blood membrane			
+ nitrous oxide	vol 8	E226, E227,	241
Human serum			
+ oxygen	vol 7	380, 381,	383
Human umbilical cord			
+ nitrous oxide	vol 8	E226, E227,	244
Human uterine muscle			
+ nitrous oxide	vol 8	E226, E227,	244

Hydrabamine pencillin G
see 4-Thia-1-azabicyclo[3,2,0]heptane-2-carboxylic acid, 3,3-dimethyl-7-oxo-6-[(phenylacetyl)amino]- complexed with N,N´-bis[(1,2,3,4,4a,9,10,10a-octahydro-1,4a-dimethyl-7-(1-methylethyl)-1-phenanthrenyl)methyl]-1,2-ethanediamine (1:2)

Hydrabamine penicillin V
see 4-Thia-1-azabicyclo[3,2,0]heptane-2-carboxylic acid, 3,3-dimethyl-7-oxo-6-[(phenoxyacetyl)amino]- complexed with N,N´-bis[(1,2,3,4,4a,9,10,10a-octahydro-1,4a-dimethyl-7-(1-methylethyl)-1-phenanthrenyl)methyl]-1,2-ethanediamine (1:2)

Hydrated aluminium oxide
see Aluminium oxide (aqueous)

Hydrated ferric oxide
see Ferric oxide, hydrated

Hydrated iron oxide
see Iron oxide (aqueous)

Hydrazine			
+ argon	vol 4		232
+ helium	vol 1		113
+ nitrogen	vol 10	270,	273
Hydrazine (1,1-dimethylhydrazine solution)			
+ argon	vol 4		236
Hydrazine 1,1-dimethyl-			
+ argon	vol 4		234
+ helium	vol 1		101
+ nitrogen	vol 10	272,	273
Hydrazine 1,2-dimethyl-			
+ argon	vol 4		235
+ helium	vol 1		102
+ nitrogen	vol 10		274

Hydrazine dinitrate		
+ water	vol 13	88, 89
Hydrazine dinitrate (aqueous)		
+ nitric acid, lanthanum salt	vol 13	88, 89
+ nitric acid, praseodymium salt	vol 13	257, 258
Hydrazine, methyl-		
+ argon	vol 4	233
+ helium	vol 1	100
+ nitrogen	vol 10	271
Hydrazine mononitrate		
+ water	vol 13	86, 87, 201, 202, 320 - 321
Hydrazine mononitrate (aqueous)		
+ nitric acid, cerium salt	vol 13	201, 202
+ nitric acid, lanthanum salt	vol 13	86, 87
+ nitric acid, neodymium salt	vol 13	320, 321
Hydrazine,tetrafluoro-		
+ nitrogen	vol 10	487
Hydrobromic acid (ternary)		
+ nitric oxide	vol 8	292
Hydrocarbon blend		
see Gas oil, Heavy naphtha		
Hydrocarbon oil		
+ ethane	vol 9	E232, E233, 236, 237
Hydrochloric acid		
+ chlorine	vol 12	107, E344, 345 - 348, E349
+ helium	vol 1	E20, 23
+ hydrogen	vol 5/6	E27, 36
+ krypton	vol 2	E12, E13, 16
+ neon	vol 1	E141, 144
+ nitrous oxide	vol 8	E28, E40, 41
+ oxygen	vol 7	E59, 77
+ sulfur dioxide	vol 12	40, 41
Hydrochloric acid (aqueous)		
+ iodic acid, barium salt	vol 14	E246, 295
+ iodic acid, calcium salt + sodium chloride	vol 14	98
Hydrochloric acid (multicomponent)		
+ nitric oxide	vol 8	287, 290, 291,
+ sulfur dioxide	vol 12	107
Hydrocortisone sodium succinate		
see pregn-4-ene-3,20-dione,21(3-carboxy-1-oxopropoxy) -11,17-dihydroxy-monosodium salt, (11-beta)-		
Hydrocyanic acid		
see Hydrogen cyanide		
Hydrogen		
+ helium-3	vol 1	E307, 308
+ helium-4	vol 1	E310, 311, 334, 335
p-Hydrogen		
+ helium	vol 1	322
Hydrogen cyanide		
+ silver cyanide + nitric acid + sulfuric acid	vol 3	E86
Hydrogen cyanide (aqueous)		
+ silver cyanide	vol 3	E46 - 47, E53 - 54, 70 - 71

Hydrogen fluoride
+ chlorine vol 12 448
+ oxygen vol 7 430
+ silver cyanide vol 3 87
+ sulfur dioxide vol 12 E1, E2
Hydrogen sulfide
+ ethane vol 9 E232, E233, 244
+ hydrogen vol 5/6 619, 620
+ hydrogen deuteride vol 5/6 299
+ nitrogen vol 10 468 - 470
Hydrogen sulfide (multicomponent)
+ hydrogen vol 5/6 565, 566
+ nitrogen vol 10 495
Hydrogen sulfide (ternary)
+ hydrogen vol 5/6 562 - 564, 567, 568
Hydrogenated cotton seed oil
see oil, hydrogenated cotton seed
Hydrogenated fuel
+ helium vol 1 114
Hydroquinone, dimethyl ether
see benzene, 1,4-dimethoxy-
Hydroterpin
see Hydroterpinol
Hydroterpinol
+ sulfur dioxide vol 12 171
Hydroxide
see ammonium hydroxide, potassium hydroxide, sodium hydroxide
2-Hydroxy-N,N,N-tris(2-Hydroxyethyl)-ethanaminium bromide)
see Ethanamine, 2-hydroxy-, N,N,N-tris (2-hydroxyethyl)-, bromide
2-Hydroxy ethyl ether
see Ethanol, 2,2´-oxybis-
4-Hydroxyproline
see proline, 4-hydroxy-

Inuline
see aconitane-7,8-diol,4-[[(2-aminobenzoyl)oxy]-methyl]-20-ethyl-1,6,14,16-tetramethoxy-, (1b,6b,14a,16b)-
Iodic acid
+ water vol 14 204, 205
Iodic acid (aqueous)
+ iodic acid, barium salt vol 14 E246, 296
+ iodic acid, magnesium salt vol 14 E17, 21 - 24
+ iodic acid, strontium salt vol 14 204, 205
Iodic acid, aluminium salt
+ water vol 14 42, 43
Iodic acid, aluminium salt (aqueous)
+ iodic acid, magnesium salt vol 14 42, 43
Iodic acid, ammonium salt
+ water vol 14 25, 26
Iodic acid, ammonium salt (aqueous)
+ iodic acid, magnesium salt vol 14 E17, 25, 26
Iodic acid, barium salt
+ water vol 14 E242 -E255, 256 - 272, 284
Iodic acid, barium salt (aqueous)
+ acetic acid, bromo-, sodium salt vol 14 263
+ acetic acid, ethyl ester vol 14 E251, 252, 309
+ alanine vol 14 E252, E253, 285, 287
+ alanine + iodic acid, potassium salt vol 14 E252, E253, 287
+ alanine + sodium hydroxide vol 14 E252, E253, 286
+ ammonia vol 14 E247, 293
+ ammonium hydroxide vol 14 E247, 294
+ barium bromide vol 14 291
+ barium chloride vol 14 289, 290
+ bromic acid, potassium salt vol 14 269
+ calcium chloride vol 14 277
+ 1,4-dioxane vol 14 E251, E252, 307, 308
+ 1,2-ethanediol vol 14 E251, E252, 303
+ chloric acid, potassium salt vol 14 E253, 267
+ ethanol vol 14 E251, E252, 299, 300
+ ethanol + nitric acid, sodium salt + sodium hydroxide vol 14 301
+ formamide, N,N-dimethyl- vol 14 E251, E252, 310
+ foran, tetrahydro- vol 14 E251, E252, 306
+ glycine vol 14 E252, 281, 284
+ glycine + iodic acid, potassium salt vol 14 E252, 284
+ glycine + potassium chloride vol 14 E252, 283
+ glycine + sodium hydroxide vol 14 E252, 282
+ glycine, N-glycyl vol 14 288
+ hydrochloric acid vol 14 E246, 295
+ iodic acid vol 14 E246, 296
+ iodic acid, magnesium salt vol 14 E17, 41
+ iodic acid, potassium salt vol 14 270
+ magnesium chloride vol 14 276
+ methanol vol 14 E251, E252, 298
+ nitric acid vol 14 E246, 297
+ nitric acid, barium salt vol 14 E250, E251, 279, 280, 292
+ nitric acid, calcium salt vol 14 278
+ nitric acid, potassium salt vol 14 E248 -E250, 271 - 275

Iodic acid, barium salt (aqueous)
+ perchloric acid, lithium salt + nitraic acid, lithium salt vol 14 261
+ perchloric acid, potassium salt vol 14 E253, 268
+ potassium chloride vol 14 264 - 266
+ 1,2,3-propanetriol vol 14 E251, E252, 304
+ 1-propanol vol 14 E251, E252, 302
+ 2-propanone vol 14 E251, E252, 305
+ thiosulfuric acid, sodium salt vol 14 262
+ urea vol 14 311
Iodic acid, calcium salt
+ water vol 14 40, E59 - E82, 83 - 178
Iodic acid, calcium salt (aqueous)
+ acetic acid, cyano-, sodium salt vol 14 E75, E76, 114
+ acetic acid, ethyl ester vol 14 E74, 174
+ acetic acid, hydroxy-, monosodium salt vol 14 E75, 76, 106 - 108
+ acetic acid, methoxy-, sodium salt vol 14 E75, E76, 113
+ alanine vol 14 E75, 142, 178
+ alanyl glycine, monosodium salt vol 14 E76, E77, 123
+ DL-alanine, monosodium salt (aqueous) vol 14 E76, E77, 115
+ ammonia vol 14 E69, 87, 88
+ ammonium chloride vol 14 E70, E73, 89, 90
+ benzoic acid, 2-hydroxy-, monosodium salt vol 14 E75, E76, 124
+ butanedioic acid, hydroxy-, monosodium salt vol 14 E75, E76, 117, 118
+ butanoic acid, 3-hydroxy-, monosodium salt vol 14 E75, E76, 119
+ calcium chloride vol 14 E70, 73, 155 - 157
+ calcium hydroxide vol 14 E69, 102, 154
+ chloric acid, lithium salt + nitric acid, lithium salt vol 14 93
+ 1,4-dioxane vol 14 E74, 170, 171
+ 1,2-ethanediol vol 14 E74, 168
+ ethanol vol 14 E74, 166
+ ferrate (4-) hexakis (cyano-O)-, tetrapotassium vol 14 E70, E74, 147
+ formamide, N,N-dimethyl- vol 14 E74, 174
+ β-D-fructofuranosyl-α-D-glucopyranoside + calcium hydroxide vol 14 159
+ furan, tetrahydro- vol 14 E74, 173
+ DL-glutamic acid, disodium salt vol 14 E76, E77, 121
+ DL-glutamic acid, monosodium salt vol 14 E76, E77, 122
+ glycine vol 14 E75, 141, 177
+ glycine, N-glycyl- vol 14 E75
+ glycine, N-glycyl- + iodic acid, potassium salt vol 14 E77, 143
+ glycine, N-glycyl-, sodium salt vol 14 E76, E77, 120, 143
+ glycine + calcium chloride vol 14 E77, 158
+ glycine + iodic acid, potassium salt vol 14 E77, 141
+ glycine, monosodium salt vol 14 E77, 109, 110
+ hippuric acid, sodium salt vol 14 E76, E77, 130
+ hydrochloric acid vol 14 E68, 98

Iodic acid, calcium salt (aqueous)
+ hydrochloric acid + sodium chloride vol 14 98
+ iodic acid, lithium salt vol 14 E69, 161 - 163
+ iodic acid, magnesium salt vol 14 E69, 151
+ iodic acid, potassium salt vol 14 140
+ iodic acid, potassium + alanine vol 14 E77, 142
+ iodic acid, sodium salt vol 14 E69, 164
+ leucyl glycine, monosodium salt vol 14 E76, E77, 128
+ lithium chloride vol 14 E70 - E72, 91, 92
+ magnesium chloride vol 14 E70, E73, 148 - 150
+ mandelic acid, monosodium salt vol 14 E75, E76, 125 - 127
+ methanol vol 14 165
+ nitric acid, potassium salt vol 14 E70, E74, 146
+ potassium bromide vol 14 E70, E74, 138
+ potassium chloride vol 14 E70, E72, 133 - 136
+ potassium chloride + glycine vol 14 137
+ potassium hydroxide vol 14 E68, E69, 132
+ potassium iodide vol 14 E70, E74, 139
+ 1,2,3-propanetriol vol 14 E74, 169
+ propanoic acid, 2-hydroxy-, calcium salt vol 14 160
+ propanoic acid, 2-hydroxy-, monosodium salt vol 14 E75, E76, 112
+ propanoic acid, 2-oxo-, sodium salt vol 14 E75, E76, 111
+ 1-propanol vol 14 E74, 167
+ 2-propanone vol 14 E74, 172
+ serine, monosodium salt vol 14 E76, E77, 116
+ sodium chloride vol 14 E70, E72, 94 - 99
+ sodium chloride + calcium hydroxide vol 14 102
+ sodium chloride + potassium chloride vol 14 100, 101
+ sodium chloride + sodium hydroxide vol 14 99
+ sodium chloride + sucrose vol 14 103
+ sodium hydroxide vol 14 E68, 99
+ sulfuric acid, magnesium salt vol 14 E70, E74, 152, 153
+ sulfuric acid, potassium salt vol 14 E70, 144, 145
+ sulfuric acid, sodium salt vol 14 E70, E74, 102
+ thiosulfuric acid, disodium salt vol 14 E70, E74, 105
+ tyrosine, 3,5-diiodo-, monosodium salt vol 14 E76, E77, 129
+ tyrosine, monosodium salt vol 14 E76, E77, 131
+ urea vol 14 175

Iodic acid, cesium salt (aqueous)
+ iodic acid, magnesium salt vol 14 E17

Iodic acid, lithium salt
+ water 27 - 30, 161 - 163

Iodic acid, lithium salt (aqueous)
+ iodic acid, calcium salt vol 14 E69, 161 - 163
+ iodic acid, magnesium salt vol 14 E17, 27 - 30

Iodic acid, magnesium salt
+ water vol 14 E11 - E18, 19 - 43

Iodic acid, magnesium salt (aqueous)
+ iodic acid vol 14 E17, 21 - 24
+ iodic acid, aluminium salt vol 14 42, 43
+ iodic acid, ammonium salt vol 14 E17, 25, 26
+ iodic acid, barium salt vol 14 E17, 41
+ iodic acid, calcium salt vol 14 E17, 40, E69, 151
+ iodic acid, cesium salt vol 14 E17
+ iodic acid, lithium salt vol 14 E17, 27 - 30
+ iodic acid, potassium salt vol 14 E17, 33, 34
+ iodic acid, rubidium salt vol 14 E17, 35, 36
+ iodic acid, sodium salt vol 14 E17, 31, 32
+ nitric acid, magnesium salt vol 14 E17, 37 - 39
Iodic acid, potassium salt
+ water vol 14 33, 34, 51, 52
Iodic acid, potassium salt (aqueous)
+ alanine + iodic acid, barium salt vol 14 E252, E253, 287
+ glycine + iodic acid, barium salt vol 14 E252, 284
+ iodic acid, barium salt vol 14 270
+ iodic acid, calcium salt vol 14 140
+ iodic acid, calcium salt + alanine vol 14 E77, 142
+ iodic acid, calcium salt + glycine vol 14 E77, 141
+ iodic acid, calcium salt + glycine, N-glycyl- vol 14 E77, 143
+ iodic acid, magnesium salt vol 14 E17, 33, 34
+ iodic acid, strontium salt + perchloric acid vol 14 203
Iodic acid, rubidium salt
+ water vol 14 35, 36
Iodic acid, rubidium salt (aqueous)
+ iodic acid, magnesium salt vol 14 E17, 35, 36
Iodic acid, sodium salt
+ water vol 14 31, 32, 164, 216
Iodic acid, sodium salt (aqueous)
+ iodic acid, calcium salt vol 14 E69, 164
+ iodic acid, magnesium salt vol 14 E17, 31, 32
Iodic acid, strontium salt
+ water vol 14 E191 -E197, 198 - 205
Iodic acid, strontium salt (aqueous)
+ furan, tetrahydro- vol 14 206
+ iodic acid vol 14 204, 205
+ iodic acid, potassium salt + perchloric acid vol 14 203
+ perchloric acid vol 14 203
+ perchloric acid, lithium salt + nitric acid, lithium salt vol 14 202
+ formamide, N,N-dimethyl- vol 14 207
+ sodium chloride vol 14 E195, 199, 200
+ sodium hydroxide vol 14 E195, 201
Iodide
see ammonium iodide, barium iodide, calcium iodide, cesium iodide,lithium iodide, potassium ,iodide, sodium iodide, magnesium, iodide, methanaminium,N,N,N-trimethyl iodide
Iodobenzene
see benzene,iodo-
Iron bromide (aqueous)
+ nitric oxide vol 8 E265, E266, 274
Iron bromide (in aqueous hydrobromic acid)
+ nitric oxide vol 8 E265, E266, 292

Iron bromide (in ethanol)		
+ nitric oxide	vol 8	E265, E266, 306
Iron chloride (aqueous)		
+ neon	vol 1	E141, 146
+ nitric oxide	vol 8	E265, E266, 268, 269, 271, 280, 283, 285, 304
Iron chloride (in aqueous hydrochloric acid)		
+ nitric oxide	vol 8	E265, E266, 287, 291
Iron chloride (in organic liquids)		
+ nitric oxide	vol 8	E265, E266, 307, 309, 310, 312, 313
Iron hydroxide (aqueous)		
+ nitrous oxide	vol 8	145 - 147
Iron iodide (aqueous)		
+ argon	vol 4	251
+ hydrogen	vol 5/6	E29, 51
+ nitric oxide	vol 8	E226, E265, 275
+ nitrogen	vol 10	E45 - E47, 61
+ oxygen	vol 7	E64, 105, 389
Iron iodide (in ethanol)		
+ nitric oxide	vol 8	E265, E266, 308
Iron oxide (aqueous)		
+ argon	vol 4	78
Iron sulfate		
see sulfuric acid, iron (2+) salt, sulfuric acid, iron (3+) salt		
3-Isoxazolidinone,4-amino-		
+ acetic acid, ethyl ester	vol 16/17	E740, 752
+ benzene	vol 16/17	E740, 747
+ benzenemethanol	vol 16/17	E740, 764
+ benzene, methyl-	vol 16/17	E740, 748
+ 1-butanol,3-methyl-	vol 16/17	E740, 745
+ 1-butanol,3-methyl- acetate	vol 16/17	E740, 753
+ 2-butanone	vol 16/17	E740, 755
+ carbon disulfide	vol 16/17	E740, 760
+ cyclohexane	vol 16/17	E740, 746
+ 1,4-dioxane	vol 16/17	E740, 758
+ ethane,dichloro-	vol 16/17	E740, 757
+ ethane, 1,1´-oxybis-	vol 16/17	E740, 756
+ ethanol	vol 16/17	E740, 743, 765
+ ethanol (quaternary)	vol 16/17	E740, 766
+ ethanol, 2-methoxy-	vol 16/17	E740, 763
+ formamide	vol 16/17	E740, 762
+ methane, tetrachloride	vol 16/17	E740, 751
+ methane, trichloro-	vol 16/17	E740, 759
+ methanol	vol 16/17	E740, 742
+ pentane, 2,2,4-trimethyl-	vol 16/17	E749, 750
+ petroleum ether	vol 16/17	E740, 749
+ 2-propanol	vol 16/17	E740, 744
+ 2-propanol (quaternary)	vol 16/17	E740, 766
+ 2-propanone	vol 16/17	E740, 754
+ pyridine	vol 16/17	E740, 761
+ water	vol 16/17	E740, 741, 766

Jojoba oil			
+ hydrogen	vol 5/6		527
Jet fuel			
see fuel ,jet			
JP-1, JP-4			
see fuel			
Kaiserol			
+ radon-222	vol 2		325
Kerosene			
+ air	vol 10	541,	542
+ oxygen	vol 7		445
+ sulfur dioxide	vol 12		172
Kidney			
see Rabbit kidney			
Koppers emulsion K-900			
+ xenon	vol 2		191
Krebs solution			
+ oxygen	vol 7	373,	385
Krypton			
+ argon	vol 4		240
+ helium	vol 1	E254,	323
+ neon	vol 1	E369,	370

Lactose
see 4-O-β-D-glucopyranosyl-D-glucose
Lanolin
+ radon-222 vol 2 331
Lanthanum acetate
see Acetic acid, lanthanum salt
Lanthanum ammonium nitrate
see Nitric acid, lanthanum ammonium salt
Lanthanum chloride (aqueous)
+ hydrogen vol 5/6 E30, 55
+ nitric acid, lanthanum salt vol 13 69
Lanthanum cobalt nitrate
see Nitric acid, cobalt lanthanum salt
Lanthanum manganese nitrate
see Nitric acid, lanthanum manganese salt
Lanthanum nickel nitrate
see Nitric acid, lanthanum nickel salt
Lanthanum nitrate
see Nitric acid, lanthanum salt
Lard
+ air vol 10 533, 534
+ hydrogen vol 5/6 274
+ nitrogen vol 10 276, 277, 280
+ oxygen vol 7 366 - 368
+ radon-222 vol 2 331
Large intestine
see Guinea pig large intestine
Lauric acid
see Dodecanoic acid
Lead nitrate
see Nitric acid, lead salt
Lecithin
+ krypton-85 vol 2 120
+ nitrogen vol 10 301
+ radon-222 vol 2 338
+ xenon-133 vol 2 206
Lecithin (benzene solution)
+ argon vol 4 250
+ oxygen vol 7 395
Leucyl glycine, monosodium salt (aqueous)
+ iodic acid, calcium salt vol 14 E76, E77, 128
Levulose
see D-fructose
Linoleic acid
see (ZZ)-9-12-Octadecadienoic acid
Lipoidal suspension
+ nitrogen vol 10 303
Lithium aluminium chloride (in 4-methyl-1,3-dioxolan-2-one solution)
+ oxygen vol 7 184
Lithium bromide
+ acetamide,N,N-dimethyl- vol 11 E313, 314
+ formamide vol 11 E4, 5, 6
+ formamide,N,N-dimethyl- vol 11 E156, 157, 158
+ formamide,N-methyl- vol 11 E93, 94, 95
Lithium bromide (aqueous)
+ oxygen vol 7 E66, E67, 134
Lithium carbonate
see Carbonic acid, lithium salt
Lithium chloride
+ acetamide, N-methyl- vol 11 E255, 256, 257
+ acetamide, N,N-dimethyl- vol 11 E309, E310, 311, 312
+ butyltrisopentylammonium tetraphenylborate + ethanol vol 18 87

Lithium chloride			
+ cesium tetraphenylborate + ethanol			
	vol 18		69
+ formamide	vol 11	E1,	2, 3
+ formamide, N-methyl-	vol 11	E88,	89 - 92
+ formamide, N,N-dimethyl-	vol 11	E151,	152 - 155
+ potassium tetraphenylborate + ethanol			
	vol 18		25, 26
+ rubidium tetraphenylborate + ethanol			
	vol 18		51
Lithium chloride (aqueous)			
+ argon	vol 4		E33 - 36,
		44 -	48, 69
+ butyltrisopentylammonium tetraphenylborate + ethanol			
	vol 18		85
+ cesium tetraphenylborate + sodium hydroxide + methanol			
	vol 18		61, 62
+ chlorine	vol 12		E349
+ ethane	vol 9		E30, E31,
			44, 45
+ helium	vol 1		E20, 29
+ hydrogen	vol 5/6	44,	61 - 63
+ iodic acid, calcium salt	vol 14		E70 - E72,
			91, 92
+ krypton	vol 2		
+ neon	vol 1	E141,	161, 173,
			175, 178
+ nitric oxide	vol 8		333
+ nitrogen	vol 10		E45 - E47,
			E49, 64
+ nitrous oxide	vol 8		E32, E39,
			76 - 78
+ oxygen	vol 7	E66 -	E67, 129,
			130, 133
+ potassium tetraphenylborate + sodium hydroxide + methanol			
	vol 18		27 - 29
Lithium chloride (aqueous, N,N-dimethyl formamdie)			
+ argon	vol 4		E34 - 36
Lithium chloride (in 4-methyl-1,3-dioxolan-2-one solution)			
+ oxygen	vol 7		183
Lithium fluoride			
+ formamide, N,N-dimethyl-	vol 11	E148,	149, 150
Lithium fluoride (aqueous)			
+ oxygen	vol 7	E66,	E67, 132
Lithium hydroxide (aqueous)			
+ oxygen	vol 7	E66,	E67, 131
Lithium iodate			
see Iodic acid, lithium salt			
Lithium iodide			
+ acetamide, N,N-dimethyl-	vol 11		E315, 316
Lithium iodide (aqueous)			
+ helium	vol 1		E20, 29
+ neon	vol 1	E141,	176, 178
+ oxygen	vol 7	E66,	E67, 135
Lithium nitrate			
see Nitric acid, lithium salt			
Lithium perchlorate			
see Perchloric acid, lithium salt			
Lithium sulfate			
see Sulfuric acid, lithium salt			
Lithium tetrachloro aluminate (1-) (in thionyl chloride solution)			
+ sulfur dioxide	vol 12		330, 331

Lithium tetraphenylborate			
+ 2-propanone	vol 18		2
+ water	vol 18		1
Lithium thiocyanate			
see Thiocyanic acid, lithium salt			
Liver			
see Human fetal liver, Rabbit liver,Guinea pig liver			
Lung			
see Guinea pig lung, human fetal lung, human lung homogenate			
Lung homogenate, human			
see Human lung homogenate			
Lung tissue (dog)			
+ ethane	vol 9	E195 -E199,	231
Lutetium nitrate			
see Nitric acid, lutetium salt			
Lymph nodes			
see Guinea pig lymph nodes			
Lysine			
+ oxygen	vol 7		407
L-1822 (mixture of mostly 10-carbon fluorocarbons)			
+ air	vol 10		529
+ nitrogen	vol 10		251

Mackerel oil			
+ oxygen	vol 7		362, 363
Magnesium bisulfite			
see Sulfurous acid, magnesium salt			
Magnesium bromate			
see Bromic acid, magnesium salt			
Magnesium chlorate			
see Chloric acid, magnesium salt			
Magnesium chloride (aqueous)			
+ argon	vol 4		E33, 43
+ iodic acid, barium salt	vol 14		276
+ iodic acid, calcium salt	vol 14		E70, E73, 148 - 150
+ neon	vol 1		E141, 147
+ nitrous oxide	vol 8		E30, 64
+ oxygen	vol 7	E65,	119 - 121
Magnesium chloride			
+ acetamide, N,N-dimethyl-	vol 11		E333, 334
+ formamide	vol 11		E72, 73
+ formamide, N,N-dimethyl-	vol 11	E235,	236, 237
+ formamide, N-methyl-	vol 11		E142, 143
Magnesium iodate			
see Iodic acid, magnesium salt			
Magnesium iodide			
+ acetamide, N,N-dimethyl-	vol 11		E335, 336
+ formamide, N,N-dimethyl-	vol 11		E238, 239
Magnesium lanthanum nitrate			
see Nitric acid, magnesium lanthanum salt			
Magnesium nitrate			
see Nitric acid, magnesium salt			
Magnesium sulfate			
see Sulfuric acid, magnesium salt			
Maltose			
see 4-O-α-D-glucopyranoxyl-D-glucose			
Mandelic acid, monosodium salt (aqueous)			
+ iodic acid, calcium salt	vol 14		E75, E76, 125 - 127
Manganese acetate			
see acetic acid, manganese salt			
Manganese chloride (aqueous)			
+ nitric oxide	vol 8	E266,	330, 331
+ oxygen	vol 7		E64, 106
Manganese nitrate			
see Nitric acid, manganese salt			
Manganese sulfate			
see Sulfate acid, manganese salt			
Manganous sulfate			
see Sulfuric acid, manganese (2+) salt (1:1)			
D-Mannitol (aqueous)			
+ hydrogen	vol 5/6		314
Mercury acetate			
see acetic acid, mercury salt			
Mercury chloride (aqueous)			
+ radon-222	vol 2	E238 - 241,	247
Mesitylene			
see Benzene, 1,3,5-trimethyl-			
Methanamine			
+ hydrogen	vol 5/6	E481,	484, 485
Methanamine, N,N-dimethyl- (aqueous)			
+ nitric acid, neodymium salt	vol 13		322
+ nitric acid, neodymium salt + nitric acid, tetramethylammonium salt	vol 13		323

Methanamine, N-methyl-		
+ sulfur dioxide	vol 12	297
+ water	vol 13	207
Methanamine, N-methyl- (aqueous)		
+ nitric acid, cerium salt	vol 13	207
Methanamine, N-methyl-, nitrate		
+ water	vol 13	90, 91
Methanamine, N-methyl-, nitrate (aqueous)		
+ nitric acid, lanthanum salt	vol 13	90, 91
Methanamine, N-methyl-N-nitroso-		
+ oxygen	vol 7	350
Methanamine nitrate		
+ water	vol 13	205, 206
Methanamine nitrate (aqueous)		
+ nitric acid, cerium salt	vol 13	205, 206
Methanaminium, N,N,N-trimethyl-, bromide (aqueous)		
+ argon	vol 4	E33 - 34, 39 - 40
+ ethane	vol 9	E29, E30, 37
Methanaminium, N,N,N-trimethyl-, hydroxide (aqueous)		
+ hydrogen	vol 5/6	100, 101
Methanaminium, N,N,N-trimethyl-, iodide (aqueous)		
+ helium	vol 1	E20, 25
+ krypton	vol 2	E12, E13, 15
Methanaminium, N,N,N-trimethyl-, nitrate (aqueous)		
+ nitric acid, neodymium salt	vol 13	323
Methane		
+ argon	vol 4	E261, 262 - 268
+ argon + nitrogen	vol 4	312 - 313
+ helium	vol 1	E263, 264
+ hydrogen	vol 5/6	E321, 322 - 331
+ hydrogen deuteride	vol 5/6	E299
+ neon	vol 1	E255, 357
+ nitrogen	vol 10	E361, E362, 363 - 375
Methane (multicomponent)		
+ hydrogen	vol 5/6	552, 553, 565
Methane (ternary)		
+ hydrogen	vol 5/6	530 - 542, 549 - 551, 556, 557
+ nitrogen	vol 10	376 - 386, 482, 482, 498 - 501
Methane, chloro-		
+ sulfur dioxide	vol 12	E1, E2
Methane, chlorodifluoro-		
+ argon	vol 4	E253 - 255, 308
+ hydrogen	vol 5/6	515, 516
+ ozone	vol 7	489
+ nitrogen	vol 10	455, 456
Methane, chlorotrifluoro-		
+ ozone	vol 7	489
+ nitrogen	vol 10	234, 460, 461
Methane, dichloro-		
+ chlorine	vol 12	E354 -E366, 393
+ oxygen	vol 7	452
+ sulfur dioxide	vol 12	E1, E2
Methane, dichlorodifluoro-		
+ helium	vol 1	E254, 280
+ krypton	vol 2	87
+ ozone	vol 7	489
+ nitrogen	vol 10	235, 457
+ xenon	vol 2	173

Methane, dimethoxy-
+ oxygen vol 7 305
+ sulfur dioxide vol 12 234
Methane, nitro-
+ argon vol 4 227
+ helium vol 1 104
+ krypton vol 2 103
+ neon vol 1 245
+ nitrogen vol 10 260
+ silver cyanide + tetraethylammonium cyanide + perchloric acid, tetraethylammonium salt vol 3 85
+ silver tetraphenylborate + perchloric acid, sodium salt + sodium tetraphenylborate vol 18 171
+ sulfur dioxide vol 12 264, 265
+ tetraphenylarsonium tetraphenylborate (1-) vol 18 222
+ thiocyanic acid, silver salt + thiocyanic acid, tetraethylammonium salt + perchloric acid, tetraethylammonium salt vol 3 227
+ xenon vol 2 187
Methane, nitro- (aqueous)
+ argon vol 4 E80 - 84, 106
+ helium vol 1 39
+ oxygen vol 7 349
Methane, 1,1-sulfinylbis-
+ deuterium vol 5/6 E281, 292, 293
+ ethane vol 9 E195 -E199, 221, 222
+ helium vol 1 99
+ hydrogen vol 5/6 E259, 260, 261
+ krypton vol 2 101
+ neon vol 1 244
+ nitrogen vol 10 E253, 258
+ silver azide + tetraethylammonium azide + nitric acid, tetraethylammonium salt vol 3 E23, 24
+ silver cyanide + tetraethylammonium cyanide + perchloric acid, tetraethylammonium salt vol 3 84
+ silver tetraphenylborate (1-) vol 18 E158, 159, 162
+ silver tetraphenylborate + perchloric acid, sodium salt + sodium tetraphenylborate vol 18 160
+ silver tetraphenylborate + sodium tetraphenylborate vol 18 161
+ sulfur dioxide vol 12 E318, 319 - 322
+ tetraphenylarsonium tetraphenylborate (1-) vol 18 E203, 204 - 207
+ thallium (1) tetraphenylborate (1-) vol 18 181
+ thiocyanic acid, silver salt + thiocyanic acid, tetraethylammonium salt + perchloric acid, tetraethylammonium salt vol 3 E213, 216
+ xenon vol 2 184
Methane, 1,1-sulfinybis- (aqueous)
+ argon vol 4 226
+ ethane vol 9 E64, 67
+ potassium tetraphenylborate vol 18 24
+ oxygen vol 7 346 - 348
+ silver azide + tetraethylammonium azide + perchloric acid, tetraethylammonium salt vol 3 E23, 25
+ silver cyanide + tetraethylammonium cyanide + perchloric acid, tetraethylammonium salt vol 3 18
+ sulfur dioxide vol 12 113
+ thiocyanic acid, silver salt + nitric acid, silver salt + perchloric acid, tetraethylammonium salt vol 3 E213, 215

Methane, 1,1-sulfinybis- (aqueous)
+ thiocyanic acid, silver salt + thiocyanic acid, potassium salt + nitric acid, potassium salt vol 3 E213, 214
+ thiocyanic acid, silver salt + thiocyanic acid, potassium salt + perchloric acid, potassium salt vol 3 E185, 187
+ thiocyanic acid, silver salt + thiocyanic acid, tetraethylammonium salt + perchloric acid, tetraethylammonium salt vol 3 E185, 186
Methane, 1,1-sulfinylbis- (quaternary)
+ hydrogen vol 5/6 100, 101
Methane, tetrachloro-
+ argon vol 4 213, E214, E253 - 255, 309
+ chlorine vol 12 E354 -E366, E395 -E397, 398 - 409, 411
+ deuterium vol 5/6 E281, 290
+ ethane vol 9 E138 -E141, 152, 153
+ hydrogen vol 5/6 E238, 245, 246
+ krypton vol 2 88
+ nitric oxide vol 8 E336, 346, 347
+ nitrogen vol 10 236, 252, 458
+ nitrous oxide vol 8 E161, 218 - 220
+ oxygen vol 7 304, 306 - 308, 452
+ ozone vol 7 486 - 488, 492
+ sulfur dioxide vol 12 E1, E2, E298, 300, 303, 304
+ xenon vol 2 176
Methane, tetrachloro- (ternary)
+ sulfur dioxide vol 12 316
Methane, tetrachloro- (quaternary)
+ chlorine vol 12 411
Methane, tetrafluoro-
+ hydrogen vol 5/6 E238, 247
+ nitrogen vol 10 234, 459 - 461
Methane, tribromo-
+ krypton vol 2 88
Methane, trichloro-
+ argon vol 4 212
+ chlorine vol 12 E354 -E366, 394
+ hydrogen vol 5/6 E185, 191 - 193
+ krypton vol 2 88
+ nitrogen vol 10 232, 233
+ nitrous oxide vol 8 E161, 220 - 222
+ oxygen vol 7 452
+ ozone vol 7 484, 492
+ radon-222 vol 2 318
+ sulfur dioxide vol 12 E1, E2, E298, 299 - 302
+ tris(0-phenanthroline)ruthenium (11) tetraphenylborate (1-) vol 18 123
+ xenon vol 2 175
Methane, trichloro- (aqueous)
+ sulfur dioxide vol 12 106
Methane, trichlorofluoro-
+ ozone vol 7 485, 489
Methanesulfenyl chloride trichloro-
+ chlorine vol 12 E354 -E366, 410
+ sulfur dioxide vol 12 E174 175 - 177, 180

Methanesulfenyl chloride, trichloro- (quaternary)
+ chlorine vol 12 411
Methanol
+ argon vol 4 E169 - 171, 172 - 175, E253 - 255, 282
+ butyltriisopentylammonium tetraphenylborate vol 18 88
+ cesium tetraphenylborate vol 18 71
+ ethane vol 9 E166, E167, 168 - 171, E232, E233, 247, 248
+ helium vol 1 81, E255
+ hydrogen vol 5/6 E185, 191 - 193, 435, 436 - 440
+ krypton vol 2 E71, 72
+ neon vol 1 226
+ nitric oxide vol 8 E336, 342
+ nitrogen vol 10 E174, E177, E178, 179 - 183, 446
+ nitrous oxide vol 8 E160, 183 - 187
+ oxygen vol 7 E262, 268, 269, 304, 452
+ potassium tetraphenylborate vol 18 38
+ radon-222 vol 2 272
+ silver azide + sodium azide + nitric acid, tetraethylammonium salt vol 3 28
+ silver tetraphenylborate (1-) vol 18 166
+ silver tetraphenylborate + sodium tetraphenylborate + nitric acid, sodium salt vol 18 167
+ silver tetraphenylborate + sodium tetraphenylborate + sodium bromide vol 18 168, 169
+ sulfur dioxide vol 12 E174, 175 - 177, 180
+ tetra-N-butylammonium tetraphenylborate vol 18 101
+ tetraphenylarsonium tetraphenylborate (1-) vol 18 E217, 218 - 220
+ thiocyanic acid, silver salt + thiocyanic acid, potassium salt + nitric acid, potassium salt vol 3 E223, 226
+ thiocyanic acid, silver salt + thiocyanic acid, sodium salt vol 3 E223, 224 - 225
+ thiocyanic acid, silver salt + thiocyanic acid, tetraethylammonium salt + nitric acid, tetraethylammonium salt vol 3 E223, 226
+ tris(O-phenanthroline)ruthenium (11) tetraphenylborate (1-) vol 18 130
Methanol (aqueous)
+ argon vol 4 E80 - 84, 85
+ cesium tetraphenylborate + lithium chloride + sodium hydroxide vol 18 61, 62
+ iodic acid, barium salt vol 14 E251, E252, 298
+ iodic acid, calcium salt vol 14 165
+ nitrogen vol 10 E83, 86, 87
+ neon vol 1 179
+ nitrous oxide vol 8 117
+ oxygen vol 7 E190, 193, 194
+ potassium tetraphenylborate + lithium chloride + sodium hydroxide vol 18 27 - 29
+ thiocyanic acid, silver salt + thiocyanic acid, potassium salt + nitric acid, potassium salt vol 3 E200, 201 - 202

Methanol (aqueous)
+ thiocyanic acid, silver salt + thiocyanic acid, potassium salt + perchloric acid, potassium salt vol 3 E200, 203 - 204
+ sulfur dioxide vol 12 108, 109
Methanol (aqueous 1,2-ethanediol)
+ argon vol 4 97
Methanol (aqueous 1,2,3-propanetriol)
+ argon vol 4 99
Methanol (ternary and multicomponent)
+ hydrogen vol 5/6 567, 568
+ nitric oxide vol 8 319
+ nitrogen vol 10 495
+ oxygen vol 7 188, 189, 192, 419
+ sulfur dioxide vol 12 178, 179
Methicillin sodium
see 4-thia-1-azabicyclo[3,2,0]heptane-2-carboxylic acid, 6-[(2,6-dimethyoxybenzoyl)amino]-3,3-dimethyl-7-oxo, monosodium salt
2-Methoxyethanol
see Ethanol, 2-methoxy-
3-Methyoxysulfolane
see Thiophene, tetrahydro-3-methoxy-, 1,1-dioxide
Methylacetate
see Acetic acid, methyl ester
N-Methylacetamide
see acetamide, N-methyl-
Methylal
see Methane, dimethoxy-
2-Methyl aminobenzene
see Benzenamine, 2-methyl-
Methylammonium tetraphenylborate
+ water vol 18 94
N-Methylaniline
see Benzenamine,N-methyl-
2-Methylaniline
see Benzenamine, 2-methyl-
3-Methylaniline
see Benzenamine, 3-methyl-
Methylbenzene
see Benzene, methyl-
Methyl benzoate
see Benzoic acid, methyl ester
2-Methyl-1-butanol
see 1-Butanol, 2-methyl-
3-Methyl-1-butanol
see 1-Butanol, 3-methyl-
2-Methyl-2-butanol
see 2-Butanol, 2-methyl-
3-Methyl-2-butanol
see 2-Butanol, 3-methyl-
3-Methyl-1-butene
see 1-Butene, 3-methyl-
Methyl-butyl ethanoate
see 1-Butanol, 3-methyl-, acetate
Methyl butyl ketone
see 2-Hexanone
Methyl isobutyl ketone
see 2-Butanone, 3,3-dimethyl-
Methyl cellosolve
see Ethanol, 1-methoxy-
Methylcyclohexane
see Cyclohexane, methyl-

Methylcyclohexane, tetrafluorodeca-
see Cyclohexane, undecafluoro (trifluoromethyl)-
4-Methyl-1,3-dioxolan-2-one
see 1,3-dioxolane-2-one, 4-methyl-
Methylene dichloride
see Methane, dichloro-
1-Methylethylcyclohexane
see Cyclohexane, 1-methylethyl-
Methylethylketone
see 2-Butanone
Methyl formate
see Formic acid, methyl ester
2-Methylheptane
see Heptane, 2-methyl-
3-Methylheptane
see Heptane, 3-methyl-
2-Methyl-2-heptanol
see 2-Heptanol, 2-methyl-
3-Methyl-3-heptanol
see 3-Heptanol, 3-methyl-
2-Methyl-2-hexanol
see 2-Hexanol, 2-methyl-
5-Methyl-2-hexanol
see 2-Hexanol, 5-methyl-
2-Methyl-3-hexanol
see 3-Hexanol, 2-methyl-
3-Methyl-3-hexanol
see 3-Hexanol, 3-methyl-
Methylhexylketone,
see 2-Octanone
Methylhydrazine
see Hydrazine, methyl-
1-Methylnaphthalene
see Naphthalene, 1-methyl-
1-Methyl-2-nitrobenzene
see Benzene,1-methyl-2-nitro-
N-Methyl-N-nitrosomethanamine
see methanamine, N-methyl-N-nitroso-
7-Methyl-1-octanol
see 1-Octanol, 7-methyl-
5-Methyl-N-methylpyrrolidinone
see 2-pyrrolidinone, 1,5-dimethyl-
4-Methyl-1-pentanol
see 1-Pentanol, 4-methyl-
2-Methyl-1-pentanol
see 1-Pentanol, 2-methyl-
2-Methyl-2-pentanol
see 2-Pentanol, 2-methyl-
3-Methyl-2-pentanol
see 2-Pentanol, 3-methyl-
4-Methyl-2-pentanol
see 2-Pentanol, 4-methyl-
2-Methyl-3-pentanol
see 3-Pentanol, 2-methyl-
3-Methyl-3-pentanol
see 3-Pentanol, 3-methyl-
4-Methyl-2-pentanone
see 2-Pentanone, 4-methyl-
4-Methyl-1-penten-3-ol
see 1-Penten-3-ol, 4-methyl-
Methyl phenyl ketone
see Ethanone, 1-phenyl-
1-Methylpiperidine
see Piperidine, 1-methyl-

Methylprednisolone sodium succinate
see Pregn-1,4-diene-3,20-dione,21-(3-carboxy-1-oxopropoxy) -11,17-dihydroxy-6-methyl-,monosodium salt,(6α,11β)-
2-Methylpropane
see Propane, 2-methyl-
2-Methylpropanoic acid
see Propanoic acid, 2-methyl-
2-Methyl-1-propanol
see 1-Propanol, 2-methyl-
2-Methyl-2-propanol
see 2-Propanol, 2-methyl-
2-Methyl-1-propene
see 1-Propene, 2-methyl-
(1-Methylpropyl) benzene
see Benzene, (1-methylpropyl)-
1-Methyl-4-propylbenzene
see Benzene, 1-methyl-4-propyl-
Methyl-N-propyl carbinol
see 2-pentanol
Methylisopropyl carbinol
see 2-Butanol, 3-methyl-
Methyl-propyl ethanoate
see Acetic acid, 2-methylpropyl ester
Methylpropylketone
see 2-Pentanone
Methyl pyridine
see Pyridine, methyl-
1-Methylpyrrolidine
see Pyrrolidine, 1-methyl-
1-Methyl-2-pyrrolidinone
see 2-Pyrrolidinone, 1-methyl-
N-Methyl-2-pyrrolidinone
see 2-Pyrrolidinone, 1-methyl-
Methyl salicylate
see Benzoic acid, 2-hyroxy-, methyl ester
Methylstyrene
see Benzene, (1-methylethenyl)-
3-Methylsulfolane
see Thiophene, tetrahydro-3-methyl-, 1,1-dioxide
Methyl tricyanamide argentate
+ water vol 3 36 - 37
Meticortelone
see Pregn-1,4-diene-3,20-dione,21-(3-carboxy-1-oxopropoxy) -11,17-dihydroxy-,monosodium salt, (11α)-
Mineral oil
see Oil, mineral
Molasses
+ oxygen vol 7 422
Monoethanolamine
see Ethanol, 2-amino-
Monosodium Glutamate
see DL-glutamic acid, monosodium salt
Monosodium glycinate
see Glycine, monosodium salt
Muscle
see Guinea pig muscle and Rabbit muscle
Myocardium
see Dog myocardium
Myoglobin
see Horse heart myoglobin

Nafcillin sodium
see 4-thia-1-azabicyclo[3,2,0]heptane-2-carboxylic acid, 6-[[(2-ethoxy-1-naphthalenyl)carbonyl]amino]-3,3-dimethyl-7-oxo-, monosodium salt

Naphta
+ nitrogen vol 10 503

Naphthalene, decahydro-
+ ethane vol 9 E138 -E141, 151
+ hydrogen vol 5/6 393
+ sulfur dioxide vol 12 129, 140, 143, 144, E147

Naphthalene, decahydro-,cis-
+ hydrogen vol 5/6 156, 157

Naphthalene, decahydro-,trans-
+ hydrogen vol 5/6 156, 158

Naphthalene, heptadecafluorodecahydro(trifluoromethyl)-
+ oxygen vol 7 340

Naphthalene, 1-methyl-
+ ethane vol 9 E138 -E141, 160
+ hydrogen vol 5/6 427 - 430
+ nitrogen vol 10 172

Naphthalene, octadecafluorodecahydro-
+ air vol 10 527
+ oxygen vol 7 336

Naphthalene, 1,2,3,4-tetrahydro-
+ hydrogen vol 5/6 185, 426
+ krypton vol 2 61
+ nitrogen vol 10 280
+ oxygen vol 7 368
+ sulfur dioxide vol 12 129, 140, E147

Naphthalene, 1,2,3,4-tetrahydro- (ternary)
+ hydrogen vol 5/6 556 - 561

Neodymium cobalt nitrate
see Nitric acid, cobalt neodymium salt

Neodymium manganese nitrate
see Nitric acid, manganese neodymium salt

Neodymium nickel nitrate
see Nitric acid, neodymium nickel salt

Neodymium nitrate
see Nitric acid, neodymium salt

Neon
+ deuterium vol 5/6 E581, 582
+ helium vol 1 E324, 325
+ hydrogen vol 5/6 E581, 583 - 589

Nickel acetate
see Acetic acid, nickel salt

Nickel nitrate
see Nitric acid, nickel salt

Nickel sulfate
see Sulfuric acid, nickel salt

Nickelous sulfate
see Sulfuric acid, nickel salt

Nitrate barium
see Nitric acid, barium salt

Nitrate calcium
see Nitric acid, calcium salt

Nitrate cesium
see Nitric acid, cesium salt

Nitrate lithium
see Nitric acid, lithium salt

Nitrate magnesium
see Nitric acid, magnesium salt

Nitrate sodium
see Nitric acid, sodium salt
Nitrate, lead
see Nitric acid, lead salt
Nitrate, silver
see Nitric acid, silver salt
Nitrate, sodium
see Nitric acid, sodium salt
Nitric acid
+ helium vol 1 E20, 23
+ hydrogen vol 5/6 E28, 41
+ krypton vol 2 E12, E13, 16
+ nitric oxide vol 8 E28, E40, 43
+ nitrogen vol 10 531
+ oxygen vol 7 E61, E62, 84 - 96
+ ozone vol 7 487, 488
+ silver cyanide + sulfuric acid + hydrogen cyanide
vol 3 E86
+ silver cyanamide + nitric acid, potassium salt + water
vol 3 33
Nitric acid (aqueous)
+ iodic acid, barium salt vol 14 E246, 297
Nitric acid (multicomponent)
+ ozone vol 7 487, 488
Nitric acid, aluminium salt
+ water vol 13 31, 319
Nitric acid, aluminium salt (aqueous)
+ nitrous oxide vol 8 E29, E30, 63
Nitric acid, ammonium salt (aqueous)
+ hydrogen vol 5/6 E29, 46
+ nitric acid vol 13 390
+ nitric acid, cerium salt vol 13 185, 186, 203, 204
+ nitric acid, cerium salt + nitric acid, magnesium salt
vol 13 185, 186
+ nitric acid, gadolinium salt vol 13 390
+ nitric acid, lanthanum salt vol 13 85
+ nitric acid, magnesium salt vol 13 185, 186
+ nitric acid, yttrium salt vol 13 32, 33, 34
+ nitric acid, yttrium salt + nitric acid
vol 13 34
+ nitrous oxide vol 8 E29, 50, 51
+ radon-222 vol 2 E238 - E241, 243
+ sulfur dioxide vol 12 E37 - E39, 62
+ water vol 13 32, 33, 85, 203, 204
Nitric acid, ammonium cerium IV salt
+ phosphoric acid tributyl ester vol 13 240
+ water vol 13 236, 237
Nitric acid, ammonium cerium IV salt (aqueous)
+ nitric acid vol 13 238, 239
Nitric acid, ammonium cerium salt
+ water vol 13 E221, 234
Nitric acid, ammonium cerium salt (aqueous)
+ nitric acid vol 13 E221, 235
Nitric acid, ammonium lanthanum salt
+ water vol 13 E148, 155
Nitric acid, barium salt
+ water vol 14 219, 220
+ water vol 13 191

Nitric acid, barium salt (aqueous)
+ bromic acid, barium salt vol 14 234, 241
+ chloric acid, barium salt vol 14 E210, 219, 220
+ chloric acid, barium salt + barium bromide vol 14 E210, 221, 222
+ iodic acid, barium salt vol 14 E250, E251, 279, 280, 292
+ neon vol 1 E141, 160
+ nitric acid, cerium salt vol 13 191
+ nitric acid, lanthanum salt vol 13 65
+ radon-222 vol 2 E238 -E241, 250
+ silver cyanide + silver iodide + potassium cyanide + potassium iodide + nitric acid, potassium salt vol 3 81

Nitric acid, beryllium salt
+ water vol 13 180, 181

Nitric acid, beryllium salt (aqueous)
+ nitric acid, cerium salt vol 13 180, 181

Nitric acid, cadmium salt
+ water vol 13 199, 200

Nitric acid, cadmium salt (aqueous)
+ nitric acid, cerium salt vol 13 199, 200
+ nitrous oxide vol 8 E29, E30, 54

Nitric acid, calcium salt
+ water vol 13 187, 188

Nitric acid, calcium salt (aqueous)
+ iodic acid, barium salt vol 14 278
+ neon vol 1 E141, 154
+ nitric acid, cerium salt vol 13 187, 188
+ nitrous oxide vol 8 E31, 73
+ oxygen vol 7 E66, 124

Nitric acid, cerium salt
+ 1,4-dioxane vol 13 E221, 224
+ ethane, 1,1´-oxybis- vol 13 E221, 222, 223
+ phosphoric acid, tributyl ester vol 13 E221, 225
+ water vol 13 E156 -E162, 163 - 182, 187 - 202

Nitric acid, cerium salt (aqueous)
+ acetamide, N-(aminocarbonyl)- vol 13 220
+ aniline nitrate vol 13 215
+ cerium chloride vol 13 192
+ ethanamine, N-ethyl- vol 13 208
+ 1,2-ethanediamine, dinitrate vol 13 209, 210
+ hydrazine mononitrate vol 13 201, 202
+ methanamine nitrate vol 13 205, 206
+ methanamine, N-methyl- vol 13 207
+ nitric acid vol 13 E161, E162, 167 - 170
+ nitric acid, ammonium salt vol 13 185, 186 203, 204
+ nitric acid, barium salt vol 13 191
+ nitric acid, beryllium salt vol 13 180, 181
+ nitric acid, cadmium salt vol 13 199, 200
+ nitric acid, calcium salt vol 13 187, 188
+ nitric acid, cesium salt vol 13 E161, E162, 178, 179
+ nitric acid, copper salt vol 13 196 - 198
+ nitric acid, lithium salt vol 13 171, 172
+ nitric acid, magnesium salt vol 13 E162, 182 - 186
+ nitric acid, magnesium salt + nitric acid, ammonium salt vol 13 185, 186
+ nitric acid, magnesium salt + nitric acid, nickel salt vol 13 183, 184

Nitric acid, cerium salt (aqueous)
+ nitric acid, manganese salt vol 13 193, 194
+ nitric acid, nickel salt vol 13 183, 184, 195
+ nitric acid, potassium salt vol 13 E161, E162, 174, 175
+ nitric acid, rubidium salt vol 13 E161, E162, 176, 177
+ nitric acid, sodium salt vol 13 173
+ nitric acid, strontium salt vol 13 189, 190
+ piperidine nitrate vol 13 211, 212
+ pyridine nitrate vol 13 213, 214
+ thiourea vol 13 219
+ urea vol 13 216, 217
+ urea mononitrate vol 13 218

Nitric acid, cerium cobalt salt (aqueous)
+ nitric acid vol 13 E221, 231

Nitric acid, cerium magnesium salt (aqueous)
+ nitric acid vol 13 229

Nitric acid, cerium magnesium salt
+ water vol 13 E226, E227, 228

Nitric acid, cerium manganese salt (aqueous)
+ nitric acid vol 13 E221, 230

Nitric acid, cerium nickel salt (aqueous)
+ nitric acid vol 13 E221, 232

Nitric acid, cerium zinc salt (aqueous)
+ nitric acid vol 13 E221, 233

Nitric acid, cesium salt
+ water vol 13 61 178, 179

Nitric acid, cesium salt (aqueous)
+ neon vol 1 E141, 172
+ nitric acid, cerium salt vol 13 E161, E162, 178, 179
+ nitric acid, lanthanum salt vol 13 61
+ oxygen vol 7 E75, 138

Nitric acid, cobalt salt
+ water vol 13 30, 80

Nitric acid, cobalt salt (aqueous)
+ nitric acid, lanthanum salt vol 13 80
+ nitric acid, yttrium salt vol 13 30

Nitric acid, cobalt gadolinium salt (aqueous)
+ nitric acid vol 13 406

Nitric acid, cobalt lanthanum salt (aqueous)
+ nitric acid vol 13 E109, 152

Nitric acid, cobalt neodymium salt
+ water vol 13 350

Nitric acid, cobalt neodymium salt (aqueous)
+ nitric acid vol 13 349

Nitric acid, cobalt praseodymium salt
+ water vol 13 295

Nitric acid, cobalt praseodymium salt (aqueous)
+ nitric acid vol 13 294

Nitric acid, cobalt samarium salt (aqueous)
+ nitric acid vol 13 373

Nitric acid, copper salt
+ water vol 13 196 - 198

Nitric acid, copper salt (aqueous)
+ nitric acid, cerium salt vol 13 196 - 198
+ nitrous oxide vol 8 E29, E30, 55

Nitric acid, copper praseodymium salt
+ water vol 13 298

Nitric acid, dysprosium salt
+ water vol 13 E414 -E417, 418 - 428

Nitric acid, dysprosium salt (aqueous)
+ aniline nitrate vol 13 E416, E417, 424
+ 1,2-ethanediamine dinitrate vol 13 421
+ 1,6-hexanediamine dinitrate vol 13 422
+ nitric acid vol 13 E415 -E417, 420
+ pyridine nitrate vol 13 E416, E417, 423
+ quinoline nitrate vol 13 E416, E417, 425
+ quinoline, 8-methyl-,nitrate vol 13 E416, E417, 426
+ urea vol 13 E416, E417, 426
428

Nitric acid, enodymium salt
+ nitric acid, praseodymium salt
+ propane, 2,2´-oxybis- vol 13 284

Nitric acid, erbium salt
+ ethane, 1,1´-oxybis- vol 13 448
+ water vol 13 E434 -E436,
437 - 443,
445 - 447

Nitric acid, erbium salt (aqueous)
+ acetamide vol 13 447
+ acetamide, N-(aminocarbonyl)- vol 13 446
+ nitric acid vol 13 439, 440
+ pyridine nitrate vol 13 441
+ quinoline nitrate vol 13 442
+ quinoline, 6-methyl-, nitrate vol 13 443
+ urea vol 13 444, 445

Nitric acid, europium salt
+ water vol 13 E376, E377, 378
380

Nitric acid, europium salt (aqueous)
+ nitric acid vol 13 379
+ urea vol 13 380

Nitric acid, gadolinium salt
+ water vol 13 E381 -E385,
386 - 389,
391 - 404

Nitric acid, gadolinium salt (aqueous)
+ diethylenetriamine trintrate vol 13 395
+ 1,2-ethanediamine, dinitrate vol 13 394
+ ethanol, 2-amino, nitrate vol 13 391
+ ethanol, 2,2´-iminobis-, nitrate vol 13 392
+ ethanol, 2,2´,2´´-nitrilotris-,
nitrate vol 13 393
+ 1,6-hexanediamine dinitrate vol 13 396
+ nitric acid vol 13 29, E384, E385,
389, 390
+ nitric acid, ammonium salt
+ nitric acid vol 13 390
+ nitric acid, yttrium salt + nitric acid
vol 13 29
+ piperazine dinitrate vol 13 402
+ piperidine nitrate vol 13 401
+ pyridine nitrate vol 13 397
+ quinoline nitrate vol 13 398
+ quinoline, 8-methyl-, nitrate vol 13 399
+ quinolinol, nitrate vol 13 400
+ urea vol 13 403
+ urea mononitrate vol 13 404

Nitric acid, gadolinium magnesium salt (aqueous)
+ nitric acid vol 13 405

Nitric acid, gadolinium nickel salt (aqueous)
+ nitric acid vol 13 407

Nitric acid, gadolinium zinc nitrate (aqueous)
+ nitric acid vol 13 408
Nitric acid, holmium salt
+ water vol 13 E429, 430 - 433
Nitric acid, holmium salt (aqueous)
+ nitric acid vol 13 431, 432
+ urea vol 13 433
Nitric acid, iron (2+) salt (aqueous)
+ nitric oxide vol 8 E265, E266, 286
Nitric acid, lanthanum salt
+ 1,4-dioxane vol 13 E109, 135
+ ethane, 1,1´-oxybis- vol 13 E109, 132, 133
+ ethanol, 2-amino vol 13 E109, 115
+ morpholine vol 13 E109, 144
+ water vol 13 E38 - E47, 48 - 72, 75, 76, 79 - 108

Nitric acid, lanthanum salt, hexahydrate
+ acetic acid, ethyl ester vol 13 E109, 141
+ acetic acid, methyl ester vol 13 E109, 140
+ acetonitrile vol 13 E109, 142
+ benzenamine vol 13 E109, 146
+ benzenamine, 2-methyl- vol 13 E109, 145
+ benzenemethanol vol 13 E109, 131
+ 1-butanol vol 13 E109, 120
+ 2-butanol vol 13 E109, 121
+ 1-butanol, 3-methyl- vol 13 E109, 126
+ 2-butanol, 2-methyl- vol 13 E109, 127
+ cyclohexanol vol 13 E109, 130
+ cyclohexanone vol 13 E109, 138
+ 1,4-dioxane vol 13 E109, 136
+ ethane, 1,1´-oxybis- vol 13 E109, 134
+ 1,2-ethanediol vol 13 E109, 112
+ ethanol vol 13 E109, 111
+ ethanol, 2-ethoxy- vol 13 E109, 114
+ ethanol, 2-methoxy- vol 13 E109, 113
+ formic acid, ethyl ester vol 13 E109, 139
+ 1-hexanol vol 13 E109, 128, 129
+ methanol vol 13 E109, 110
+ 1-pentanol vol 13 E109, 124
+ 3-pentanol vol 13 E109, 125
+ phosphoric acid, tributyl ester vol 13 E109, 147
+ 1,2,3-propanetriol vol 13 E109, 119
+ 1-propanol vol 13 E109, 116
+ 2-propanol vol 13 E109, 117
+ 1-propanol, 2-methyl- vol 13 E109, 122
+ 2-propanol, 2-methyl- vol 13 E109, 123
+ 2-propanone vol 13 E109, 137
+ 2-propene-1-ol vol 13 E109, 118
Nitric acid, lanthanum salt (aqueous)
+ acetic acid, lanthanum salt vol 13 66
+ aniline nitrate vol 13 104, 105
+ benzamide vol 13 108
+ ethanamine, N,N-diethyl- vol 13 95
+ ethanamine, N,N-diethyl-, nitrate vol 13 96, 97
+ guanidine mononitrate vol 13 92
+ 1-hexanol vol 13 E39, E47
+ hydrazine dinitrate vol 13 88, 89
+ hydrazine mononitrate vol 13 86, 87
+ lanthanum chloride vol 13 69
+ methanamine, N-methyl-, nitrate vol 13 90, 91

Nitric acid, lanthanum salt (aqueous)
+ nitric acid vol 13 28, E39, E45, E47 53 - 55
+ nitric acid, ammonium salt vol 13 E45, E47, 85
+ nitric acid, barium salt vol 13 65
+ nitric acid, cesium salt vol 13 E45, E47, 61
+ nitric acid, cobalt salt vol 13 80
+ nitric acid, magnesium salt vol 13 62 - 64
+ nitric acid, manganese salt vol 13 79
+ nitric acid, nickel salt vol 13 81
+ nitric acid, neodynium salt vol 13 E39, E47, 71, 72
+ nitric acid, neodynium salt + phosphoric acid, tributyl ester vol 13 73, 74
+ nitric acid, potassium salt vol 13 E45, E47 58, 59
+ nitric acid, praseodymium salt vol 13 70
+ nitric acid, rubidium salt vol 13 E45, E47, 60
+ nitric acid, samarium salt vol 13 75, 76
+ nitric acid, samarium salt??? vol 13 77, 78
+ nitric acid, sodium salt vol 13 57
+ nitric acid, yttrium salt + nitric acid vol 13 28
+ nitric acid, zinc salt vol 13 82 - 84
+ oxalic acid, lanthanum salt vol 13 67, 68
+ piperidine nitrate vol 13 100, 101
+ 2-propanone vol 13 56
+ pyridine nitrate vol 13 102, 103
+ quinoline nitrate vol 13 106, 107
Nitric acid, lanthanum maganese salt (aqueous)
+ nitric acid vol 13 E109, 151
Nitric acid, lanthanum magnesium salt
+ water vol 13 E148, 150
Nitric acid, lanthanum magnesium salt (aqueous)
+ nitric acid vol 13 E148, 149
Nitric acid, lanthanum nickel salt (aqueous)
+ nitric acid vol 13 E109, 153
Nitric acid, lanthanum zinc salt (aqueous)
+ nitric acid vol 13 E109, 154
Nitric acid, lead salt (aqueous)
+ radon-222 vol 2 E238 -E241, 248
Nitric acid, lithium salt
+ water vol 13 171, 172
Nitric acid, lithium salt (aqueous)
+ iodic acid, calcium salt + chloric acid, lithium salt vol 14 93
+ neon vol 1 E141, 162
+ nitric acid, cerium salt vol 13 171, 172
+ oxygen vol 7 E66, E67, 138
+ perchloric acid, lithium salt + iodic acid, barium salt vol 14 261
+ perchloric acid, lithium salt + iodic acid, strontium salt vol 14 199
Nitric acid, lutetium salt
+ water vol 13 E461, 462 - 464
Nitric acid, lutetium salt (aqueous)
+ nitric acid vol 13 463
+ urea vol 13 464
Nitric acid, magnesium salt
+ water vol 13 62 - 64, 182
Nitric acid, magnesium salt
+ water vol 14 37 - 39
Nitric acid, magnesium salt (aqueous)
+ bromic acid, barium salt vol 14 233

Nitric acid, magnesium salt
+ bromic acid, magnesium salt vol 14 10
+ iodic acid, magnesium salt vol 14 E17, 37 - 39
+ neon vol 1 E141, 149
+ nitric acid, ammonium salt vol 13 185, 186
+ nitric acid, cerium salt vol 13 E162, 182 - 186
+ nitric acid, cerium salt + nitric acid, ammonium salt
vol 13 185, 186
+ nitric acid, cerium salt + nitric acid, nickel salt
vol 13 183, 184
+ nitric acid, lanthanum salt vol 13 62 - 64
+ nitric acid, nickel salt vol 13 183, 184
+ nitrous oxide vol 8 E30, E31, 68, 69

Nitric acid, magnesium neodymium salt
+ water vol 13 E339 -E341, 343, 344

Nitric acid, magnesium neodymium salt (aqueous)
+ nitric acid vol 13 E339 -E441, 442, 445
+ nitric acid, magnesium salt vol 13 346

Nitric acid, magnesium praseodymium salt
+ water vol 13 E286 -E285, 290, 291

Nitric acid, magnesium praseodymium salt (aqueous)
+ nitric acid vol 13 E286 - 288, 289

Nitric acid, magnesium samarium salt (aqueous)
+ nitric acid vol 13 371, 372

Nitric acid, manganese salt
+ water vol 13 79, 193, 194

Nitric acid, manganese salt (aqueous)
+ nitric acid, cerium salt vol 13 193, 194
+ nitric acid, lanthanum salt vol 13 79

Nitric acid, manganese neodymium salt
+ water vol 13 348

Nitric acid, manganese neodymium salt (aqueous)
+ nitric acid vol 13 347

Nitric acid, manganese praseodymium salt
+ water vol 13 293

Nitric acid, manganese praseodymium salt (aqueous)
+ nitric acid vol 13 E286 -E288, 292

Nitric acid, neodymium salt
+ 1,4-dioxane vol 13 E335, 338
+ ethane, 1,1′-oxybis- vol 13 E335, 336, 337
+ ethanol, 2-ethoxy- + nitric acid praseodymium salt vol 13 280
+ nitric acid, lanthanum salt + phosphoric acid, tributyl ester
vol 13 73, 74
+ nitric acid, praseodymium salt + ethane, 1,1′-oxybis- vol 13 282
+ nitric acid, praseodymium salt + ethanol, 2-ethoxy- vol 13 280
+ nitric acid, praseodymium salt + 2-methoxyethanol vol 13 279
+ water vol 13 E301 -E314, 316 - 334

Nitric acid, neodymium salt (aqueous)
+ 2-butanamine nitrate vol 13 326
+ cyclohexanamine nitrate vol 13 328
+ ethanamine, N,N-diethyl-, nitrate vol 13 325
+ ethanamine, N-ethyl-, nitrate vol 13 324
+ 1,2-ethanedioic acid, neodymium salt
vol 13 316

Nitric acid, neodymium salt (aqueous)
+ 1,6-hexanediamine dinitrate vol 13 327
+ hydrazine mononitrate vol 13 320, 321
+ methanamine, N,N-dimethyl-, nitrate vol 13 322
+ nitric acid vol 13 E304 -E308, 313 - 314
+ nitric acid, aluminium salt vol 13 319
+ nitric acid, lanthanum salt vol 13 71, 72
+ nitric acid, praseodymium salt vol 13 255
+ nitric acid, rubidium salt + nitric acid vol 13 315
+ nitric acid, samarium salt vol 13 317
+ nitric acid, zinc salt vol 13 318
+ piperazine dinitrate vol 13 330
+ piperidine nitrate vol 13 329
+ pyridine nitrate vol 13 331
+ quinoline nitrate vol 13 332, 333
+ urea vol 13 334
Nitric acid, neodymium nickel salt (aqueous)
+ nitric acid vol 13 351
Nitric acid, neodymium zinc salt
+ water vol 13 353
Nitric acid, neodymium zinc salt (aqueous)
+ nitric acid vol 13 352
Nitric acid, nickel salt
+ water vol 13 81, 195
Nitric acid, nickel salt (aqueous)
+ nitric acid, cerium salt vol 13 183, 184, 195
+ nitric acid, lanthanum salt vol 13 81
+ nitric acid, magnesium salt vol 13 183, 184
Nitric acid, nickel praseodymium salt
+ water vol 13 297
Nitric acid, nickel praseodymium salt (aqueous)
+ nitric acid vol 13 296
Nitric acid, nickel samarium salt (aqueous)
+ nitric acid vol 13 374
Nitric acid, potassium salt
+ thiocyanic acid, silver salt + thiocyanic acid, potassium salt + acetonitrile vol 3 208
+ thiocyanic acid, silver salt + thiocyanic acid, potassium salt + N,N-dimethylacetamide vol 3 211
+ thiocyanic acid, silver salt + thiocyanic acid, potassium salt + N,N-dimethylformamide vol 3 212
+ thiocyanic acid, silver salt + thiocyanic acid, potassium salt + dimethyl sulfoxide vol 3 E213, 214
+ thiocyanic acid, silver salt + thiocyanic acid, potassium salt + formamide vol 3 221
+ thiocyanic acid, silver salt + thiocyanic acid, potassium salt + hexamethylphosphorotriamide vol 3 222
+ thiocyanic acid, silver salt + thiocyanic acid, potassium salt + methanol vol 3 E223, 226
+ thiocyanic acid, silver salt + thiocyanic acid, potassium salt + N-methyl-2-pyrrolidinone vol 3 229
+ water vol 13 58, 59, 174, 175
Nitric acid, potassium salt (aqueous)
+ bromic acid, barium salt vol 14 232
+ hydrogen vol 5/6 E35, 44, 87, 88
+ iodic acid, barium salt vol 14 E248 -E250, 271 - 275
+ iodic acid, calcium salt vol 14 E70, E74, 146
+ krypton vol 2

Nitric acid, potassium salt (aqueous)
+ nitric acid, cerium salt vol 13 E161, E162, 174, 175
+ nitric acid, lanthanum salt vol 13 58, 59
+ nitrous oxide vol 8 E37, 107 - 110
+ oxygen vol 7 E74, 138, 180
+ selenocyanic acid, silver salt + selenocyanic acid, potassium salt vol 3 E90, 96 - 97
+ silver azide + sodium azide vol 3 9
+ silver cyanamide + nitric acid, potassium salt vol 3 E31, 33
+ silver cyanide + silver iodide + nitric acid, barium salt + potassium cyanide + potassium iodide vol 3 81
+ sulfur dioxide vol 12 E37 - E39, 103
+ thiocyanic acid, silver salt + nitric acid, silver salt vol 3 147
+ thiocyanic acid, silver salt + thiocyanic acid, potassium salt vol 3 E103, 145 - 146, 150 - 151, 152 - 153, 165 - 166, 169 - 170
+ thiocyanic acid, silver salt + thiocyanic acid, potassium salt + 1,4-dioxane vol 3 E188, 189
+ thiocyanic acid, silver salt + thiocyanic acid, potassium salt + ethanol vol 3 E193, 194 - 195
+ thiocyanic acid, silver salt + thiocyanic acid, potassium salt + methanol vol 3 E200, 201 - 202
+ thiocyanic acid, silver salt + thiocyanic acid, potassium salt + 1,2,3-propanetriol vol 3 199
+ thiocyanic acid, silver salt + thiocyanic acid, potassium salt + 2-propanone vol 3 E177 - 178, 179 - 180
+ thiocyanic acid, silver salt + thiocyanic acid, potassium salt + 1-propanol vol 3 205
+ thiocyanic acid, silver salt + thiocyanic acid, potassium salt + 2-propanol vol 3 206

Nitric acid, praseodymium salt
+ 1,4-dioxane vol 13 285
+ ethane, 1,1´-oxybis- vol 13 281, 283
+ ethanol, 2-ethoxy- + nitric acid, neodymium salt vol 13 280
+ nitric acid, neodymium salt + ethane, 1,1´-oxybis- vol 13 282
+ nitric acid, neodymium salt + ethanol, 2-ethoxy- vol 13 280
+ nitric acid, neodymium salt + 2-methoxyethanol vol 13 279
+ nitric acid, neodymium salt + propane, 2,2´-oxybis- vol 13 284
+ water vol 13 E241 -E246, 247 - 253

Nitric acid, praseodymium salt (aqueous)
+ aniline, nitrate vol 13 271
+ cyclohexylamine nitrate vol 13 266, 267
+ ethanamine, N,N-diethyl- vol 13 261
+ ethanamine, N-ethyl-, nitrate vol 13 260
+ 1,2-ethanediamine, dinitrate vol 13 262, 263
+ guanidine mononitrate vol 13 259
+ hexamethyleneimine nitrate vol 13 264, 265
+ hydrazine dinitrate vol 13 257, 258
+ nitric acid vol 13 252 - 254
+ nitric acid, lanthanum salt vol 13 70

Nitric acid, praseodymium salt (aqueous)
+ nitric acid, neodymium salt vol 13 255
+ nitric acid, rubidium salt + nitric acid vol 13 254
+ nitric acid, zinc salt vol 13 256
+ piperidine nitrate vol 13 268
+ pyridine nitrate vol 13 269, 270
+ quinoline nitrate vol 13 272, 273
+ quinoline, 8-methyl-, nitrate vol 13 274, 275
+ urea vol 13 276
+ urea mononitrate vol 13 277
Nitric acid, praseodymium zinc salt
+ water vol 13 300
Nitric acid, praseodymium zinc salt (aqueous)
+ nitric acid vol 13 299
Nitric acid, propyl ester
+ air vol 10 531
+ nitrogen vol 10 262
+ oxygen vol 7 351
Nitric acid, rubidium salt
+ water vol 13 60, 176, 177
Nitric acid, rubidium salt (aqueous)
+ nitric acid vol 13 25, 254, 315
+ nitric acid, cerium salt vol 13 E161, E162, 176, 177
+ nitric acid, lanthanum salt vol 13 60
+ nitric acid, neodymium salt + nitric acid vol 13 315
+ nitric acid, yttrium salt + nitric acid vol 13 25
+ oxygen vol 7 E74, E75, 138
+ nitric acid, praeseodymium salt + nitric acid vol 13 254
Nitric acid, samarium salt
+ water vol 13 75, 76, E354 -E359, 360 - 370
Nitric acid, samarium salt (aqueous)
+ acetamide, N-(aminocarbonyl)- vol 13 370
+ nitric acid vol 13 E354 -E359, 364 - 367
+ nitric acid, lanthanum salt vol 13 75, 76
+ nitric acid, lanthanum salt + phosphoric acid, tributyl ester vol 13 77, 78
+ nitric acid, neodymium salt vol 13 317
+ nitric acid, zinc salt vol 13 E358, 368, 369
Nitric acid, samarium zinc salt (aqueous)
+ nitric acid vol 13 375
Nitric acid, scandium salt
+ water vol 13 E1, 2 - 14
Nitric acid, scandium salt (aqueous)
+ butane, 1,1´-oxybis- vol 13 10
+ 1-butanol, 3-methyl- vol 13 6
+ ethanone, 1-phenyl- vol 13 13
+ 2-hexanone vol 13 12
+ 1-heptanol vol 13 7
+ 2-octanol vol 13 8
+ nitric acid vol 13 3
+ 2-pentanone vol 13 11
+ propanoic acid, butyl ester vol 13 14
+ propane, 2,2´-oxybis- vol 13 9
+ urea vol 13 4, 5

Nitric acid, silver salt
+ thiocyanic acid, silver salt
+ perchloric acid, tetraethylammonium salt
+ dimethysulfoxide vol 3 E213, 215
Nitric acid, silver salt (aqueous)
+ radon-222 vol 2 E238 -E241, 245
+ silver cyanide vol 3 E47, 60
+ silver cyanide + potassium silver cyanide + ammonia vol 3 66
+ thiocyanic acid, silver salt vol 3 E104, 115
+ thiocyanic acid, silver salt
+ nitric acid, potassium salt vol 3 147
+ selenocyanic acid, silver salt vol 3 E90 - 91, 98 - 99
Nitric acid, sodium salt
+ silver azide + sodium azide + dimethylacetamide vol 3 21
+ silver azide + sodium azide + dimethylformamide vol 3 22
+ silver azide + sodium azide + formamide vol 3 26
+ silver azide + sodium azide + hexamethylphosphorotriamide vol 3 27
+ silver azide + sodium azide + methanol vol 3 28
+ silver azide + sodium azide + N-methyl-2-pyrrolidinone vol 3 30
+ silver azide + tetraethylammonium azide + methanol vol 3 28
+ silver tetraphenylborate
+ nitric acid, sodium salt
+ N,N-dimethylacetamide vol 18 153
+ silver tetraphenylborate
+ sodium tetraphenylborate
+ acetonitrile vol 18 150
+ silver tetraphenylborate
+ sodium tetraphenylborate
+ nitromethane vol 18 169
+ silver tetraphenylborate
+ sodium tetraphenylborate
+ methanol vol 18 167
+ silver tetraphenylborate
+ sodium tetraphenylborate
+ N-methylacetamide vol 18 155
+ water vol 13 57, 173
Nitric acid, sodium salt (aqueous)
+ argon vol 4 E33, 59
+ ethanol + iodic acid, barium salt
+ sodium hydroxide vol 14 301
+ helium vol 1 E20, 33
+ hydrogen vol 5/6 E33, 44, 74, 75
+ neon vol 1 E141, 166
+ nitric acid, lanthanum salt vol 13 57
+ nitric acid, cerium salt vol 13 173
+ nitrous oxide vol 8 E35, 90, 91
+ oxygen vol 7 E71, 138, 153, 160
+ silver tetraphenylborate
+ sodium tetraphenylborate vol 18 143
Nitric acid, strontium salt
+ water vol 13 189, 190

Nitric acid, strontium salt (aqueous)
+ nitric acid vol 13 412, 413
+ nitric acid, cerium salt vol 13 189, 190
Nitric acid, tetraethylammonium salt
+ silver azide + sodium azide + methanol vol 3 28
+ silver azide + tetraethylammonium azide + acetonitrile vol 3 20
+ silver azide + tetraethylammonium azide + dimethyl sulfoxide vol 3 E23, 24
+ silver azide + tetraethylammonium azide + formamide vol 3 26
+ thiocyanic acid, silver salt + thiocyanic acid, tetraethylammonium salt + formamide vol 3 221
+ thiocyanic acid, silver salt + thiocyanic acid, tetraethylammonium salt + methanol vol 3 E223, 226
Nitric acid, terbium salt
+ water vol 13 E409, 410 - 412
Nitric acid, thulium salt
+ water vol 13 E449, 450 - 453
Nitric acid, thulium salt (aqueous)
+ nitric acid vol 13 451, 452
+ urea vol 13 453
Nitric acid, ytterbium salt
+ water vol 13 E454 -E456, 457 - 460
Nitric acid, ytterbium salt (aqueous)
+ nitric acid vol 13 459
+ urea vol 13 460
Nitric acid, yttrium salt
+ water vol 13 E15 - E19, 20, 21
+ 1,1´-oxybisethane vol 13 37
Nitric acid, yttrium salt (aqueous)
+ chromic acid, potassium salt vol 13 24
+ nitric acid vol 13 22, 25, 28 - 33
+ nitric acid, aluminium salt vol 13 31
+ nitric acid, ammonium salt vol 13 32, 33, 34
+ nitric acid, ammonium salt + nitric acid vol 13 34
+ nitric acid, cobalt salt vol 13 30
+ nitric acid, gadolinium salt + nitric acid vol 13 29
+ nitric acid, lanthanum salt + nitric acid vol 13 28
+ nitric acid, rubidium salt + nitric acid vol 13 25
+ urea vol 13 35, 36
+ yttrium hydroxide vol 13 26, 27
Nitric acid, zinc salt
+ water vol 13 82 - 84, 318 E358, 368, 369
Nitric acid, zinc salt (aqueous)
+ nitric acid, lanthanum salt vol 13 82 - 84
+ nitric acid, praseodymium salt vol 13 256
+ nitrous oxide vol 8 E29, E30, 53
Nitric oxide
+ hydrogen vol 5/6 609
+ hydrogen deuteride vol 5/6 E299
Nitrilotriethanol nitrate
+ water vol 13 393
Nitrobenzene
see Benzene, nitro-

Nitrogen
+ deuterium vol 5/6 597
+ helium vol 1 E328, 329
+ hydrogen vol 5/6 E589, E590, 591 - 600
+ hydrogen deuteride vol 5/6 E299
+ neon vol 1 E373, 374

Nitrogen + methane
+ argon vol 4 312 - 313

Nitrogen (quaternary)
+ hydrogen vol 5/6 551

Nitrogen (ternary)
+ hydrogen vol 5/6 311, 530, 531, 631 - 634
+ oxygen vol 7 434, 435

Nitrogen oxide (N2O)
+ helium vol 1 E255, 346
+ krypton vol 2 102
+ nitrogen vol 10 484
+ oxygen vol 7 466, 467
+ xenon vol 2 186

Nitrogen oxide (N2O4)
+ argon vol 4 237 - 238
+ helium vol 1 111
+ nitrogen vol 10 317, 318
+ nitrous oxide vol 8 E350
+ oxygen vol 7 432, 433

Nitrogen oxide (N2O4) (multicomponent)
+ oxygen vol 7 E62, 85, 96

Nitrogen trifluoride
+ nitrogen vol 10 486

Nitromethane
see Methane, nitro-

Nitropropane
see Propane, nitro-

N-Nitroso-dimethylamine
see methanamine,N-methyl-N-nitroso-

Nitrosyl chloride
+ nitric oxide vol 8 E350

2-Nitrotoluene
see Benzene, 1-methyl-2-nitro-

Nitrous acid, sodium salt (aqueous)
+ hydrogen vol 5/6 E33, 73

Nitrous oxide
see Nitrogen oxide (N2O)

Nitryl fluoride
+ nitrogen vol 10 485

1,1,2,2,3,3,4,4,4-Nonafluoro-N,N-bis (nonafluorobutyl)-1-butanamine
see 1-Butanamine, 1,1,2,2,3,3,4,4,4-nona-fluoro-N,N-bis (nonafluorobutyl)-

Nonane
+ argon vol 4 E125, 126 - 127
+ ethane vol 9 E77, E78, 92, 93
+ helium vol 1 50
+ hydrogen vol 5/6 E122, 137, 138
+ krypton vol 2 39
+ neon vol 1 195
+ nitrogen vol 10 E119, E120, 137, 138
+ nitrous oxide vol 8 E160, 171
+ oxygen vol 7 E214, E215, 228, 229, 239, 452
+ sulfur dioxide
+ xenon vol 2 159

Nonane (ternary)
+ hydrogen vol 5/6 561
Nonane, eicosafluoro-
+ oxygen vol 7 332
Nonane, 2,2,4,4,6,8,8-heptamethyl-
+ ethane vol 9 E77, E78, 107
Nonanoic acid
+ radon-222 vol 2 205
1-Nonanol
+ hydrogen vol 5/6 207
+ nitrogen vol 10 E177, E178, 213
+ nitrous oxide vol 8 204
+ oxygen vol 7 E266, 287, 453
+ water vol 15 E392, E393, 394 - 397
2-Nonanol
+ water vol 15 398
3-Nonanol
+ water vol 15 399
4-Nonanol
+ water vol 15 400
5-Nonanol
+ water vol 15 401
3,6,9,12,15,21,24,27-Nonaoxatriacontane,1,1,1,2,4,4,5,7,7,8,10,10,11,13,13,14,16,16,17,19,20,22,22,23,25,25,26,28,28,29,29,30,30,30-pentacontafluoro-5,8,14,17,20,23,26-octakis(trifluoromethyl)-
+ oxygen vol 7 344
Non-aqueous liquids
+ radon-222 vol 2 E259 - 261

(Z,Z)-9,12-Octadecadienoic acid
+ radon-222 vol 2 310
Octadecafluorodecahydronaphthalene
see naphthalene, octadecafluorodecahydro-
Octadecafluorooctane
see Octane,octadecafluoro-
Octadecane
+ ethane vol 9 E77, E78, 99, 101
+ oxygen vol 7 E214, E215, 239
(Z)-9-Octadecanoic acid
+ nitrogen vol 10 280
+ radon-222 vol 2 311
Octadecanol
+ water vol 15 427
Octamethycyclotetrasiloxane
see Cyclotetrasiloxane, octamethyl-
Octane
+ air vol 10 517
+ argon vol 4 E114 - 115, 116 - 119, E253 - 255, 274
+ chlorine vol 12 E354 -E366, 370
+ deuterium vol 5/6 E281, 285, E300, E355
+ ethane vol 9 E77, E78, 89 - 91, E110 -E112, 127, 128
+ helium vol 1 E43, 44
+ hydrogen vol 5/6 E121, 130, 134, 135, E355, 365 - 367
+ krypton vol 2 E35, 36 - 38
+ neon vol 1 E188, 189
+ nitrogen vol 10 E119, E120, 131-134, E419, 431, 432
+ nitrous oxide vol 8 E160, 169
+ oxygen vol 7 E214, E215, 222 - 224, 239, 452
+ sulfur dioxide vol 12 E116, E117, 123, 124
+ xenon vol 2 159
iso-Octane
see Pentane, 2,2,4-trimethyl-
Octane, octadecafluoro-
+ air vol 10 526
+ oxygen vol 7 326
Octane, 1-bromo-1,1,2,2,3,3,4,4,5,5,6,6,7,7,8,8,8-heptadecafluoro-
+ air vol 10 525
+ oxygen vol 7 325
Octane, 1-chloro-
+ sulfur dioxide vol 12 314
Octane, 1-chloro-1,1,2,2,3,3,4,4,5,5,6,6,7,8,8,8-hexadecafluoro-7-(trifluoromethyl)-
+ oxygen vol 7 331
Octane, 1-bromo-
+ sulfur dioxide vol 12 315
Octane, 1,1,1,2,2,3,3-heptafluoro-4,4-bis(trifluoromethyl)-
+ oxygen vol 7 334

Octanoic acid, sodium salt (aqueous)
+ oxygen vol 7 451
Octanoic acid
+ radon-222 vol 2 304
Octanoic acid, 1,2,3-propanetriyl ester
+ radon-222 vol 2 317
1-Octanol
+ air vol 10 520
+ hydrogen vol 5/6 E189, 206, 207
+ ethane vol 9 E166, E167, 188
+ helium vol 1 86
+ krypton vol 2 79
+ nitrogen vol 10 E176 - E178, 209-212
+ nitrous oxide vol 8 203
+ oxygen vol 7 E266, 283 - 285, 453
+ sulfur dioxide vol 12 E189, E191, 193, 195
+ water vol 15 E364 -E366, 367 - 381
2-Octanol
+ water vol 15 E382, 383 - 385
1-Octanol
+ neon vol 1 231
+ argon vol 4 E169 - 170, 196 - 197, E253 - 255, 290, 291
1-Octanol, 7-methyl-
+ water vol 15 391
2-Octanone
+ sulfur dioxide vol 12 250
2-Octanone
+ argon vol 4 E253 - 255, 303
1-Octene
+ hydrogen vol 5/6 E400, 405
Octyl bromide
see Octane, 1-bromo-
Octyl chloride
see Octane, 1-chloro-
Octyl formate
see Formic acid, octyl ester
Oil
+ oxygen vol 7 437, 442
Oil, apiezon GW
see Apiezon oil GW
Oil, animal

Oil, asphaltic
+ air vol 10 538
Oil, butter
+ air vol 10 534
+ nitrogen vol 10 277
Oil, cod liver
+ air vol 10 535, 536
Oil, corn
+ air vol 10 533, 536
+ nitrogen vol 10 276
Oil, cotton seed
+ air vol 10 533, 534, 536
+ nitrogen vol 10 276, 277
+ oxygen vol 7 366, 367

Oil, gas		
+ hydrogen	vol 5/6	521
+ nitrogen	vol 10	504
Oil, herring		
+ air	vol 10	536
Oil, hydrocarbon		
+ air	vol 10	544
+ nitrogen	vol 10	328
Oil, hydrogenated cotton seed		
+ air	vol 10	533
+ nitrogen	vol 10	276
+ oxygen	vol 7	366
Oil, linseed		
+ air	vol 10	536
Oil, lubricating		
+ air	vol 10	539, 545
Oil, mineral		
+ air	vol 10	536, 537 - 539
+ nitrogen	vol 10	319 - 321, 331
+ oxygen	vol 7	436
Oil, olive		
+ air	vol 10	535, 536
+ argon	vol 4	E241, 242 - 244
+ hydrogen	vol 5/6	275
+ helium	vol 1	118
+ krypton	vol 2	E108, E109, 110 - 113
+ neon	vol 1	250
+ nitrogen	vol 10	E279, 280 - 283
+ nitrous oxide	vol 8	256
+ oxygen	vol 7	368 - 371, 438
+ radon-222	vol	E328, 329 - 331
+ xenon	vol 2	E193 - 194, 195 - 198
Oil, soybean		
+ nitrogen	vol 10	278
Oil, transformer type		
+ air	vol 10	538
Oil, vegetable		
+ hydrogen	vol 5/6	272, 274
Oils		
+ hydrogen	vol 5/6	270
Oil		
see Colza oil, Creozote oil, Harp seal oil, Hydrocarbon oil, Jojoba oil, Paraffin oil, Silicone oil, Solvent refined coal oil, Sperm whale oil and Vaseline oil		
Oleic acid		
see (Z)-9-Octadecenoic acid		
Oleum		
see Disulfuric acid		
Omental fat		
see Guinea pig omental fat		
Ovalbumen (aqueous)		
+ argon	vol 4	246
+ oxygen	vol 7	391, 392, 397
Ovaries		
see guinea pig ovaries		
3-Oxabutanoic acid, ethyl ester		
+ sulfur dioxide	vol 12	242
Oxacillin sodium		
see 4-thia-1-azabicyclo[3,2,0]heptane-2-carboxylic acid, 3,3-dimethyl-6[[(5-methyl-3-phenyl-4-isoxazoyl)carbonyl]amino]-7-oxo-, monosodium salt		

Oxalic acid
see 1,2-Ethanedioic acid
Oxide
see Aluminium oxide, Iron oxide
Oxirane
+ ethane vol 9 E195 -E199, 213, 214
2-Oxopropanoic acid, sodium salt
see Propanoic acid, 2-oxo, sodium salt
1,1´-Oxybisbutane
see Butane, 1,1´-oxybis-
1,1´-Oxybis-2-chloroethane
see Ethane, 1,1´-oxybis(2-chloro-
1,1´-Oxybisethane
see ethane, 1,1´-oxybis-
2,2´-Oxybisethanol
see Ethanol, 2,2´-oxybis-
2,2´-Oxybispropane
see Propane, 2,2´-oxybis-
Oxygen
+ helium vol 1 E348, 349
+ neon vol 11 E380, 381
Oxygen (multicomponent)
+ nitrogen vol 10 76, 353, 354, 466
Oxygen fluoride
+ nitrogen vol 10 467
Ox blood
see Bovine blood
Ox erythrocytes
+ nitrogen vol 10 290, 291, 295

Paraffin
see kerosene in addition to entries given below

Paraffin oil		
+ air	vol 10	538
+ argon	vol 4	143
+ hydrogen	vol 5/6	271
+ krypton-85	vol 2	120
+ krypton	vol 2	68
+ nitrogen	vol 10	322 - 324
+ oxygen	vol 7	439, 440, 446
+ radon-222	vol 2	E322, 325, 326, 331
+ xenon-133	vol 2	205, 206
Paraffin, soft		
+ sulfur dioxide	vol 12	173
Paraffin, soft (ternary)		
+ sulfur dioxide	vol 12	317
Paraffin wax		
+ argon	vol 4	144
+ nitrogen	vol 10	325
+ oxygen	vol 7	447, 448

Penicillin V
see 4-thia-1-azabicyclo[3,2,0]heptane-2-carboxylic acid, 3,3-dimethyl-7-oxo-6-[(phenoxyacetyl)amino]-

Pentadecane		
+ argon	vol 4	140
+ helium	vol 1	58
+ hydrogen	vol 5/6	144
+ krypton	vol 2	49
+ neon	vol 1	203
+ nitrogen	vol 10	E119, 145
+ nitrous oxide	vol 8	E160, 177
+ oxygen	vol 7	E214, E215, 236, 239
+ sulfur dioxide	vol 12	E116, E117, 135
+ xenon	vol 2	161
1-Pentadecanol		
+ water	vol 15	421
Pentanal		
+ argon	vol 4	E169 - 170
Pentane		
+ argon	vol 4	107, 271
+ ethane	vol 9	E77, E78, 79, E110 -E112, 119, 120
+ helium	vol 1	40
+ hydrogen	vol 5/6	E119, 125, E354, 358
+ krypton	vol 2	25
+ neon	vol 1	185
+ nitrogen	vol 10	123, E419, 420
+ nitrous oxide	vol 8	E160, 163, 164
+ oxygen	vol 7	E214, E215, 216, 452
+ sulfur dioxide	vol 12	E116, E117, 118
+ xenon	vol 2	155
Pentane (ternary)		
+ ethane	vol 9	137
Pentane, 1-chloro-		
+ sulfur dioxide	vol 12	E1, E2
Pentane, 1,1,1,2,2,3,3,5,5,5-decafluoro-4-methoxy-(trifluoromethyl)-		
+ oxygen	vol 7	316
Pentane, 1,1,1,2,2,3,3,5,5,5-decafluoro-4-propoxy-4-(trifluoromethyl)-		
+ oxygen	vol 7	330
Pentane, 2-ethoxy-1,1,1,3,3,4,4,5,5,5-decafluoro-2-(trifluoromethyl)-		
+ oxygen	vol 7	323

Pentane, 2,2,4-trimethyl-		
+ air	vol 10	517, 518
+ argon	vol 4	E123, 124, E253 - 255, 275
+ chlorine	vol 12	E354 -E366, 371
+ deuterium	vol 5/6	E281, 286
+ ethane	vol 9	E77, E78, 106
+ helium	vol 1	49
+ hydrogen	vol 5/6	136, E356, 376 - 379
+ krypton	vol 2	34
+ neon	vol 1	194
+ nitrogen	vol 10	E121, 131, 135, 136, 433, 434
+ nitrous oxide	vol 8	E160, 170
+ oxygen	vol 7	225 - 227, 238, 304
+ sulfur dioxide	vol 12	124 - 126
+ xenon	vol 2	158
Pentane, 2,2,4-trimethyl- (ternary)		
+ nitrogen	vol 10	E104, 113 115 - 118
+ oxygen	vol 7	213
1,5-Pentanediamine, bis-tetraphenylborate		
+ water	vol 18	95
Pentanedioic acid, dimethyl ester		
+ hydrogen	vol 5/6	479
+ nitrogen	vol 10	451
1,5-Pentanediol (aqueous)		
+ nitrous oxide	vol 8	127
Pentanoic acid		
+ radon-219	vol 2	E339, 342
+ radon-222	vol 2	301
Pentanoic acid, sodium salt (aqueous)		
+ oxygen	vol 7	451
1-Pentanol		
+ argon	vol 4	E169 - 170, 193
+ ethane	vol 9	E166, E167, 182 - 184
+ hydrogen	vol 5/6	E189, 204, 205, E442, 464
+ krypton	vol 2	78
+ nitrogen	vol 10	E176 -E178, 202 - 205
+ nitrous oxide	vol 8	200
+ oxygen	vol 7	E265, 282, 453
+ radon-219	vol 2	E339, 342
+ radon-222	vol 2	284, 285
+ water	vol 15	E162 -E167, 168 - 195
1-Pentanol (multicomponent)		
+ ethane	vol 9	E61, 62, 63
2-Pentanol		
+ water	vol 15	E196 -E197, 198 - 202
3-Pentanol		
+ water	vol 15	E203, E204, 205 - 209
iso-Pentanol		
see 1-butanol, 3-methyl-		
tert-Pentanol		
see 2-butanol, 2-methyl-		
2-Pentanol, 1,1,1,3,3,4,4,5,5,5-decafluoro-2-(trifluorormethyl)-		
+ oxygen	vol 7	310
1-Pentanol, 2,2-diethyl-		
+ water	vol 15	386

1-Pentanol, 2,2-dimethyl-
+ water vol 15 300
1-Pentanol, 2,4-dimethyl-
+ water vol 15 310
1-Pentanol, 4,4-dimethyl-
+ water vol 15 317
2-Pentanol, 2,3-dimethyl-
+ water vol 15 E304, 305, 306
2-Pentanol, 2,4-dimethyl-
+ water vol 15 E311, 312, 313
3-Pentanol, 2,2-dimethyl-
+ water vol 15 E301, 302, 303
3-Pentanol, 2,3-dimethyl-
+ water vol 15 E307, 308, 309
3-Pentanol, 2,4-dimethyl-
+ water vol 15 E314, 315, 316
3-Pentanol, 3-ethyl-
+ water vol 15 E318, 319, 320
1-Pentanol, 2-methyl-
+ water vol 15 E240, 241, 242
1-Pentanol, 4-methyl-
+ water vol 15 256
2-Pentanol, 2-methyl-
+ water vol 15 E243, 244 - 246
2-Pentanol, 3-methyl-
+ water vol 15 E250, 251, 252
2-Pentanol, 4-methyl-
+ water vol 15 E257, E258, E259 -E258
3-Pentanol, 2-methyl-
+ water vol 15 E247, 248, 249
3-Pentanol, 3-methyl-
+ water vol 15 E253, 254, 255
3-Pentanol, 2,2,3-trimethyl-
+ water vol 15 E258, 259 - 262
2-Pentanone
+ argon vol 4 E169 -E170, 253 - 255, 300
+ oxygen vol 7 295, 453
+ sulfur dioxide vol 12 210
3-Pentanone
+ argon vol 4 E169 -E170, 253 - 255, 301
+ oxygen vol 7 453
2-Pentanone, 4-methyl-
+ sulfur dioxide vol 12 210
3,6,9,12,15-Pentaoxaoctadecane, 1,1,1,2,4,4,5,7,7,8,10,10,11,13,13,14,16,16,17,17,18,18,18-tricosfluoro-5,8,11,14-tetrakis(trifluoromethyl)-
+ oxygen vol 7 343
2,3,8,11,14-Pentaoxapentadecane
+ sulfur dioxide vol 12 258
1-Penten-3-ol
+ water vol 15 120
3-Penten-2-ol
+ water vol 15 121
4-Penten-1-ol
+ water vol 15 122
1-Penten-3-ol, 4-methyl-
+ water vol 15 225
Pentyl acetate
see acetic acid, pentyl ester
iso-Pentyl alcohol
see 1-butanol, 3-methyl-
tert-Pentyl carbinol
see 1-butanol, 2,2-dimethyl-
Pentylmethyl carbinol
see 2-heptanol

Perchloric acid (aqueous)
+ argon vol 4 E33, 37
+ chlorine vol 12 E349
+ helium vol 1 E20, 24
+ iodic acid, potassium salt + iodic acid, strontium salt vol 14 203
+ iodic acid, strontium salt vol 14 203
+ nitric oxide vol 8 334, 335
+ silver cyanide + sodium cyanide + perchloric acid, sodium salt vol 3 79 - 80
+ thiocyanic acid, silver salt + thiocyanic acid, potassium salt vol 3 164, 171 - 172
Perchloric acid, lithium salt (aqueous)
+ iodic acid, barium salt + nitric acid, lithium salt vol 14 261
+ iodic acid, strontium salt + nitric acid, lithium salt vol 14 199
+ nitric oxide vol 8 333
+ oxygen vol 7 E66, E67, 136
Perchloric acid, lithium salt (in 4-methyl-1,3-dioxolan-2-one soln.)
+ oxygen vol 7 185
Perchloric acid, potassium salt
+ thiocyanic acid, silver salt + thiocyanic acid, potassium salt + dimethylsulfoxide vol 3 E185, 187
+ thiocyanic acid, silver salt + thiocyanic acid, potassium salt + 1,2-ethanediol vol 3 220
Perchloric acid, potassium salt (aqueous)
+ chlorine vol 12 E349
+ iodic acid, barium salt vol 14 E253, 268
+ oxygen vol 7 E73, 136
+ thiocyanic acid, silver salt + thiocyanic acid, potassium salt vol 3 196 - 197, 203 - 204
+ thiocyanic acid, silver salt + thiocyanic acid, potassium salt + dimethylformamide vol 3 184
+ thiocyanic acid, silver salt + thiocyanic acid, potassium salt + 1,4-dioxane vol 3 E188, 190 - 192
+ thiocyanic acid, silver salt + thiocyanic acid, potassium salt + ethanol vol 3 E193, 196 - 198
+ thiocyanic acid, silver salt + thiocyanic acid, potassium salt + methanol vol 3 E200, 203 - 204
+ thiocyanic acid, silver salt + thiocyanic acid, potassium salt + 2-propanone vol 3 E177 - 178, 181 - 183
Perchloric acid, rubidium salt (aqueous)
+ oxygen vol 7 E74, E75, 136
Perchloric acid, silver salt (aqueous)
+ selenocyanic acid, silver salt vol 3 E90 - 91, 100 - 101
Perchloric acid,sodium salt
+ silver tetraphenylborate + sodium tetraphenylborate + N,N-dimethylformamide vol 18 156
+ silver tetraphenylborate + sodium tetraphenylborate + dimethylsulfoxide vol 18 160
+ silver tetraphenylborate + sodium tetraphenylborate + ethanol vol 18 163
Perchloric acid, sodium salt (aqueous)
+ chlorine vol 12 E349
+ nitric oxide vol 8 333
+ oxygen vol 7 136
+ silver azide + sodium azide vol 3 E1, 15 - 16
+ silver tetraphenylborate + sodium tetraphenylborate + ethanol vol 18 151
+ silver tetraphenylborate + sodium tetraphenylborate + 4-methyl-1,3-dioxolan-2-one vol 18 176

Perchloric acid, sodium salt (aqueous)
+ silver tetraphenylborate + sodium tetraphenylborate + N-methyl-2-pyrrolidinone vol 18 170
+ silver tetraphenylborate + sodium tetraphenylborate + nitromethane vol 18 171
+ silver tetraphenylborate + sodium tetraphenylborate + 2-propanone vol 18 173
+ silver tetraphenylborate + sodium tetraphenylborate + tetrahydrothiophene-1,1-dioxide vol 18 176
+ sodium dithiocyanoargentate + thiocyanic acid, sodium salt vol 3 E104, 157 - 158
+ thiocyanic acid, silver salt vol 3 175 - 176
+ thiocyanic acid, silver salt + thiocyanic acid, sodium salt vol 3 E104, 154 - 156, 159 - 163, 172
+ thiocyanic acid, silver salt + thiocyanic acid, sodium salt + sodium dithiocyanoargentate vol 3 E104, 157 - 158
+ thiocyanic acid, silver salt + thiosulfuric acid, sodium salt vol 3 173 - 174

Perchloric acid, tetrabutylammonium salt
+ thiocyanic acid, silver salt + thiocyanic acid, tetrabutylammonium salt + 1,2-dimethoxyethane vol 3 209 - 210
+ thiocyanic acid, silver salt + thiocyanic acid, tetrabutylammonium salt + tetrahydrofuran vol 3 230 - 231

Perchloric acid, tetraethylammonium salt
+ silver azide + tetraethylammonium azide + dimethyl sulfoxide vol 3 E23, 25
+ silver cyanide + tetraethylammonium cyanide + dimethyl sulfoxide vol 3 84
+ silver cyanide + tetraethylammonium cyanide + nitromethane vol 3 85
+ thiocyanic acid, silver salt + nitric acid, silver salt + dimethyl sulfoxide vol 3 E213, 215
+ thiocyanic acid, silver salt + thiocyanic acid, potassium salt + tetrahydrothiophene-1,1-dioxide vol 3 232
+ thiocyanic acid, silver salt + thiocyanic acid, tetraethylammonium salt + dimethyl sulfoxide vol 3 E213, 216
+ thiocyanic acid, silver salt + thiocyanic acid, tetraethylammonium salt + nitromethane vol 3 227
+ thiocyanic acid, silver salt + thiocyanic acid, tetraethylammonium salt + 4-methyl-1,3-dioxolan-2-one vol 3 228

Perchloric acid, tetraethylammonium salt (aqueous)
+ silver azide + tetraethylammonium azide + dimethyl sulfoxide + vol 3 18

Perchloryl fluoride
see chlorine, hexafluoride

Perfluorobenzene
see benzene, hexafluoro-

Perfluorobutyltetrahydrofuran
see furan, heptafluorotetrahydro(nonafluorobutyl)-

Perfluorodecalin
see naphthalene, octadecafluorodecahydro-

Perfluorodimethylcyclohexane
see cyclohexane, decafluorobis(trifluoromethyl)-

Perfluoro-1,4-diisopropoxybutane
see butane, 1,1,2,2,3,3,4,4-octafluoro-1,4-bis(1,2,2,2-tetrafluoro-1-(trifluoromethyl)-ethoxy)-

Perfluoroheptane
see heptane, hexadecafluoro-

Perfluoromethylcyclohexane
see cyclohexane, undecafluoro(trifluoromethyl)-
Perfluoromethyldecalin
see naphthalene, heptadecafluorodecahydro(trifluoromethyl)-
Perfluorononane
see nonane, eicosafluoro-
Perfluorooctane
see octane, octadecafluoro-
Perfluoro-1-isopropoxyhexane
see hexane, 1,1,1,2,2,3,3,4,4,5,5,6,6-tridecafluoro-6-(1,2,2,2-tetrafluoro-1-(trifluoromethyl11)ethoxy)-
Perfluorotributylamine
see 1-butanamine, 1,1,2,2,3,3,4,4,4-nonafluoro-N,N-bis(nonafluorobutyl)-

Periodic acid		
+ nitrous oxide	vol 8	E40
Periodic acid, potassium salt (aqueous)		
+ nitrous oxide	vol 8	E38, 111
Permanganate, potassium		
see permanganic acid, potassium salt		
Permanganic acid, potassium salt (acqueous)		
+ radon-222	vol 2	E238 - 241, 254
Petroleum		
+ hydrogen	vol 5/6	269
+ krypton	vol 2	68
+ nitrogen	vol 10	329
+ nitrous oxide	vol 8	E29, 257
+ oxygen	vol 7	444
+ radon-219	vol 2	E339, 341
+ radon-220	vol 2	E343, 345
+ radon-221	vol 2	E322, 323 - 325
Petroleum ether		
+ oxygen	vol 7	304
+ radon-222	vol 2	E322, 325
Petrowax A		
+ nitrogen	vol 10	326
+ oxygen	vol 7	448
Phenanthrene, 9,10-dihydro-		
+ hydrogen	vol 5/6	432, 433

Phenethicillin potassium
see 4-thia-1-azabicyclo[3,2,0]heptane-2-carboxylic acid, 3,3-dimethyl-7-oxo-6-[(1-oxo-2-phenoxypropyl)amino], monopotassium salt
Phenetole
see benzene, ethoxy-

Phenol		
+ hydrogen	vol 5/6	226
+ ethane	vol 9	E166, E167, 191
+ sulfur dioxide	vol 12	235, 236
Phenol, 2-chloro-		
+ sulfur dioxide	vol 12	313
Phenol, 2-methoxy-		
+ sulfur dioxide	vol 12	236
Phenol, methyl-		
+ hydrogen	vol 5/6	475

Phenoxymethyl penicillin
see 4-thia-1-azabicyclo[3,2,0]heptane-2-carboxylic acid, 3,3-dimethyl-7-oxo-6-[(phenoxyacetyl)amino]-
7-Phenylacetamidodeacetoxycephalosporanic acid
see 5-thia-1-azabicyclo[4,2,0]oct-2-ene-2-carboxylic acid, 3-methyl-8-oxo-7-(phenylacetamido)-
Phenyl cellosolve
see ethanol,2-phenoxy-

Phosphate buffer		
+ hydrogen	vol 5/6	118
+ nitric oxide	vol 8	333

Phosphate, tricresyl-
see phosphoric acid, tris(methylphenyl) ester
Phosphoric acid
+ nitrous oxide vol 8 E40, 44, 45
Phosphoric acid, potassium salt (aqueous)
+ oxygen vol 7 E74, 164
Phosphoric acid, sodium salt (aqueous)
+ nitrous oxide vol 8 E35, 92
+ oxygen vol 7 E71, 164
Phosphoric acid, sodium salt (phosphate buffer in water)
+ krypton vol 2 E12, E13, 23
+ xenon vol 2 E149, 154
Phosphoric acid, tributyl ester
+ argon vol 4 E253 -E255, 310
+ ethane vol 9 E195 -E199, 230
+ hydrogen vol 5 519
+ nitric acid, cerium salt vol 13 E221, 225
+ sulfur dioxide vol 12 324, 326, 327
Phosphoric acid, triethyl ester
+ ethane vol 9 E195 -E199, 230
+ sulfur dioxide vol 12 325
Phosphoric acid, trimethyl ester
+ ethane vol 9 E195 -E199, 230
Phosphoric acid, tripropyl ester
+ ethane vol 9 E195 -E199, 230
+ sulfur dioxide vol 12 325
Phosphoric acid, trisodium salt
+ nitrous oxide vol 9 E35, 92
Phosphoric acid, tris (methylphenyl) ester
+ krypton vol 2 106
Phosphoric acid, tris (2-methyl propyl) ester
+ ethane vol 9 E195 -E199, 230
+ sulfur dioxide vol 12 324, 325
Phosphoric triamide, hexamethyl-
+ ethane vol 9 E195 -E199, 218
+ silver azide + sodium azide + nitric acid, sodium salt vol 3 27
+ silver tetraphenylborate (1-) vol 18 165
+ tetraphenylarsonium tetraphenylborate (1-) vol 18 E213, 214 - 216
+ thiocyanic acid, silver salt + thiocyanic acid, potassium salt + nitric acid, potassium salt vol 3 222
Physiologic salt solution
+ oxygen vol 7 386
Picoline
see pyridine,methyl-
Pine oil
+ xenon vol 2 191
Pine oil (ternary)
+ xenon vol 2 191
Pinene
+ sulfur dioxide vol 12 E147, 169
Piperazine dinitrate
+ water vol 13 330, 402
Piperazine dinitrate (aqueous)
+ nitric acid, gadolinium salt vol 13 402
+ nitric acid, neodymium salt vol 13 330
Piperidine
+ hydrogen vol 5/6 265, E482, 500
+ nitrogen vol 10 263
+ oxygen vol 7 352
Piperidine nitrate
+ water vol 13 100, 101, 211, 212, 329, 401

Piperidine nitrate (aqueous)
+ nitric acid, cerium salt vol 13 211, 212
+ nitric acid, gadolinium salt vol 13 401
+ nitric acid, lanthanum salt vol 13 100, 101
+ nitric acid, neodymium salt vol 13 329
+ nitric acid, praseodymium salt vol 13 268
Piperidine, 1-methyl-
+ hydrogen vol 5/6 E482, 502
Placenta
see human placenta
Plasma
see also human plasma, human hyperlipidemic plasma
Plasma
+ krypton -85 vol 2 121, 124, 127
+ nitrogen vol 10 294, 295
+ xenon-133 vol 2 207, 213 - 215, 218
Polyethylene glycol
+ sulfur dioxide vol 12 260
Polydimethylsiloxane oil
+ oxygen vol 7 441
Poppy oil
+ radon-222 vol 2 333
Potassium amide (ternary)
+ deuterium vol 5/6 296
Potassium azide (ternary)
+ silver azide vol 3 14
Potassium bicarbonate
see carbonic acid, monopotassium salt
Potassium bromide
+ acetamide vol 11 E249, 250
+ acetamide, N,N-dimethyl- vol 11 E327, 328
+ acetamide, N-methyl- vol 11 E274, E275, 276, 277
+ formamide vol 11 E35, 36 - 38
+ formamide, N,N-dimethyl- vol 11 E194, 197
+ formamide, N-methyl- vol 11 E118, 119 - 122
Potassium bromide (aqueous)
+ argon vol 4 E33 - E36, 62 - 63, 72
+ iodic acid, calcium salt vol 14 E70, E74, 138
+ krypton vol 2 E12, E13, 21
+ neon vol 1 E141, 169
+ nitrous oxide vol 8 E36, 101, 102
+ oxygen vol 7 E73, 134, 176
+ silver cyanide + silver bromide + bis(cyano-C-)-potassium argentate vol 3 E45, 64
+ thiocyanic acid, silver salt + silver bromide + thiocyanic acid, potassium salt vol 3 E102, 127
+ thiocyanic acid, silver salt + thiocyanic acid, potassium salt + sulfuric acid vol 3 121
+ sulfur dioxide vol 12 E37 - E39, 100
Potassium carbonate
see carbonic acid, dipotassium salt
Potassium chlorate
see chloric acid, potassium salt
Potassium chloride
+ acetamide vol 11 E247, 248
+ acetamide, N,N-dimethyl- vol 11 E325, 326
+ acetamide, N-methyl- vol 11 E271, 272, 273
+ formamide vol 11 E28, E29, 30-34
+ formamide, N,N-dimethyl- vol 11 E186, 187 - 193
+ formamide, N-methyl- vol 11 E113, 114 - 117
Potassium chloride(aqueous)
+ argon vol 4 E33 - E36, 47, 49, 61

Potassium chloride (aqueous)
+ bromic acid, barium salt vol 14 230
+ chloric acid, calcium salt + calcium chloride
vol 14 57, 58
+ chlorine vol 12 E349, 353
+ ethane vol 9 E32, 57
+ glycine + iodic acid, barium salt
vol 14 E252, 283
+ helium vol 1 E20, 29
+ hydrogen vol 5/6 E34, 44 ,83 - 85
+ iodic acid, barium salt vol 14 E247, E248, 264 - 266
+ iodic acid, calcium salt vol 14 E70, E72, 133 - 136
+ iodic acid, calcium salt + glycine
vol 14 137
+ krypton vol 2 E12, E13, 21
+ neon vol 1 E141, 168, 177, 178
+ nitrogen vol 10 E45 - E47, E54, E55, 78, 79
+ nitrous oxide vol 8 E35, E36, E39 96 - 100
+ oxygen vol 7 E72, E73, 133, 170 - 175
+ radon-219 vol 2 E339, 340
+ radon-222 vol 2 E238 - E241, 255
+ silver cyanide + silver chloride + bis(cyano-C-)-potassium argentate vol 3 E45, 63
+ sodium chloride + iodic acid, calcium salt
vol 14 100, 101
+ sulfur dioxide vol 12 E37 - E39, 97 - 99
+ thiocyanic acid + silver chloride + thiocyanic acid potassium salt vol 3 E102, 126
Potassium chromate
see chromic acid, potassium salt
Potassium cyanate
+ formamide, N,N-dimethyl- vol 11 E206, 207
Potassium cyanate (aqueous)
+ silver cyanide vol 3 E38 - E39, 43 - 44
Potassium cyanide
+ formamide vol 11 E46, 47
+ formamide, N,N-dimethyl- vol 11 E204, 205
Potassium cyanide (aqueous)
+ silver cyanide vol 3 E45, 68 - 69
+ silver cyanide + nitric acid, potassium salt
vol 3 72 - 73, 77, 78
+ silver cyanide + silver iodide + nitric acid, barium salt + potassium iodide + nitric acid, potassium salt
vol 3 81
Potassium dicyanoargentate
see argentate(1)-, bis(cyano-C)-potassium
Potassium ferrocyanide
see ferrate (4-), hexakis(cyano-C-), tetrapotassium
Potassium fluoride (aqueous)
+ neon vol 1 E141, 174
+ nitrous oxide vol 8 E35, 95
+ oxygen vol 7 E72, 132
Potassium hydroxide (aqueous)
+ argon vol 4 E33, E34, 60
+ helium vol 1 E20, 28
+ hydrogen vol 5/6 E33, E34, 77-82
+ iodic acid, calcium salt vol 14 E68, E69, 132
+ neon vol 1 E141, 167
+ nitrous oxide vol 8 E35, 94

Potassium hydroxide (aqueous)
+ oxygen vol 7 E71, E72, 131, 162, 165 - 169
Potassium hydroxide (aqueous methanol soln)
+ oxygen vol 7 185, 189, E192
Potassium iodide
+ acetamide vol 11 E251, E252, 253, 254
+ acetamide, N,N-dimethyl- vol 11 E329, 330
+ acetamide, N-methyl- vol 11 E278, E279, 280, 281
+ formamide vol 11 E39, E40, 41 - 45
+ formamide, N,N-dimethyl- vol 11 E198, 199 - 203
+ formamide, N-methyl- vol 11 E123, 124 - 127
Potassium iodide (aqueous)
+ argon vol 4 E33 - E36, 48, 64 - 65, 73
+ ethane vol 9 E32, 33, 58
+ helium vol 1 E20, 30
+ hydrogen vol 5/6 E34, 44, 86
+ iodic acid, calcium salt vol 14 E70, E74, 139
+ krypton vol 2 E12, E13, 21
+ neon vol 1 E141, 170, 173, 177
+ nitrogen vol 10 E45 - E47 E55, 80
+ nitrous oxide vol 8 E37, 103, 104
+ oxygen vol 7 E73, 135, 177
+ silver cyanide + siver iodide + nitric acid, barium salt + potassium cyanide + nitric acid, potassium salt vol 3 81
+ sulfur dioxide vol 12 E37 - E39, 101
+ xenon vol 2 E149, 151
Potassium nitrate
see nitric acid,potassium salt
Potassium penicillin G
see 4-thia-1-azabicyclo[3,2,0]heptane-2-carboxylic acid, 3,3-dimethyl-7-oxo-6-[(phenylacetyl)amino], monopotassium salt
Potassium penicillin V
see 4-thia-1-azabicyclo[3,2,0]heptane-2-carboxylic acid, 3,3-dimethyl-7-oxo-6-[(phenoxyacety)amino], monopotassium salt
Potassium perchlorate
see perchloric acid, potassium salt
Potassium periodate
see periodic acid, potassium salt
Potassium permanganate
see permanganic acid, potassium salt
Potassium phosphate
see phosphoric acid, potasium salt
Potassium selenocyanate (aqueous)
+ silver selenocyanate vol 3 E90, 94 - 95
+ silver selenocyanate + nitric acid, potassium salt vol 3 E90, 96 - 97
Potassium sulfate
see sulfuric acid, potassium salt
Potassium tetraphenylborate
+ acetonitrile vol 18 E31, E32, 33-36
+ formamide vol 18 37
+ methanol vol 18 38
+ 1-methyl-2-pyrrolidinone vol 18 39
+ 2-propanone vol 18 E40, E41, 42
+ water vol 18 E7 - E10, 11-14, 20

Potassium tetraphenylborate (aqueous)
+ lithium chloride + ethanol vol 18 25, 26
+ lithium chloride + methanol + sodium hydroxide
vol 18 27 - 29
+ 2-propanone vol 18 15, 16
E21, 22, 23
+ 1,1´-sulfinylbismethane vol 18 24
+ sulfuric acid, sodium salt vol 18 17, 18
+ tris(hydroxymethyl)aminomethane + acetic acid
vol 18 19
+ urea vol 18 30
Potassium thiocyanate
see thiocyanic acid, potassium salt
Praseodymium cobalt nitrate
see nitric acid, cobalt praseodymium nitrate
Praseodymium copper nitrate
see nitric acid copper praseodymium nitrate
Praseodymium manganese nitrate
see nitric acid, praseodymium manganese salt
Praseodymium nickel nitrate
see ntric acid, nickel praseodymium nitrate
Praseodymium nitrate
see nitric acid, praseodymium salt
Praseodymium zinc nitrate
see nitric acid praseodymium zinc salt
Prednisolone sodium succinate
see pregn-1,4-diene-3,20-dione,21-(3-carboxy-1-oxopropoxy)-11,17-dihydroxy-,monosodium salt,(11β)-
Pregna-1,4-diene-3,20-dione,9-fluoro-11,17-dihydroxy-16-methyl-21-(phosphonoxy)-, disodium salt, (11β,16β)-
+ oxygen vol 7 380, 381
Pregna-1,4-diene-3,20-dione,9-fluoro-11,17-dihydroxy-16-methyl-21-(phosphonoxy)-, disodium salt, (11β,16α)-
+ oxygen vol 7 380, 381
Pregna-1,4-diene-3,20-dione,21-(3-carboxy-1-oxopropoxy)-11,17-dihydroxy-, monosodium salt, (11β)-
+ oxygen vol 7 380, 381
Pregna-1,4-diene-3,20-dione,21-(3-carboxy-1-oxypropoxy)-11,17-dihydroxy-6-methyl- ,monosodium salt,(6α,11β)-
+ oxygen vol 7 380, 381
Pregn-4-ene-3,11,20-trione,21-(acetyloxy)-17-hydroxy-
+ oxygen vol 7 380, 381
Pregn-4-ene-3,20-dione,21-(3-carboxy-1-oxopropoxy)-11,17-dihydroxy-, monosodium salt,
+ oxygen vol 7 380, 381
Procaine hydrochloride
see benzoic acid, 4-amino-2-(diethylamino)- ethyl ester, monohydrochloride
Procaine penicillin G monohydrate
see 4-thia-1-azabicyclo[3,2,0]heptane-2-carboxylic acid, 3,3-dimethyl-7-oxo-6-[(phenylacetyl)amino], complexed with 2-(diethylamino)ethyl-4-aminobenzoate (1:1) monohydrate
Procaine penicillin G
see 4-thia-1-azabicyclo[3,2,0]heptane-2-carboxylic acid, 3,3-dimethyl-7-oxo-6-[(phenylacetyl)amino], complexed with 2-(diethylamino)ethyl-4-aminobenzoate (1:1)
Proline, 4-hydroxy-
+ oxygen vol 7 397, 400
Propanal
+ argon vol 4 E169 - 170
1-Propanamine
+ hydrogen vol 5/6 E482, 491 - 493
+ oxygen vol 7 453
2-Propanamine
+ hydrogen vol 5 E482, 494

1-Propanaminium, N,N,N-tripropyl-, bromide (aqueous)		
+ argon	vol 4	E33,E34, 39 - 40
+ ethane	vol 9	E29, E30, 39
Propane		
+ ethane	vol 9	E110- E112, 113 - 115
+ helium	vol 1	E254, 278
+ hydrogen	vol 5/6	E339, 340 - 345
+ nitrogen	vol 10	E402, 403 - 408
Propane (multi component)		
+ hydrogen	vol 5/6	553
Propane (ternary)		
+ hydrogen	vol 5/6	528, 529, 540 - 542, 545 - 548
+ nitrogen	vol 10	383, 384
Propane, 1-chloro-		
+ sulfur dioxide	vol 12	E1, E2
Propane, 2-chloro-		
+ sulfur dioxide	vol 12	E1, E2
Propane, 1-chloro-2-methyl-		
+ sulfur dioxide	vol 12	E1, E2
Propane, 2-chloro-2-methyl-		
+ sulfur dioxide	vol 12	E1, E2
Propane, 1,2-dibromo-1,1,2,3,3,3-hexafluoro-		
+ oxygen	vol 7	309
Propane, 1,2-dichloro-		
+ sulfur dioxide	vol 12	E1, E2, 306
Propane, 2,2-dimethyl-		
+ argon	vol 4	E253 - E255, 272
Propane, 1,1,1,2,3,3-hexafluoro-2-(heptafluoropropoxy)-3-(1,2,2-trifluoro-2-(1,2,2,2-tetrafluoroethoxy)-1-(trifluoromethyl)ethoxy)-		
+ oxygen	vol 7	388
Propane, 2-methyl-		
+ ethane	vol 9	E110 - E112, 117, 118
+ hydrogen	vol 5/6	E356, 375
+ nitrogen	vol 10	416, 417
Propane, 2-methyl-(ternary)		
+ nitrogen	vol 10	414, 415
Propane, nitro-		
+ tris(o-phenanthroline)ruthenium (11) tetraphenylborate (1-)	vol 18	134
Propane, 1,1´-oxybis-		
+ hydrogen	vol 5/6	E214, 218
+ nitrogen	vol 10	230
+ oxygen	vol 7	230
Propane, 2,2´-oxybis-		
+ sulfur dioxide	vol 12	234
+ tris(o-phenanthroline)ruthenium (11)tetraphenylborate (1-)	vol 18	129
Propane, 1,2,3-trichloro-		
+ sulfur dioxide	vol 12	E1, E2, 306
1,3-Propanediamine		
+ hydrogen	vol 5/6	E482, 497
1,2-Propanediamine		
+ hydrogen	vol 5/6	E482, 495, 496
1,3-Propanediamine, N-(2-aminoethyl)-N-(3-aminopropyl)-		
+ hydrogen	vol 5/6	E483, 508
1,3-Propanediamine, N-(3-aminopropyl)-		
+ hydrogen	vol 5/6	E483, 505
1,3-Propanediamine, N,N-dimethyl-		
+ hydrogen	vol 5/6	E482, 501
1,3-Propanediol		
+ hydrogen	vol 5/6	E442, 451
1,2-Propanediol		
+ sulfur dioxide	vol 12	204

1,3-Propanediol, 2-amino-2-(hydroxymethyl)-, (aqueous)
+ cesium tetraphenylborate + acetic acid vol 18 60
+ silver tetraphenylborate + acetic acid vol 18 146
+ ammonium tetraphenylborate + acetic acid vol 18 78
+ potassium tetraphenylborate + acetic acid vol 18 19
+ rubidium tetraphenylborate + acetic acid vol 18 47
+ thallium tetraphenylborate + acetic acid vol 18 179
1,2,3-Propanetricarboxylic acid, 2-hydroxy-
+ oxygen vol 7 410
1,2,3-Propanetriol
+ krypton vol 2 81
+ nitrogen vol 10 E84, 100 - 102
+ nitrous oxide vol 8 185
+ oxygen vol 7 305
+ radon -222 vol 2 286, 287
+ sulfur dioxide vol 12 204, 219
1,2,3-Propanetriol (aqueous)
+ argon vol 4 E80 - E84, 98
+ hydrogen vol 5/6 104, 105
+ iodic acid, barium salt vol 14 E251, E252, 304
+ iodic acid, calcium salt vol 14 E74, 169
+ nitrous oxide vol 8 123 - 125
+ oxygen vol 7 207, 409
+ thiocyanic acid, silver salt + thiocyanic acid, potassium salt + nitric acid, nitric acid vol 3 199
1,2,3-Propanetriol (aqueous 1,2-ethanediol)
+ argon vol 4 E80 - E84, 100
1,2,3-Propanetriol (aqueous methanol)
+ argon vol 4 E80 - E84, 99
1,2,3-Propanetriol, triacetate
+ argon vol 4 E253 -E255, 306
+ hydrogen vol 5/6 480
+ nitrogen vol 10 452
+ radon -222 vol 2 314
Propanoic acid
+ argon vol 4 E169, 170
+ nitrogen vol 10 E84, 94, 95
+ nitrous oxide vol 8 E116, 133, 134
+ oxygen vol 7 453
+ ozone vol 7 492
+ radon -222 vol 2 E294, 295, 296
+ sulfur dioxide vol 12 213
Propanoic acid (aqueous)
+ ozone vol 7 493
Propanoic acid,sodium salt (aqueous)
+ oxygen vol 7 451
Propanoic acid, 2-hydroxy-, monosodium salt (aqueous)
+ iodic acid, calcium salt vol 14 E75, E76, 112
Propanoic acid, 2-methyl-
+ nitrogen vol 10 222
+ radon-222 vol 2 298
Propanoic acid, 2-methyl- (aqueous)
+ nitrogen vol 10 E83, 103
Propanoic acid, 2-oxo-, sodium salt (aqueous)
+ iodic acid, calcium salt vol 14 E75, E76, 111
1-Propanol
+ argon vol 4 E169 -E170, E180 -E181, 182 - 184, E253 -E255, 284

1-Propanol
+ ethane vol 9 E166, E167, 176 - 178
+ hydrogen vol 5/6 E187, 200, 201, E441, 446
+ krypton vol 2 75
+ nitrogen vol 10 E174, E175, E177, 178, 191 - 194
+ nitrous oxide vol 8 186, 192
+ oxygen vol 7 E263, 277, 304, 453
+ radon-222 vol 2 278
+ sodium tetraphenylborate vol 18 6
+ tetraethylammonium tetraphenylborate vol 18 105
+ tetramethylammonium tetraphenylborate(1-) vol 18 111
+ tetrapropylammonium tetraphenylborate vol 18 114
+ water vol 15 1
+ xenon vol 2 170

1-Propanol (aqueous)
+ argon vol 4 E80 - E84, 92
+ ethane vol 9 E64, 65
+ iodic acid, barium salt vol 14 E251, E252, 302
+ iodic acid, calcium salt vol 14 E74, 167
+ nitrogen vol 10 E84, 96, 97
+ nitrous oxide vol 8 119, 120
+ oxygen vol 7 E191, 203, 204
+ sulfur dioxide vol 12 114, 115
+ thiocyanic acid, silver salt + nitric acid, potassium salt + thiocyanic acid, potassium salt vol 3 205

1-Propanol (ternary)
+ nitrogen vol 10 E104, 115, 116

2-Propanol
+ hydrogen vol 5/6 E188, E441, 447 - 449
+ nitrogen vol 10 E175, 195, 196
+ nitrous oxide vol 8 186, 192
+ oxygen vol 7 E263, 264, 278, 304
+ radon-222 vol 2 279
+ sulfur dioxide vol 12 192
+ tris(o-phenanthroline)ruthenium (11)tetraphenylborate (1-) vol 18 128
+ water vol 15 2

2-Propanol (aqueous)
+ argon vol 4 E80 - E84, 93
+ nitrogen vol 10 E84, 98, 99
+ nitrous oxide vol 8 121
+ oxygen vol 7 E191, 205, 206
+ silver tetraphenylborate + methylbenzene vol 18 147, 148
+ thiocyanic acid, silver salt + nitric acid, potassium salt + thiocyanic acid, potassium salt vol 3 206
+ triisoamylbutylammonium tetraphenylborate + methylbenzene vol 18 86

2-Propanol (ternary)
+ hydrogen vol 5/6 106 - 108

iso-Propanol
see 2-propanol

2-Propanol, 1-amino-
+ nitrous oxide vol 8 140

1-Propanol, 2,2-dimethyl-
+ water vol 15 E123, 124, 125

2-Propanol, 1,1´-iminobis- (aqueous)
+ nitrous oxide vol 8 142
1-Propanol, 2-methyl-
+ argon vol 4 E169,E170, 190 - 191, E253-E255, 287
+ helium vol 1 85
+ hydrogen vol 5/6 E442, 455 - 459
+ krypton vol 2 77
+ neon vol 1 230
+ nitrogen vol 10 E175, E176, 201
+ oxygen vol 7 E265, 281
+ radon-222 vol 2 282
+ water vol 15 E3 - E7, 8 - 28
1-Propanol, 2-methyl- (aqueous)-
+ argon vol 4 E81
2-Propanol, 2-methyl-
+ argon vol 4 E169, E170, 190
+ hydrogen vol 5/6 106 - 108
+ oxygen vol 7 E192, 208 - 210
+ sulfur dioxide vol 12 192
+ water vol 15 29 - 31
2-Propanol, 2-methyl- (aqueous)
+ argon vol 4 E80-E84, 94 - 95
+ helium vol 1 37
+ oxygen vol 7 E192, 208 - 210
2-Propanol, 2-methyl- (aqueous potassium chloride)
+ argon vol 4 E34, E35
1-Propanol, oxybis-
+ ethane vol 9 E166, E167, 193
iso-Propanolamine
see 2-propanol, 1-amino-
2-Propanone
+ argon vol 4 E169, E170, 203, E253 - 255, 297-298
+ ethane vol 9 E195- E199, 206, E232, E233, 246
+ helium vol 1 89
+ hydrogen vol 5/6 E221, 223, 224
+ krypton vol 2 83
+ neon vol 1 234
+ nitrogen vol 10 E216, 217, 219
+ nitrous oxide vol 8 E160, 193 - 196
+ oxygen vol 7 E265, 281, 290 - 293
+ potassium tetraphenylborate vol 18 E40 - E41, 42
+ radon -222 vol 2 289
+ silver tetraphenylborate + perchloric acid, sodium salt + sodium tetraphenylborate vol 18 173
+ sulfur dioxide vol 12 E189, E191 207 - 212
+ tetraphenylarsonium tetraphenylborate (1-) vol 18 224
+ thiocyanic acid, silver salt + thiocyanic acid, potassium salt vol 3 207
+ tris(o-phenanthroline)ruthenium (11)tetraphenylborate (1-) vol 18 135, 136
2-Propanone (aqueous)
+ ammonium tetraphenylborate vol 18 79
+ argon vol 4 E82, E83
+ cesium tetraphenylborate vol 18 63
+ iodic acid, barium salt vol 14 E251, E252, 305
+ iodic acid, calcium salt vol 14 E74, 172
+ potassium tetraphenylborate vol 18 15, 16 E21, 22, 23

2-Propanone
+ rubidium tetraphenylborate vol 18 48
+ sodium tetraphenylborate vol 18 4
+ thiocyanic acid, silver salt + thiocyanic acid, potassium salt + nitric acid, potassium salt vol 3 E177, 179 - 180
+ thiocyanic acid, silver salt + thiocyanic acid, potassium salt + perchloric acid, potassium salt vol 3 E177, 181 - 182, 183
2-Propanone (aqueous potassium chloride)
+ argon vol 4 E34, E35
2-Propanone (ternary)
+ hydrogen vol 5/6 153, 168, 212
+ nitric oxide vol 8 310, 324
+ nitrogen vol 10 E105, 106, 114
+ sulfur dioxide vol 12 178
1-Propene
+ argon vol4 E253-E255, 276
+ ethane vol 9 E232, E233, 235
+ hydrogen vol 5/6 E400, 401, 402
+ krypton vol 2 50, 51
+ nitrogen vol 10 400, 401
1-Propene (multi component)
+ hydrogen vol 5/6 553
1-Propene (ternary)
+ hydrogen vol 5/6 547 - 551
1-Propene,2-methyl-
+ nitrogen vol 10 418
+ sulfur dioxide vol 12 E1, E2
2-Propenoic acid
+ radon -222 vol 2 297
Propicillin potassium
see 4-thia-1-azabicyclo[3,2,0]heptane-2-carboxylic acid, 3,3-dimethyl-7-oxo-6-[(1-oxo-2-phenoxybutyl)amino]-monopotassium salt
Propionic acid
see propanoic acid
Propyl acetate
see Acetic acid, propyl ester
iso-Propyl alcohol
see 2-propanol
Propylammonium tetraphenylborate
+ water vol 18 96
Propylbenzene
see benzene, propyl-
iso-Propylbenzene
see benzene, (1-methylethyl)-
Propylene
see 1-Propene
Propylene carbonate
see 1,3-dioxolan-2-one, 4-methyl-
Propylene glycol
see 1,2-Propanediol
iso-Propyl ether
see propane, 2,2′-oxybis-
Propylethyl carbinol
see 3-Hexanol
iso-Propylethyl carbinol
see 3-pentanol, 2-methyl-
Propyl nitrate
see nitric acid, propyl ester
Pullulan
+ oxygen vol 7 435
Pyran, 2,3-dihydro-
+ hydrogen vol 5/6 E215, 219
+ nitrogen vol 10 229

2H-Pyran, tetrahydro-
+ hydrogen vol 5/6 E215, 219
+ nitrogen vol 10 229
+ oxygen vol 7 301
Pyridine
+ hydrogen vol 5/6 265
+ nitrogen vol 10 263
+ nitrous oxide vol 8 216, 217
+ oxygen vol 7 352
+ sulfur dioxide vol 12 E261, E262, 278 - 280
Pyridine (ternary)
+ nitric oxide vol 8 310
Pyridine (aqueous)
+ silver azide vol 3 19
+ oxygen vol 7 352
+ silver azide vol 3 29
Pyridine, 2-chloro-6-(trichloromethyl)-
+ chlorine vol 12 437
Pyridine, 5-chloro-2-(trichloromethyl)- (ternary)
+ chlorine vol 12 438
Pyridine, 3,5-dichloro-2-(dichloromethyl)- (ternary)
+ chlorine vol 12 438
Pyridine, 3,5-dichloro-2-(trichloromethyl)- (ternary)
+ chlorine vol 12 438
Pyridine, methyl-
+ sulfur dioxide vol 12 278
Pyridine nitrate
+ nitric acid, erbium salt vol 13 441
+ water vol 13 102, 103, 213, 214, 331, 397, 423, 441
Pyridine nitrate (aqueous)
+ nitric acid, cerium salt vol 13 213, 214
+ nitric acid, dysprosium salt vol 13 E416, E417, 423
+ nitric acid, gadolinium salt vol 13 397
+ nitric acid, lanthanum salt vol 13 102, 103
+ nitric acid, neodymium salt vol 13 331
+ nitric acid, praseodymium salt vol 13 269, 270
Pyridine, 3,4,5-trichloro-2-(dichloromethyl)-
+ chlorine vol 12 433
Pyridine, 2-(trichloromethyl)-
+ chlorine vol 12 442
Pyridine, 3,4,5-trichloro-2-(trichloromethyl)-
+ chlorine vol 12 432
Pyridine, trimethyl-
+ sulfur dioxide vol 12 297
Pyridinium tetraphenylborate
+ water vol 18 97
Pyridinium, 4-(aminocarbonyl)-1[[[2-carboxy-8-oxo-7-(phenylsulfoacetyl)-amino]-5-thia-1-azobicyclo[4,2,0]oct-2-en-3-yl]methyl]-hydroxide, inner salt, monosodium salt
+ acetic acid, ethyl ester vol 16/17 E733, 737
+ hexane vol 16/17 E733, 738
+ methane, trichloro- vol 16/17 E733, 739
+ methanol vol 16/17 E733, 735
+ phosphoric acid, disodium salt (multicomponent) vol 16/17 E733, 734
+ water vol 16/17 E733, 734
+ 1,2,3-propanetricarboxylic acid, 2-hydroxy vol 16/17 E733, 734
+ 2-propanone vol 16/17 E733, 736
Pyridinium,1-[[2-carboxy-8-oxo-7-[2-thienylacetyl]amino-5-thia-1-aza bicyclo[4,2,0]oct-2-en-3-yl]methyl]- hydroxide, inner salt
+ acetic acid, ethyl ester vol 16/17 E706, 719
+ benzene vol 16/17 E706, 715
+ 1-butanol, 3-methyl- vol 16/17 E706, 713

Pyridinium,1-[[2-carboxy-8-oxo-7-[2-thienylacetyl]amino-5-thia-1-aza bicyclo[4,2,0]oct-2-en-3-yl]methyl]- hydroxide, inner salt
+ 1-butanol, 3-methyl- acetate vol 16/17 E706, 720
+ 2-butanone vol 16/17 E706, 722
+ carbon disulfide vol 16/17 E706, 727
+ cyclohexane vol 16/17 E706, 714
+ 1,4-dioxane vol 16/17 E706, 725
+ ethane, dichloro- vol 16/17 E706, 724
+ 1,2-ethanediol vol 16/17 E706, 730
+ ethane, 1,1´-oxybis- vol 16/17 E706, 723
+ ethanol vol 16/17 E706, 711
+ formamide vol 16/17 E706, 729
+ hydrochloric acid vol 16/17 E706, 708
+ methane, tetrachloride vol 16/17 E706, 718
+ methane, sulfinylbis- vol 16/17 E706, 732
+ methane, trichloro- vol 16/17 E706, 726
+ methanol vol 16/17 E706, 710
+ pentane, 2,2,4-trimethyl- vol 16/17 E706, 717
+ petroleum ether vol 16/17 E706, 716
+ 1,2-propanediol vol 16/17 E706, 731
+ 2-propanol vol 16/17 E706, 712
+ 2-propanone vol 16/17 E706, 721
+ pyridine vol 16/17 E706, 728
+ sodium hydroxide (aqueous) vol 16/17 E706, 709
+ water vol 16/17 E706, 707 - 709

Pyrrolidine
+ hydrogen vol 5/6 265, E482, 499
+ nitrogen vol 10 263
+ oxygen vol 7 352

Pyrrolidine, 1-methyl-
+ hydrogen vol 5/6 E482, 498

Pyrrolidine, 2-methyl-
+ hydrogen vol 5/6 511

2-Pyrrolidinone, 1-methyl-
+ ethane vol 9 E195 -E199, 216, 217
+ hydrogen vol 5/6 511
+ nitrogen vol 10 462
+ potassium tetraphenylborate vol 18 39
+ sodium tetraphenylborate vol 18 5
+ silver tetraphenylborate + sodium tetraphenylborate + perchloric acid, sodium salt vol 18 170
+ sulfur dioxide vol 12 281
+ tetrabutylammonium tetraphenylborate vol 18 102
+ tetraphenylarsonium tetraphenylborate (1-) vol 18 221
+ thallium (1) tetraphenylborate (1-) vol 18 180

2-Pyrrolidinone, N-methyl-
+ silver azide + sodium azide + nitric acid, sodium salt vol 3 30
+ thiocyanic acid, silver salt + thiocyanic acid, potassium salt + nitric acid, potassium salt vol 3 229

2-Pyrrolidinone, 1,5-dimethyl-
+ argon vol 4 E253 - 255, 311
+ hydrogen vol 5/6 512
+ nitrogen vol 10 463

Quinoline
+ hydrogen vol 5/6 510
+ sulfur dioxide vol 12 278
Quinoline, 6-methyl-, nitrate
+ water vol 13 443
Quinoline, 6-methyl-, nitrate (aqueous)
+ nitric acid, erbium salt vol 13 443
Quinoline, 8-methyl-, nitrate
+ water vol 13 399, E416, E417, 426, 428
Quinoline, 8-methyl-, nitrate (aqueous)
+ nitric acid, dysprosium salt vol 13 E416, E417, 426
+ nitric acid, gadolinium salt vol 13 399
Quinoline nitrate
+ water vol 13 106, 107, 332, 333, 398, 425, 442
Quinoline nitrate (aqueous)
+ nitric acid, dysprosium salt vol 13 E416, E417, 425
+ nitric acid, erbium salt vol 13 442
+ nitric acid, gadolinium salt vol 13 398
+ nitric acid, lanthanum salt vol 13 106, 107
+ nitric acid, neodymium salt vol 13 332, 333
+ nitric acid, praseodymium salt vol 13 272, 273
Quinolinol (aqueous)
+ nitric acid, dysprosium nitrate vol 13 E416, E417, 427
Quinolinol nitrate
+ water vol 13 400, 427
Quinolinol nitrate (aqueous)
+ nitric acid vol 13 E416, E417, 427
+ nitric acid, gadolinium salt vol 13 400

Rabbit blood			
see Blood,rabbit			
Rabbit brain			
see Brain,rabbit			
Rabbit, choroid layer			
+ krypton-85	vol 2		127
Rabbit, erythrocytes			
+ krypton-85	vol 2		127
Rabbit heart			
+ nitrous oxide	vol 8	E226, E227,	240
Rabbit kidney			
+ nitrous oxide	vol 8	E226, E227,	240
Rabbit leg muscle (saline homogenate)			
+ krypton	vol 2		128
+ xenon	vol 2		222
Rabbit liver			
+ nitrous oxide	vol 8	E226, E227,	240
Rabbit muscle			
+ nitrous oxide	vol 8	E226, E227,	240
Rabbit, plasma			
+ krypton-85	vol 2		127
Rabbit, retina			
+ krypton-85	vol 2		127
Rabbit, sclera			
+ krypton-85	vol 2		127
Rabbit, vitreous body			
+ krypton-85	vol 2		127
Rat abdominal muscle			
+ hydrogen	vol 5/6		278
+ nitrous oxide	vol 8	E226, E227,	247
+ helium	vol 1		121
Rat blood (venous)			
see Blood, rat (venous)			
Rat fatty acids (extracted)			
+ radon-222	vol 2		335
Rat fat (pooled)			
+ krypton	vol 2		
+ radon-222	vol 2		336
+ xenon	vol 2	E199,	203
Rat tissues			
+ radon-222	vol 2		336
Red blood cells			
see Human red blood cells			
Red blood cell ghosts (in phosphate buffer)			
+ nitrogen	vol 10		298
+ oxygen	vol 7		382
Red cell membrane			
see Human red cell membrane			
Redfish oil			
+ oxygen	vol 7	362,	363
Resorcinol dimethyl ester			
see Benzene, 1,3-dimethoxy-			
Retina			
see Rabbit retina			
Rubber			
+ sulfur dioxide	vol 12		332
Rubidium bromide			
+ formamide	vol 11	E52,	53
Rubidium bromide (aqueous)			
+ oxygen	vol 7	E74, E75,	134
Rubidium chloride (aqueous)			
+ argon	vol 4	E33 -	36,
		68,	76
+ neon	vol 1	E141,	174

Rubidium chloride (aqueous)
+ nitrous oxide vol 8 E38, 113
+ oxygen vol 7 E74, E75, 133, 181
Rubidium fluoride (aqueous)
+ oxygen vol 7 E74, E75, 132
Rubidium hydroxide (aqueous)
+ oxygen vol 7 E74, E75, 131
Rubidium iodate
see Iodic acid, rubidium salt
Rubidium iodide (aqueous)
+ oxygen vol 7 E74, E75, 135
Rubidium nitrate
see Nitric acid, rubidium salt
Rubidium perchlorate
see Perchloric acid, rubidium salt
Rubidium sulfate
see Sulfuric acid, rubidium salt
Rubidium tetraphenylborate
+ acetonitrile vol 18 49
+ ethane, 1,2-dichloro- vol 18 50
+ water vol 18 E43, 44 - 46
Rubidium tetraphenylborate (aqueous)
+ lithium chloride + ethanol vol 18 51
+ 2-propanone vol 18 48
+ tris(hydroxymethyl)aminoethane + acetic acid
vol 18 47

Saline		
+ nitrogen	vol 10	293
Saline solution		
see also Sodium chloride (aqueous)		
Samarium cobalt nitrate		
see Nitric acid, cobalt samarium salt		
Samarium manganese nitrate		
see Nitric acid, manganese samarium salt		
Samarium nickel nitrate		
see Nitric acid, nickel samarium salt		
Samarium nitrate trihydrate		
see Nitric acid, samarium salt (trihydrate)		
Samarium nitrate		
see Nitric acid, samarium salt		
Samarium zinc nitrate		
see Nitric acid, samarium zinc salt		
Santowax R		
+ helium	vol 1	356
+ hydrogen	vol 5/6	526
+ nitrogen	vol 10	502
Scalp		
see Human fetal scalp		
Scandium nitrate		
see Nitric acid, scandium salt		
Sclera		
see Rabbit sclera		
SDS		
see Sulfuric acid, monododecyl ester sodium salt		
Sea water		
+ argon	vol 4	E27 - 28, 29 - 32
+ argon + nitrogen + oxygen	vol 10	37, 38
+ hydrogen	vol 5/6	E17, E18, 19 - 24
+ helium	vol 1	E16, 17
+ helium-3	vol 1	19
+ krypton	vol 2	E9, 10, 11
+ neon	vol 1	E138, 139
+ nitrogen	vol 10	E31 - E33, 34 - 44
+ nitrous oxide	vol 8	E23, E24, 25, 26
+ oxygen	vol 7	E41 - E43, 44 - 55
+ sulfur dioxide	vol 12	E34, 35, 36
+ radon-222	vol 2	236, 237
+ xenon	vol 2	147, 148
Selenic acid (ternary)		
+ nitric oxide	vol 8	295
Selenic acid, iron (2+) salt (aqueous)		
+ nitric oxide	vol 8	E265, E266, 296
Selenic acid, iron (2+) salt (in aqueous ethanol solution)		
+ nitric oxide	vol 8	E265, E266, 294
Selenic acid, iron (2+) salt (in aqueous selenic acid)		
+ nitric oxide	vol 8	E265, E266, 295
Seminal vesicles		
see guinea pig seminal vesicles		
Serine, monosodium salt (aqueous)		
+ iodic acid, calcium salt	vol 14	E76, E77, 116
Serum albumen see Bovine serum albumen in addition to entries below		
Serum albumen		
+ argon	vol 4	247
+ hydrogen	vol 5/6	113
+ nitrogen	vol 10	297
+ nitrous oxide	vol 8	E226, E227, 248 - 250
+ oxygen	vol 7	388, 390

Sheep placental tissue
+ hydrogen vol 5/6 276
+ oxygen vol 7 384
Silane, tetrachloro-
+ chlorine vol 12 E354 -E366, 446
+ nitrogen vol 10 252
Silane, trichloro-
+ nitrogen vol 10 252
Silica (aqueous suspension)
+ nitrous oxide vol 8 149
Silicic acid
+ nitrous oxide vol 8 150, 151
Silicone fluid
+ oxygen vol 7 438, 441
Silicone oil
+ helium vol 1 117
+ xenon vol 2 192
Silicon tetrachloride
see Silane, tetrachloro-
Silver argentocyanide
see Silver cyanide
Silver azide
+ nitric acid, tetraethylammonium salt + nitric acid, sodium salt + methanol vol 3 28
+ pyridine vol 3 29
+ sodium azide + nitric acid, sodium salt + dimethylacetamide vol 3 21
+ sodium azide + nitric acid, sodium salt + dimethylformamide vol 3 22
+ sodium azide + nitric acid, sodium salt + formamide vol 3 26
+ sodium azide + nitric acid, sodium salt + hexmethylphosphorotriamide vol 3 27
+ sodium azide + nitric acid, sodium salt + methanol vol 3 28
+ sodium azide + nitric acid, sodium salt + N-methyl-2-pyrrolidinone vol 3 30
+ tetraethylammonium azide + nitric acid, tetraethylammonium salt + acetonitrile vol 3 20
+ tetraethylammonium azide + nitric acid, tetraethylammonium salt + dimethyl sulfoxide vol 3 E23, 24
+ tetraethylammonium azide + nitric acid, tetraethylammonium salt + formamide vol 3 26
+ tetraethylammonium azide + perchloric acid, tetraethylammonium salt + dimethyl sulfoxide vol 3 E23, 25
Silver azide (aqueous)
+ ammonia vol 3 E1, E7 - 8, 17
+ potassium azide vol 3 14
+ pyridine vol 3 19
+ methane, 1,1´-sulfinylbis- vol 3 E23
+ sodium azide vol 3 E1, E6, 10 - 11
+ sodium azide + nitric acid, potassium salt vol 3 9
+ sodium azide + perchloric acid vol 3 E1, 15 - 16
+ tetraethylammonium azide + perchloric acid, tetraethylammonium salt + dimethyl sulfoxide vol 3 18
+ thallium azide vol 3 E2, 12 - 13
Silver bromide (aqueous)
+ silver cyanide + potassium bromide + bis-(cyano-C-)potassium argentate vol 3 E45, 64

Silver bromide (aqueous)
+ thiocyanic acid, silver salt + potassium bromide + thiocyanic acid, potassium salt vol 3 E102, 127
Silver chloride (aqueous)
+ silver cyanide + potassium chloride + bis-(cyano-C-)potassium argentate vol 3 E45, 63
+ thiocyanic acid, silver salt + potassium chloride + thiocyanic acid, potassium salt vol 3 E102, 126
Silver cyanamide
+ water vol 3 E31, 32
Silver cyanamide (aqueous)
+ nitric acid, potassium salt + nitric acid vol 3 E31, 33
Silver cyanate
+ water vol 3 E38 - E39,
Silver cyanate (aqueous)
+ potassium cyanide vol 3 E38 - E39, 43 - 44
+ sodium cyanide vol 3 E38 - E39, 42 40 - 41
Silver cyanide
+ hydrogen cyanide vol 3 E86
+ hydrogen fluoride vol 3 87
+ nitric acid + sulfuric acid + hydrogen cyanide vol 3 E86
+ sulfur dioxide vol 3 88
+ tetraethylammonium cyanide + perchloric acid, tetraethylammonium salt + dimethyl sulfoxide vol 3 84
+ tetraethylammonium cyanide + perchloric acid, tetraethylammonium salt + nitromethane vol 3 85
+ water vol 3 E45, E52, 61 - 62
Silver cyanide (aqueous)
+ ammonia vol 3 E47, E54 - 55, 59, 66
+ ammonia + nitric acid, silver salt + bis(cyano-C-) potassium argentate vol 3 66
+ bis(cyano-C-) potassium argentate + sodium hydroxide vol 3 75 - 76
+ hydrogen cyanide vol 3 E46, E53 - 54, 70 - 71
+ nitric acid, silver salt vol 3 E47, 60
+ perchloric acid + sodium cyanide + perchloric acid, sodium salt vol 3 79 - 80
+ potassium cyanide vol 3 E45, 68 - 69, 78
+ potassium cyanide + nitric acid, potassium salt vol 3 72 - 73, 77, 78
+ silver bromide + potassium bromide + bis(cyano-C-) potassium argentate vol 3 E45, 64
+ silver chloride + potassium chloride + bis(cyano-C-) potassium argentate vol 3 E45, 63
+ silver iodide + nitric acid, barium salt + potassium cyanide + potassium iodide + nitric acid, potassium salt vol 3 81
+ sodium cyanide + perchloric acid, sodium salt + sodium hydroxide vol 3 E47, 79 - 80
+ thiocyanic acid, silver salt + bis(cyano-C-) potassium argentate + thiocyanic acid, potassium salt vol 3 E46, 65, E102, 122 - 123
Silver dicyanamide
+ water vol 3 34 - 35

Silver iodide
+ silver cyanide + nitric acid, barium salt + potassium cyanide + potassium iodide + nitric acid, potassium salt vol 3 81
Silver nitrate
see Nitric acid, silver salt
Selenocyanic acid, silver salt
+ water vol 3 E90, 92 - 93
Selenocyanic acid, silver salt (aqueous)
+ nitric acid, silver salt vol 3 E90 - 91, 98 - 99
+ perchloric acid, silver salt vol 3 E90 - 91, 100 - 101
+ selenocyanic acid, potassium salt vol 3 E90, 94 - 95
+ selenocyanic acid, potassium salt + nitric acid, potassium salt vol 3 E90, 96 - 97
Silver tetraphenylborate (1-)
+ acetonitrile vol 18 E149
+ formamide, N,N-dimethyl- vol 18 E154
+ methane, sulfinylbis- vol 18 E158, 159, 162
+ methanol vol 18 166
+ nitric acid, sodium salt + sodium tetraphenylborate + acetonitrile vol 18 150
+ nitric acid, sodium salt + sodium tetraphenylborate + N,N-dimethylacetamide vol 18 153
+ nitric acid, sodium salt + sodium tetraphenylborate + N,N-dimethylformamide vol 18 155
+ nitric acid, sodium salt + sodium tetraphenylborate + methanol vol 18 167
+ nitric acid, sodium salt + sodium tetraphenylborate + nitromethane vol 18 169
+ perchloric acid, sodium salt + sodium tetraphenylborate + acetonitrile vol 18 151
+ perchloric acid, sodium salt + sodium tetraphenylborate + ethanol vol 18 163
+ perchloric acid, sodium salt + sodium tetraphenylborate + N,N-dimethylformamide vol 18 156
+ perchloric acid, sodium salt + sodium tetraphenylborate + N-methyl-2-pyrolidinone vol 18 170
+ perchloric acid, sodium salt + sodium tetraphenylborate + nitromethane vol 18 171
+ perchloric acid, sodium salt + sodium tetraphenylborate + sulfinybismethane vol 18 160
+ perchloric acid, sodium salt + sodium tetraphenylborate + tetrahydrothiophene-1,1-dioxide vol 18 176
+ phosphoric triamide, hexamethyl- vol 18 165
+ sodium tetraphenylborate + acetonitrile vol 18 152
+ sodium tetraphenylborate + N,N-dimethylformamide vol 18 157
+ sodium tetraphenylborate + perchloric acid, sodium salt + 4-methyl-1,3-dioxolan-2-one vol 18 172
+ sodium tetraphenylborate + perchloric acid, sodium salt + 2-propanone vol 18 173
+ sodium tetraphenylborate + sodium bromide + methanol vol 18 168, 169
+ sodium tetraphenylborate + sulfinylbismethane vol 18 161
+ thiophene, tetrahydro-, 1,1-dioxide vol 18 174, 175
+ water vol 18 E140

Silver tetraphenylborate (1-) (aqueous)
+ nitric acid, sodium salt + sodium tetraphenylborate vol 18 143
+ 2-propanol + methylbenzene vol 18 147, 148
+ sodium tetraphenylborate vol 18 141, 142
+ sodium tetraphenylborate + sodium iodide vol 18 144, 145
+ tris(hydroxymethyl)aminomethane + acetic acid vol 18 146

Skeletal muscle
see Human fetal skeletal muscle

Skin
see Human fetal skin

Small intestine
see Guinea pig small intestine

Sodium acetate
see Acetic acid, sodium salt

Sodium aminopropionate
see DL-alanine, monosodium salt

Sodium animoacetate
see Glycine, monosodium salt

Sodium bicarbonate
see Carbonic acid, monosodium salt

Sodium bisulfate
see Sulfuric acid, monosodium salt

Sodium bisulfite
see Sulfurous acid, monosodium salt

Sodium bromide
+ acetamide, N,N-dimethyl- vol 11 E319, 320
+ acetamide, N-methyl- vol 11 E262, 263 - 265
+ formamide vol 11 E14, E15, 16 - 20
+ formamide, N,N-dimethyl- vol 11 E168, 169, 170
+ formamide, N-methyl- vol 11 E101, 102 - 105
+ silver tetraphenylborate + sodium tetraphenylborate + methanol vol 18 168, 169

Sodium bromide (aqueous)
+ argon vol 4 E33 - 36, 54
+ ethane vol 9 E32, 51
+ helium vol 1 E20, 30
+ neon vol 1 E141, 164
+ nitrous oxide vol 8 E33, 85
+ oxygen vol 7 134, 147
+ sulfur dioxide vol 12 E37 - E39, 89

Sodium bromoacetate
see Acetic acid, bromo-, sodium salt

Sodium butanoate
see Butanoic acid, sodium salt

Sodium butyrate
see Butanoic acid, sodium salt

Sodium carbonate
see Carbonic acid, disodium salt

Sodium chlorate
see Chloric acid, sodium salt

Sodium chloride
+ acetamide, N,N-dimethyl- vol 11 E317, 318
+ acetamide, N-methyl- vol 11 E258, E259, 260, 261
+ formamide vol 11 E9, 10 - 13
+ formamide, N,N-dimethyl- vol 11 E162, 163 - 167
+ formamide, N-methyl- vol 11 E96, 97 - 100

Sodium chloride (aqueous)
+ air vol 10 513, 514
+ argon vol 4 E33 - 36, 47 - 53, 70

Sodium chloride (aqueous)
+ calcium hydroxide + iodic acid, calcium salt vol 14 102
+ chloric acid, barium salt vol 14 E210, 215
+ chlorine vol 12 E349, 352
+ ethane vol 9 E31, E32, 46 - 50, 55, 56
+ helium vol 1 E20, 29, 30, 31, 261
+ hydrogen vol 5/6 E32, E33, 44, 66 - 70
+ iodic acid, calcium salt vol 14 E70, E72, 94 - 99
+ iodic acid, calcium salt + hydrochloric acid vol 14 98
+ iodic acid, calcium salt + sodium hydroxide vol 14 99
+ iodic acid, calcium salt + sucrose vol 14 103
+ iodic acid, strontium salt vol 14 E195, 200, 201
+ krypton-85 vol 2 124 - 125
+ krypton vol 2 E12, E13, 20, 21
+ neon vol 1 E141, 165, 176, 178
+ nitric oxide vol 8 E266, 332, 333
+ nitrogen vol 10 E45 - E47, E50- E52, 67 - 71 348 - 351
+ nitrous oxide vol 8 E33, 80 - 84, 241
+ oxygen vol 7 E68, E69, 133, 140 - 146, 397
+ potassiuim chloride + iodic acid, calcium salt vol 14 100, 101
+ radon-222 vol 2 E238 -E241, 251, 252
+ sulfur dioxide vol 12 E37 - E39, 88
+ xenon-133 vol 2 213, 217, 218
+ xenon vol 2 E149, 151 - 153, 213, 217, 218
Sodium chloride (ternary)
+ nitric oxide vol 8 334, 335
Sodium cyanate (aqueous)
+ cyanic acid, silver salt vol 3 E38 - 39, 42
Sodium cyanide
+ formamide, N,N-dimethyl-
Sodium cyanide (aqueous)
+ silver cyanide + perchloric acid, sodium salt + perchloric acid vol 3 79 - 80
+ silver cyanide + perchloric acid, sodium salt + sodium hydroxide vol 3 79 - 80
Sodium cyano-acetate
see Acetic acid, cyano-, sodium salt
Sodium dicyanoargentate
+ water vol 3 57
Sodium dicyanoargentate (aqueous)
+ silver cyanide + sodium hydroxide vol 3 75 - 76
+ ethanol vol 3 82
Sodium dithiocyanoargentate (aqueous)
+ thiocyanic acid, silver salt + thiocyanic acid, sodium salt + perchloric acid vol 3 E104, 162 - 163
+ thiocyanic acid, sodium salt + perchloric acid, sodium salt vol 3 E104, 157 - 158

Sodium dithionite
 see Dithionous acid, sodium salt
Sodium dodecyl sulfate
 see Sulfuric acid, monododecyl ester sodium salt
Sodium fluoride
 + formamide vol 11 E7, 8
 + formamide, N,N-dimethyl- vol 11 E159, 160, 161
Sodium fluoride (aqueous)
 + oxygen vol 7 E68, 132
Sodium gluconate
 see D-gluconic acid, monosodium salt
Sodium glutamate
 see DL-glutamic acid, disodium salt
Sodium glycolate
 see Acetic acid, hydroxy-, monosodium salt
Sodium glycyl glycinate
 see Glycine, N-glycyl-, sodium salt
Sodium hexanoate
 see Hexanoic acid, sodium salt
Sodium hydroxide (aqueous)
 + alanine + iodic acid, barium salt
 vol 14 E252, E253, 286
 + butyltriisopentylammonium tetraphenylborate
 vol 18 84

S

Sodium hydroxide (aqueous)
 + cesium tetraphenylborate + lithium chloride + methanol
 vol 18 61, 62
 + ethanol + iodic acid, barium salt + nitric acid, sodium salt
 vol 14 301
 + glycine + iodic acid, barium salt
 vol 14 E252, 282
 + hydrogen vol 5/6 E31, 64, 65
 + iodic acid, calcium salt vol 14 E68
 + iodic acid, calcium salt + sodium chloride
 vol 14 99
 + iodic acid, strontium salt vol 14 E195, 202
 + nitrogen vol 10 E45 - E47, E49, E50, 65, 66
 + oxygen vol 7 E67 - E70, 106, 131, 154 - 161
 + ozone vol 7 E474 -E450
 + potassium tetraphenylborate + lithium chloride + methanol
 vol 18 27 - 29
 + silver cyanide + bis(cyano-C-) potassium argentate
 vol 3 75 - 76
 + silver cyanide + sodium cyanide + perchloric acid, sodium salt vol 3 79 - 80
 + tetrabutylammonium tetraphenylborate
 vol 18 98
Sodium hydroxide (ternary)
 + nitric oxide vol 8 334, 335
Sodium hydroxyacetate
 see Acetic acid, hydroxy-, monosodium salt
Sodium B-hydroxybutyrate
 see Butanoic acid, 3-hydroxy-, monosodium salt
Sodium iodide
 + acetamide vol 11 E245, 246
 + acetamide, N,N-dimethyl- vol 11 E321, 322
 + acetamide, N-methyl- vol 11 E266, 267, 268
 + formamide vol 11 E21, 22 - 25
 + formamide, N,N-dimethyl- vol 11 E171, 172 - 175
 + formamide, N-methyl- vol 11 E106, 107 - 110
Sodium iodide (aqueous)
 + argon vol 4 E33 - 36, 55 - 58, 71
 + ethane vol 9 E32, 52

Sodium iodide (aqueous)
+ neon vol 1 E141, 165, 176, 178
+ oxygen vol 7 E69, 106, 135
+ silver tetraphenylborate + sodium tetraphenylborate vol 18 144 - 145
Sodium iodide (methanol)
+ helium vol 1 E20, 34
Sodium lactate
see Propanoic acid, 2-hydroxy-, monosodium salt
Sodium malate
see Butanedioic acid, hydroxy-, monosodium salt
Sodium mandelate
see Mandelic acid, monosodium salt
Sodium methoxyacetate
see Acetic acid, methoxy-, sodium salt
Sodium nitrate
see Nitric acid, sodium salt
Sodium nitrite
see Nitrous acid, sodium salt
Sodium octanoate
see Octanoic acid, sodium salt
Sodium penicillin G
see 4-thia-1-azabicyclo[3,2,0]heptane-2-carboxylic acid, 3,3-dimethyl-7-oxo-6-[(phenylacetyl)amino],monosodium salt
Sodium pentanoate
see Pentanoic acid, sodium salt
Sodium perchlorate
see Perchloric acid, sodium salt
Sodium phosphate
see Phosphoric acid, sodium salt
Sodium propanoate
see Propanoic acid, sodium salt
Sodium propionate
see Propanoic acid, sodium salt
Sodium pyruvate
see Propanoic acid, 2-oxo-, sodium salt
Sodium salycilate
see Benzoic acid, 2-hydroxy-, monosodium salt
Sodium sulfate
see Sulfuric acid, sodium salt
Sodium sulfite
see Sulfurous acid, sodium salt
Sodium tartarate
see Butanedioic acid, 2,3-dihydroxy-, disodium salt
Sodium tetraphenylborate
+ 1-propanol vol 18 6
+ 2-pyrrolidinone, 1-methyl- vol 18 5
+ silver tetraphenylborate + acetonitrile vol 18 152
+ silver tetraphenylborate + N,N-dimethylformamide vol 18 157
+ silver tetraphenylborate + nitric acid, sodium salt + N,N-dimethylacetamide vol 18 153
+ silver tetraphenylborate + nitric acid, sodium salt + N,N-dimethylformamide vol 18 155
+ silver tetraphenylborate + nitric acid, sodium salt + methanol vol 18 167
+ silver tetraphenylborate + nitric acid, sodium salt + nitromethane vol 18 169
+ silver tetraphenylborate + perchloric acid, sodium salt + acetonitrile vol 18 151
+ silver tetraphenylborate + perchloric acid, sodium salt + N,N-dimethylformamide vol 18 156
+ silver tetraphenylborate + perchloric acid, sodium salt + ethanol vol 18 163

Sodium tetraphenylborate
+ silver tetraphenylborate + perchloric acid, sodium salt + N-methyl-2-pyrrolidinone vol 18 170
+ silver tetraphenylborate + perchloric acid, sodium salt + nitromethane vol 18 171
+ silver tetraphenylborate + perchloric acid, sodium salt + 2-propanone vol 18 173
+ silver tetraphenylborate + perchloric acid, sodium salt + sulfinylbismethane vol 18 160
+ silver tetraphenylborate + perchloric acid, sodium salt + tetrahydrothiophene-1,1-dioxide vol 18 176
+ silver tetraphenylborate + sodium tetraphenylborate + perchloric acid, sodium salt vol 18 172
+ silver tetraphenylborate + sodium bromide + methanol vol 18 168, 169
+ silver tetraphenylborate + sulfinylbismethane vol 18 161
+ water vol 18 E3

Sodium tetraphenylborate (aqueous)
+ 2-propanone vol 18 4
+ silver tetraphenylborate vol 18 141, 142
+ silver tetraphenylborate + nitric acid, sodium salt vol 18 143
+ silver tetraphenylborate + sodium iodide vol 18 144 - 145

Sodium thiocyanate
see Thiocyanic acid, sodium salt

Sodium thiosulfate
see Thiosulfuric acid, sodium salt

Solu Cortef
see Pregn-4-ene-3,20-dione,21(3-carboxy-1-oxopropoxy)-11,17-dihydroxy-monosodium salt (11β)

Solu Medrol
see Pregna-1,4-diene-3,20-dione,21-(3-carboxy-1-oxopropoxy)-11,17-dihydroxy-6-methyl-, monosodium salt, (6α ,11β-)

Solvent refined coal recycle oil
+ hydrogen vol 5/6 523

Sorbitol
see D-glucitol

Soybean oil
+ oxygen vol 7 364, 365

Sperm whale oil
+ oxygen vol 7 443

Spleen
see guinea pig spleen and human fetal spleen

Squalane
see Tetracosane, 2,6,10,15,19,23-hexamethyl-

Starch (aqueous colloidal)
+ hydrogen vol 5/6 117
+ oxygen vol 7 406
+ nitrous oxide vol 8 156, 157

Steam rendered lard
+ oxygen vol 7 366

Stomach
see guinea pig stomach

Strontium bromate
see Bromic acid, strontium salt

Strontium bromide
+ acetamide, N-methyl- vol 11 E302, 303, 304
+ formamide vol 11 E82, 83

Strontium chlorate
see Chloric acid, strontium salt

Strontium chloride
+ acetamide, N,N-dimethyl- vol 11 E343, 344
+ acetamide, N-methyl- vol 11 E300, 301
+ formamide vol 11 E80, 81

Strontium chloride (aqueous)
+ argon vol 4 E33, 43
+ chlorine vol 12 E349
+ neon vol 1 E141, 155
+ nitrous oxide vol 8 E31, E32, 74
Strontium iodate
see Iodic acid, strontium salt
Strontium nitrate
see Nitric acid, strontium salt
Subcutaneous fat
see Guinea pig subcutaneous fat
Sucrose
see α-D-glucopyranoside,β-D-fructofuranosyl
Sulfate
see under sulfuric acid
Sulfinylbismethane
see Methane, sulfinylbis-
Sulfolane
see Thiophene, tetrahydro-1,1-dioxide
Sulfur hexafluoride
+ nitrogen vol 10 259
Sulfuric acid
+ air vol 10 515, 516
+ ethane vol 9 E29, E32, 34, 53
+ hydrogen vol 5/6 E27, E33, 37 - 40
+ nitric oxide vol 8 E350, 351
+ nitrogen vol 10 E45 - E48, 57, 58, 74, 76
+ nitrous oxide vol 8 E28, E40, 42
+ oxygen vol 7 E59, E60, 78 - 83
+ ozone vol 7 E474 -E480, 490, 491
+ radon-219 vol 2 E339, 340
+ silver cyanide + nitric acid + hydrogen cyanide vol 3 E86
+ sulfur dioxide vol 12 42 - 52, 350
Sulfuric acid (aqueous)
+ thiocyanic acid, silver salt + thiocyanic acid, potassium salt + potassium bromide vol 3 121
Sulfuric acid (ternary)
+ nitric oxide vol 8 289, 311, 317
+ nitrous oxide vol 8 E34, 89
Sulfuric acid (multicomponent)
+ oxygen vol 7 E64, E65, E70, 117, 118, 149, 162
Sulfuric acid, aluminium salt (aqueous)
+ nitrous oxide vol 8 E29, E30, 62
+ oxygen vol 7 E63, 100
+ sulfur dioxide vol 12 65, 66
Sulfuric acid, ammonium salt (aqueous)
+ hydrogen vol 5/6 E28, 45
+ nitric oxide vol 8 278
+ nitrous oxide vol 8 E29, 49
+ oxygen vol 7 E63, 98
+ sulfur dioxide vol 12 E37 - E39, 56
Sulfuric acid, cadmium salt (aqueous)
+ sulfur dioxide vol 12 E37 - E39, 81
Sulfuric acid, cesium salt (aqueous)
+ oxygen vol 7 E75, 137
Sulfuric acid, chromium salt (aqueous)
+ nitrous oxide vol 8 E29, E30, 61

Sulfuric acid, cobalt (aqueous)
+ hydrogen vol 5/6 E29, 50
+ nitric oxide vol 8 E266, 329, 331
+ nitrous oxide vol 8 E29, E30, 57
+ oxygen vol 7 E64, 104
Sulfuric acid, copper (2+) salt (1:1)
+ hydrogen vol 5/6 48
+ oxygen vol 7 E63, 102
+ radon-222 vol 2 E238 -E241, 244
Sulfuric acid, copper (2+) salt (ternary)
+ nitric oxide vol 8 E266, 317
Sulfuric acid, copper (2+) salt (1:1) (multicomponent)
+ oxygen vol 7 E64, E65, 117, 118
Sulfuric acid, dimethyl ester
+ sulfur dioxide vol 12 320
Sulfuric acid, iron (2+) salt (aqueous)
+ nitric oxide vol 8 E266, E267, 270, 278, 279, 281, 282, 284, 288, 293, 299 - 303, 305, 331
+ nitrous oxide vol 8 E29, E30, 58
+ radon-222 vol 2 E238 - 241, 249
Sulfuric acid, iron (3+) salt (aqueous)
+ nitrous oxide vol 8 E29, E30, 59
Sulfuric acid, iron (3+) salt (ternary)
+ nitric oxide vol 8 E266, 311
Sulfuric acid, iron (2+) ammonium salt (aqueous)
+ nitric oxide vol 8 E265, E266, 272, 273, 276, 277. 289, 290, 297, 298
Sulfuric acid, lithium salt (aqueous)
+ nitrous oxide vol 8 E32, 79
+ oxygen vol 7 E66, E67, 137
Sulfuric acid, magnesium salt (aqueous)
+ hydrogen vol 5/6 E31, 58
+ iodic acid, calcium salt vol 14 E70, E74, 152, 153
+ nitrogen vol 10 345
+ nitrous oxide vol 8 E30, 65 - 67
+ neon vol 1 E141, 148
+ oxygen vol 7 E65, 122, 123
Sulfuric acid, manganese salt (aqueous)
+ hydrogen vol 5/6 E30, 52, 58
+ nitrous oxide vol 8 E29, E30, 60
+ oxygen vol 7 E64, 106
Sulfuric acid, monododecyl ester, sodium salt (aqueous and multicomponent)
+ argon vol 4 259
+ ethane E32, 54 - 56, E61, 62, 63
+ oxygen vol 7 212
Sulfuric acid, monosodium salt (aqueous)
+ sulfur dioxide vol 12 92 - 94
Sulfuric acid, nickel salt (aqueous)
+ hydrogen vol 5/6 E29, 49
+ nitric oxide vol 8 E266, 328, 331
+ nitrous oxide vol 8 E29, E30, 56
+ oxygen vol 7 E64, 103
Sulfuric acid, potassium salt (aqueous)
+ iodic acid, calcium salt vol 14 E70, 144, 145
+ nitrous oxide vol 8 E37, 105, 106
+ oxygen vol 7 E73, E74, 137, 178, 179
+ sulfur dioxide vol 12 E37 - E39, 102

Sulfuric acid, rubidium salt (aqueous)		
+ oxygen	vol 7	E74, E75, 137
Sulfuric acid, sodium salt (aqueous)		
+ air	vol 10	345, 515, 516
+ ammonium tetraphenylborate	vol 18	76, 77
+ cesium tetraphenylborate	vol 18	55, 56
+ ethane	vol 9	E32, 53
+ helium	vol 1	E20, 32
+ hydrogen	vol 5/6	E32, E33, 71, 72
+ iodic acid, calcium salt	vol 14	E70, E74, 104
+ krypton	vol 2	E12 - E13, 22
+ nitric oxide	vol 8	E266, 278, 332
+ nitrogen	vol 10	E45 - E47, E52, E53, 73 - 76 347
+ nitrous oxide	vol 8	E33, E34, 86 - 88
+ oxygen	vol 7	E70, 137, 147 - 151, 161
+ potassium tetraphenylborate	vol 18	17, 18
+ sulfur dioxide	vol 12	E37 - E39, 90 - 91
Sulfuric acid, sodium salt (ternary)		
+ nitrous oxide	vol 8	E34, 89
Sulfuric acid, zinc salt (aqueous)		
+ hydrogen	vol 5/6	E29, 47
+ nitrous oxide	vol 8	52
+ radon-222	vol 1	E238 - 241, 246
+ sulfur dioxide	vol 12	82, 83, 84 - 87
Sulfurous acid, calcium salt (aqueous)		
+ sulfur dioxide	vol 12	E37 - E39, 67 - 75
Sulfurous acid, magnesium salt (aqueous)		
+ sulfur dioxide	vol 12	E37 - E39, 73 - 75, 77
Sulfurous acid, monoammonium salt (aqueous)		
+ sulfur dioxide	vol 12	E37 - E39, 57 - 61
Sulfurous acid, monosodium salt (aqueous)		
+ sulfur dioxide	vol 12	93 - 94
Sulfurous acid, sodium salt (aqueous)		
+ nitrogen	vol 10	E45 - E47, E52, E53, 72, 75
+ oxygen	vol 7	E70, 152
+ thiocyanic acid, silver salt	vol 3	E103, 168
Sulfurous acid, zinc salt		
+ sulfur dioxide	vol 12	84 - 87
Sulfuryl chloride		
+ chlorine	vol 12	E354 -E366, 445
+ sulfur dioxide	vol 12	E1, E2, 329
Sulfur dioxide		
+ helium	vol 1	E255
+ hydrogen	vol 5/6	622
+ hydrogen deuteride	vol 5/6	E299
+ nitrogen	vol 10	315, 475
+ oxygen	vol 7	431
+ silver cyanide	vol 3	88
+ thiocyanic acid, silver salt	vol 3	234
Sunflower seed oil		
+ oxygen	vol 7	365

Tallow, beef
  see Beef tallow
Tartaric acid
  see Butanedioic acid, 2,3-dihydooxy-
Terbium nitrate
  see Nitric acid, terbium salt
Ternary systems involving salt plus water are listed under the salt
Terpentin
  + sulfur dioxide vol 12 E147, 170
Terphenyl
  + krypton vol 2 69
Testes
  see guinea pig testes
Tetrabutylammonium bromide
  see 1-butanaminium, N,N,N-tributyl-, bromide
Tetra-N-butylammonium tetraphenylborate
  + ethane,1,1-dichloro- vol 18 99
  + ethane,1,2-dichloro- vol 18 100
  + methanol vol 18 101
  + 2-pyrrolidinone,1-methyl- vol 18 102
Tetra-N-butylammonium tetraphenylborate (aqueous)
  + sodium hydroxide vol 18 98
1,1,2,2-Tetrachloroethane
  see ethane, 1,1,2,2-tetrachloro-
1,1,2,2-Tetrachloroethene
  see ethene, 1,1,2,2-tetrachloro-
Tetrachloromethane
  see methane, tetrachloro-
Tetrachlorosilane
  see silane, tetrachloro-
Tetracosane,2,6,10,15,19,23-hexamethyl-
  + ethane vol 9 E77, E78, 108
  + hydrogen vol 5/6 E123, 147
Tetradecafluoromethylcyclohexane
  see cyclohexane, undecafluoro (trifluoromethyl)-
Tetradecane
  + argon vol 4 E137, 138 - 139
  + chlorine vol 12 E354 -E366, 373
  + helium vol 1 57
  + hydrogen vol 5/6 143, E356, 372
  + krypton vol 2 48
  + neon vol 1 202
  + nitrogen vol 10 E119, 144
  + nitrous oxide vol 8 E160, 176
  + oxygen vol 7 E214, E215, 235, 239
  + sulfur dioxide vol 12 E116, E117, 134
  + xenon vol 2 161
1-Tetradecanol
  + water vol 15 E418, 419, 420
Tetraethanolammonium bromide
  see ethanamine, 2-hydroxy-, N,N,N-tris (2-hydroxyethyl)-, bromide
Tetraethylammonium azide
  + silver azide + nitric acid, tetraethylammonium salt + acetonitrile vol 3 20
  + silver azide + nitric acid, tetraethylammonium salt + dimethyl sulfoxide vol 3 E23, 24
  + silver azide + nitric acid, tetraethylammonium salt + formamide vol 3 26
  + silver azide + perchloric acid, tetraethylammonium salt + dimethyl sulfoxide vol 3 E23, 25
Tetraethylammonium azide (aqueous)
  + silver azide + perchloric acid, tetraethylammonium salt + dimethyl sulfoxide vol 3 18

Tetraethylammonium bromide
  see 1-ethanaminium N,N,N-triethyl bromide
Tetraethylammonium tetraphenylborate
  + ethane,1,1-dichloro- vol 18 103
  + ethane,1,2-dichloro- vol 18 104
  + 1-propanol vol 18 105
Tetraethylene glycol dimethyl ether
  see 2,5,8,11,14-pentaoxapentadecane
Tetrafluorohydrazine
  see Hydrazine, tetrafluoro-
Tetrahydrofuran
  see Furan, tetrahydro-
Tetrahydrofurfuryl alcohol
  see 2-Furanmethanol, tetrahydro-
1,2,3,4-Tetrahydronaphthalene
  see Naphthalene, 1,2,3,4-tetrahydro-
Tetrahydro-2H-pyran
  see 2H-pyran, tetrahydro-
Tetralin
  see Naphthalene, 1,2,3,4-tetrahydro-
Tetramethylammonium bromide
  see methanamine, N,N,N-trimethyl-, bromide
Tetramethylammonium hydroxide
  see methanaminium, N,N,N-trimethyl-, hydroxide
Tetramethylammonium iodide,
  see methanaminium, N,N,N-trimethyl-, iodide
Tetramethylammonium tetraphenylborate(1-)
  + ethane,1,1-dichloro- vol 18 109
  + ethane,1,2-dichloro- vol 18 110
  + 1-propanol vol 18 111
  + water vol 18 E106, 107, 108
Tetramethylene glycol dimethyl ester
  see 2,5,8,11,15-pentaoxapentadecane
Tetramethylene sulfone
  see Thiophene,tetrahydro-, 1,1-dioxide
2,2,4,4-Tetramethyl-3-pentanone
  see 3-pentanone,2,2,4,4-tetramethyl-
Tetramethyl sulfone
  see Thiophene,tetrahydro-,1,1-dioxide
3,6,9,12-Tetraoxopentadecane,1,1,1,2,4,4,5,7,7,8,10,10,11,13,13,14,14,
15,15,15-eicosafluoro-5,8,11-tris(trifluoromethyl)
  + oxygen vol 7 342
Tetraphenylarsonium nitrate
  see nitric acid, tetraphenylarsonium salt
Tetraphenylarsonium tetraphenylborate (1-)
  + acetamide, N,N-dimethyl- vol 18 E194, 195 - 197
  + acetonitrile vol 18 E187, 188 - 190
  + 1,3-dioxolan-2-one,4-methyl- vol 18 223
  + ethane,1,1-dichloro- vol 18 192
  + ethane,1,2-dichloro- vol 18 193
  + ethanol vol 18 208
  + formamide vol 18 E209, 210 - 212
  + formamide, N,N-dimethyl- vol 18 E198, 199 - 202
  + methane, nitro- vol 18 222
  + methane, sulfinylbis- vol 18 E203, 204 - 207
  + methanol vol 18 E217, 218 - 220
  + nitric acid, tetraphenylarsonium salt
  vol 18 191
  + phosphoric triamide,hexamethyl- vol 18 E213, 214 - 216
  + 2-propanone vol 18 224
  + 2-pyrrolidinone,1-methyl- vol 18 221
  + thiophene,tetrahro-,1,1-dioxide vol 18 E225, 226, 227
  + water vol 18 E182, 183,
  184, 186

Tetraphenylphosphonium tetraphenylborate (1-)
+ acetonitrile vol 18 229
+ ethane,1,1-dichloro- vol 18 230
+ ethane,1,2-dichloro- vol 18 231
+ water vol 18 E228
Tetrapropylammonium bromide
see 1-propanaminium, N,N,N-tripropyl bromide
Tetrapropylammonium tetraphenylborate
+ ethane,1,1-dichloro- vol 18 112
+ ethane,1,2-dichloro- vol 18 113
+ 1-propanol vol 18 114
Tetrasulfophthalocyanine metal complexes (aqueous)
+ argon vol 4 E35 - 36
Thallium (1) tetraphenylborate (1-)
+ methane, sulfinylbis- vol 18 181
+ 2-pyrrolidinone,1-methyl- vol 18 180
+ tris(hydroxymethyl)aminoethane + acetic acid + water
vol 18 179
+ water vol 18 E177, 178
Thallium azide (aqueous)
+ silver azide vol 3 E2, 12 - 13
Thallium dicyanoargentate
+ water vol 3 E45, 58
Thallium silver cyanide
see Thallium dicyanoargentate
4-Thia-1-azabicyclo[3,2,0]heptane-2-carboxylic acid,
6-[[(1-aminocyclohexyl)carbonyl]amino]-3,3-dimethyl-7-oxo-
+ hydrochloric acid (multicomponent)
Vol 16/17 E359, 360, 362
+ phosphoric acid, trisodium salt (multicomponent)
Vol 16/17 E359, 362
+ potassium chloride (multicomponent)
Vol 16/17 E359, 360
+ potassium hydroxide (multicomponent)
Vol 16/17 E359, 360
+ sodium chloride (multicomponent) Vol 16/17 E359, 362
+ water Vol 16/17 E359, 360 - 362
4-Thia-1-azabicyclo[3,2,0]heptane-2-carboxylic acid,
6-amino-3,3-dimethyl-7-oxo-
+ acetic acid Vol 16/17 E1, 4
+ acetic acid (quaternary) Vol 16/17 E1, 5
+ acetic acid, sodium salt (quaternary)
Vol 16/17 E1, 5
+ ammonium chloride (aqueous) Vol 16/17 E1, 7
+ ammonium chloride (multicomponent) Vol 16/17 E1, 8
+ benzeneacetic acid (multicomponent)
Vol 16/17 E1, 8
+ ethanol Vol 16/17 E1, 6
+ hydrochloric acid (multicomponent) Vol 16/17 E1, 3, 8, 9
+ sodium chloride (multicomponent) Vol 16/17 E1, 8, 9
+ water Vol 16/17 E1, 2 - 9
4-Thia-1-azabicyclo[3,2,0]heptane-2-carboxylic acid,
6-[(amino-1,4-cyclohexadien-1-ylacetyl)amino]-3,3-dimethyl-
7-oxo-
+ hydrochloric acid (multicomponent) Vol 16/17 E487, 488, 489
+ potassium chloride (multicomponent)
Vol 16/17 E487, 488, 489
+ potassium hydroxide (multicomponent)
Vol 16/17 E487, 488, 489
+ water Vol 16/17 E487, 488, 489

4-Thia-1-azabicyclo[3,2,0]heptane-2-carboxylic acid,
6-[(amino(4-hydroxyphenyl)acetyl)amino]-3,3-dimethyl-
7-oxo-, trihydrate
+ hydrochloric acid (multicomponent) Vol 16/17 E479, 480, 481
+ potassium chloride (multicomponent)
Vol 16/17 E479, 480, 481
+ potassium hydroxide (multicomponent)
Vol 16/17 E479, 480, 481
+ water Vol 16/17 E479, 480, 481
4-Thia-1-azabicyclo[3,2,0]heptane-2-carboxylic acid,
6-[(aminophenylacetyl)amino]-3,3-dimethyl-7-oxo-
+ acetic acid, ethyl ester Vol 16/17 E410, 424
+ benzene Vol 16/17 E410, 420
+ 1-butanol, 3-methyl- Vol 16/17 E410, 418
+ 1-butanol, 3-methyl-, acetate Vol 16/17 E410, 425
+ 2-butanone Vol 16/17 E410, 428
+ carbon disulfide Vol 16/17 E410, 434
+ cholan-24-oic acid, 3,7,12-trihydroxy- (aqueous)
Vol 16/17 E394, 408
+ cyclohexane Vol 16/17 E410, 419
+ 1,4-dioxane Vol 16/17 E410, 431
+ 1,4-dioxane (quaternary) Vol 16/17 E410, 432
+ ethane, dichloro- Vol 16/17 E410, 430
+ ethane, 1,1´-oxybis- Vol 16/17 E410, 429
+ 1,2-ethanediol Vol 16/17 E410, 437
+ ethanol Vol 16/17 E410, 413
+ ethanol (quaternary) Vol 16/17 E410, 414
+ formamide Vol 16/17 E410, 436
+ furan, tetrahydro- (quaternary) Vol 16/17 E410, 441
+ hydrochloric acid (multicomponent) Vol 16/17 E394, 402,
403, 405, 406
+ methane, sulfinylbis- Vol 16/17 E410, 439
+ methane, sulfinylbis- (quaternary) Vol 16/17 E410, 440
+ methane, tetrachloro- Vol 16/17 E410, 423
+ methane, trichloro- Vol 16/17 E410, 433
+ methanol Vol 16/17 E410, 411
+ methanol (quaternary) Vol 16/17 E410, 412
+ pentane, 2,2,4-trimethyl- Vol 16/17 E410, 422
+ petroleum ether Vol 16/17 E410, 421
+ potassium chloride (multicomponent)
Vol 16/17 E394, 400 - 403
+ potassium chloride (quaternary) Vol 16/17 E410, 412, 414,
416, 417, 427,
432, 440, 441
+ potassium hydroxide (multicomponent)
Vol 16/17 E394, 402, 403
+ 1,2-propanediol Vol 16/17 E410, 438
+ 1-propanol (quaternary) Vol 16/17 E410, 417
+ 2-propanol Vol 16/17 E410, 415
+ 2-propanol (quaternary) Vol 16/17 E410, 416
+ 2-propanone Vol 16/17 E410, 426
+ 2-propanone (quaternary) Vol 16/17 E410, 427
+ pyridine Vol 16/17 E410, 435
+ sodium hydroxide (aqueous) Vol 16/17 E394, 404
+ sulfuric acid, monododecyl ester, sodium salt,
(aqueous) Vol 16/17 E394, 407
+ water Vol 16/17 E394, 395 - 409,
E410, 412, 414,
416, 417, 427,
432, 440, 441

4-Thia-1-azabicyclo[3,2,0]heptane-2-carboxylic acid, 6-[(aminophenylacetyl)amino]-3,3-dimethyl-7-oxo-, trihydrate
+ acetic acid, ethyl ester Vol 16/17 E457, 465
+ benzene Vol 16/17 E457, 461
+ 1-butanol, 3-methyl-, acetate Vol 16/17 E457, 466
+ 2-butanone Vol 16/17 E457, 468
+ carbon disulfide Vol 16/17 E457, 473
+ cholan-24-oic acid, 3,7,12-trihydroxy- Vol 16/17 E442, 454
+ cyclohexane Vol 16/17 E457, 470
+ 1,4-dioxane Vol 16/17 E457, 471
+ ethane, dichloro- Vol 16/17 E457, 470
+ ethane, 1,1´-oxybis- Vol 16/17 E457, 469
+ 1,2-ethanediol Vol 16/17 E457, 476
+ ethanol Vol 16/17 E457, 459
+ formamide Vol 16/17 E457, 475
+ hydrochloric acid (multicomponent) Vol 16/17 E442, 448 - 450, 452, 455
+ methane, sulfinylbis- Vol 16/17 E457, 478
+ methane, trichloro- Vol 16/17 E457, 472
+ methane, tetrachloro- Vol 16/17 E457, 464
+ methanol Vol 16/17 E457, 458
+ pentane, 2,2,4-trimethyl- Vol 16/17 E457, 463
+ petroleum ether Vol 16/17 E457, 462
+ phosphoric acid, trisodium salt (multicomponent) Vol 16/17 E442, 455
+ potassium chloride (multicomponent) Vol 16/17 E442, 448, 449
+ potassium hydroxide (multicomponent) Vol 16/17 E442, 448, 449
+ 1,2-propanediol Vol 16/17 E457, 477
+ 2-propanone Vol 16/17 E457, 467
+ pyridine Vol 16/17 E457, 474
+ sodium chloride (multicomponent) Vol 16/17 E442, 455
+ sodium hydroxide (aqueous) Vol 16/17 E442, 451
+ sulfuric acid, monododecyl ester, sodium salt (aqueous) Vol 16/17 E442, 453
+ water Vol 16/17 E442, 443 - 456

4-Thia-1-azabicyclo[3,2,0]heptane-2-carboxylic acid, 6-[(aminophenylacetyl)amino]-3,3-dimethyl-7-oxo-, monosodium salt
+ acetic acid, ethyl ester Vol 16/17 E367, 380
+ benzene Vol 16/17 E367, 376
+ 1-butanol, 3-methyl- Vol 16/17 E367, 374
+ 1-butanol, 3-methyl-, acetate Vol 16/17 E367, 381
+ 2-butanone Vol 16/17 E367, 383
+ carbon disulfide Vol 16/17 E367, 388
+ cyclohexane Vol 16/17 E367, 375
+ 1,4-dioxane Vol 16/17 E367, 386
+ ethane, dichloro- Vol 16/17 E367, 385
+ ethane, 1,1´-oxybis- Vol 16/17 E367, 384
+ 1,2-ethanediol Vol 16/17 E367, 391
+ ethanol Vol 16/17 E367, 372
+ formamide Vol 16/17 E367, 390
+ hydrochloric acid Vol 16/17 E367, 369
+ methane, sulfinylbis- Vol 16/17 E367, 393
+ methane, tetrachloro- Vol 16/17 E367, 379
+ methane, trichloro- Vol 16/17 E367, 387
+ methanol Vol 16/17 E367, 371
+ pentane, 2,2,4-trimethyl- Vol 16/17 E367, 378
+ petroleum ether Vol 16/17 E367, 377
+ 1,2-propanediol Vol 16/17 E367, 392
+ 2-propanol Vol 16/17 E367, 373
+ 2-propanone Vol 16/17 E367, 382
+ pyridine Vol 16/17 E367, 389
+ sodium hydroxide (aqueous) Vol 16/17 E367, 370
+ water Vol 16/17 E367, 368 - 370

4-Thia-1-azabicyclo[3,2,0]heptane-2-carboxylic acid,
6-[2-azido-2-phenyl-acetamido]-3,3-dimethyl-7-oxo-,
+ hydrochloric acid (multicomponent) Vol 16/17 E367, 368
+ phosphoric acid, trisodium salt (multicomponent) Vol 16/17 E367, 368
+ sodium chloride (multicomponent) Vol 16/17 E367, 368
+ water Vol 16/17 E367, 368
4-Thia-1-azabicyclo[3,2,0]heptane-2-carboxylic acid,
6-[[[3-(2-chlorophenyl)-5-methyl-4-isoxazoyl]carbonyl]amino]-3,3-dimethyl-7-oxo-, monosodium salt, monohydrate
+ acetic acid, ethyl ester Vol 16/17 E576, 587
+ benzene Vol 16/17 E576, 583
+ 1-butanol, 3-methyl- Vol 16/17 E576, 602
+ 1-butanol, 3-methyl-, acetate Vol 16/17 E576, 588
+ 2-butanone Vol 16/17 E576, 590
+ carbon disulfide Vol 16/17 E576, 595
+ cyclohexane Vol 16/17 E576, 582
+ 1,4-dioxane Vol 16/17 E576, 593
+ ethane, dichloro- Vol 16/17 E576, 592
+ ethane, 1,1´-oxybis- Vol 16/17 E576, 591
+ 1,2-ethanediol Vol 16/17 E576, 598
+ ethanol Vol 16/17 E576, 581
+ formamide Vol 16/17 E576, 597
+ hydrochloric acid Vol 16/17 E576, 578
+ methane, sulfinylbis- Vol 16/17 E576, 594
+ methane, tetrachloro- Vol 16/17 E576, 600
+ methane, trichloro- Vol 16/17 E576, 594
+ methanol Vol 16/17 E576, 580
+ pentane, 2,2,4-trimethyl- Vol 16/17 E576, 585
+ petroleum ether Vol 16/17 E576, 584
+ 1,2-propanediol Vol 16/17 E576, 599
+ 2-propanol Vol 16/17 E576, 601
+ 2-propanone Vol 16/17 E576, 589
+ pyridine Vol 16/17 E576, 596
+ sodium hydroxide (aqueous) Vol 16/17 E576, 579
+ water Vol 16/17 E576, 577 - 596
4-Thia-1-azabicyclo[3,2,0]heptane-2-carboxylic acid,
6-[[[3-(2,6-dichlorophenyl)-5-methyl-4-isoxazoyl]carbonyl]amino]-3,3-dimethyl-7-oxo-, monosodium salt, monohydrate
+ acetic acid, ethyl ester Vol 16/17 E548, 560
+ benzene Vol 16/17 E548, 556
+ 1-butanol, 3-methyl- Vol 16/17 E548, 575
+ 1-butanol, 3-methyl-, acetate Vol 16/17 E548, 561
+ 2-butanone Vol 16/17 E548, 563
+ carbon disulfide Vol 16/17 E548, 568
+ cyclohexane Vol 16/17 E548, 555
+ 1,4-dioxane Vol 16/17 E548, 566
+ ethane, dichloro- Vol 16/17 E548, 565
+ ethane, 1,1´-oxybis- Vol 16/17 E548, 564
+ 1,2-ethanediol Vol 16/17 E548, 571
+ ethanol Vol 16/17 E548, 554
+ formamide Vol 16/17 E548, 570
+ hydrochloric acid Vol 16/17 E548, 551
+ methane, sulfinylbis- Vol 16/17 E548, 573
+ methane, tetrachloro- Vol 16/17 E548, 559
+ methane, trichloro- Vol 16/17 E548, 567
+ methanol Vol 16/17 E548, 553
+ pentane, 2,2,4-trimethyl- Vol 16/17 E548, 568
+ petroleum ether Vol 16/17 E548, 557
+ 1,2-propanediol Vol 16/17 E548, 572
+ 2-propanol Vol 16/17 E548, 574
+ 2-propanone Vol 16/17 E548, 562
+ pyridine Vol 16/17 E548, 569
+ sodium hydroxide (aqueous) Vol 16/17 E548, 552
+ water Vol 16/17 E548, 549 - 552

4-Thia-1-azabicyclo[3,2,0]heptane-2-carboxylic acid,
6-[(2,6-dimethoxybenzoyl)amino]-3,3-dimethyl-7-oxo-,
monosodium salt

+ acetic acid, ethyl ester	Vol 16/17	E518, 529
+ benzene	Vol 16/17	E518, 525
+ 1-butanol, 3-methyl-	Vol 16/17	E518, 544
+ 1-butanol, 3-methyl-, acetate	Vol 16/17	E518, 530
+ 2-butanone	Vol 16/17	E518, 532
+ carbon disulfide	Vol 16/17	E518, 537
+ cyclohexane	Vol 16/17	E518, 524
+ 1,4-dioxane	Vol 16/17	E518, 535
+ ethane, dichloro-	Vol 16/17	E518, 534
+ ethane, 1,1´-oxyxbis-	Vol 16/17	E518, 533
+ 1,2-ethanediol	Vol 16/17	E518, 540
+ ethanol	Vol 16/17	E518, 523
+ formamide	Vol 16/17	E518, 539
+ hydrochloric acid	Vol 16/17	E518, 520
+ methane, sulfinylbis-	Vol 16/17	E518, 542
+ methane, tetrachloro-	Vol 16/17	E518, 528
+ methane, trichloro-	Vol 16/17	E518, 536
+ methanol	Vol 16/17	E518, 522
+ pentane, 2,2,4-trimethyl-	Vol 16/17	E518, 527
+ petroleum ether	Vol 16/17	E518, 526
+ 1,2-propanediol	Vol 16/17	E518, 541
+ 2-propanol	Vol 16/17	E518, 543
+ 2-propanone	Vol 16/17	E518, 531
+ pyridine	Vol 16/17	E518, 538
+ sodium hydroxide (aqueous)	Vol 16/17	E518, 521
+ water	Vol 16/17	E518, 519 - 521

4-Thia-1-azabicyclo[3,2,0]heptane-2-carboxylic acid,
3,3-dimethyl-6[[(5-methyl-3-phenyl-4-isoxazoyl)carbonyl]amino]-
7-oxo-, monosodium salt

+ acetic acid, ethyl ester	Vol 16/17	E603, 614
+ benzene	Vol 16/17	E603, 610
+ 1-butanol, 3-methyl-	Vol 16/17	E603, 629
+ 1-butanol, 3-methyl-, acetate	Vol 16/17	E603, 615
+ 2-butanone	Vol 16/17	E603, 617
+ carbon disulfide	Vol 16/17	E603, 622
+ cyclohexane	Vol 16/17	E603, 609
+ 1,4-dioxane	Vol 16/17	E603, 620
+ ethane, dichloro-	Vol 16/17	E603, 619
+ ethane, 1,1´-oxybis-	Vol 16/17	E603, 618
+ 1,2-ethanediol	Vol 16/17	E603, 625
+ ethanol	Vol 16/17	E603, 608
+ formamide	Vol 16/17	E603, 624
+ hydrochloric acid	Vol 16/17	E603, 605
+ methane, sulfinylbis-	Vol 16/17	E603, 627
+ methane, tetrachlaoro-	Vol 16/17	E603, 613
+ methane, trichloro-	Vol 16/17	E603, 621
+ methanol	Vol 16/17	E603, 607
+ pentane, 2,2,4-trimethyl-	Vol 16/17	E603, 612
+ petroleum ether	Vol 16/17	E603, 611
+ 1,2-propanediol	Vol 16/17	E603, 626
+ 2-propanol	Vol 16/17	E603, 628
+ 2-propanone	Vol 16/17	E603, 616
+ pyridine	Vol 16/17	E603, 623
+ sodium hydroxide (aqueous)	Vol 16/17	E603, 606
+ water	Vol 16/17	E603, 604 - 606

4-Thia-1-azabicyclo[3,2,0]heptane-2-carboxylic acid,
3,3-dimethyl-7-oxoo-6-[2-(allylthio)-amino]-ethyl-
4-amino-2-chlorobenzoate (1:1)
+ acetic acid, ethyl ester Vol 16/17 E334, 346
+ benzene Vol 16/17 E334, 341
+ benzene, methyl- Vol 16/17 E334, 342
+ benzenemethanol Vol 16/17 E334, 358
+ 1-butanol, 3-methyl- Vol 16/17 E334, 339
+ 1-butanol, 3-methyl-, acetate Vol 16/17 E334, 347
+ 2-butanone Vol 16/17 E334, 349
+ carbon disulfide Vol 16/17 E334, 354
+ cyclohexane Vol 16/17 E334, 340
+ 1,4-dioxane Vol 16/17 E334, 352
+ ethane, dichloro- Vol 16/17 E334, 351
+ ethane, 1,1´-oxybis- Vol 16/17 E334, 350
+ 1,2-ethanediol Vol 16/17 E334, 357
+ ethanol Vol 16/17 E334, 337
+ formamide Vol 16/17 E334, 356
+ methane, tetrachloro- Vol 16/17 E334, 345
+ methane, trichloro- Vol 16/17 E334, 353
+ methanol Vol 16/17 E334, 336
+ pentane, 2,2,4-trimethyl- Vol 16/17 E334, 344
+ petroleum ether Vol 16/17 E334, 343
+ 2-propanol Vol 16/17 E334, 338
+ 2-propanone Vol 16/17 E334, 348
+ pyridine Vol 16/17 E334, 355
+ water Vol 16/17 E334, 335
4-Thia-1-azabicyclo[3,2,0]heptane-2-carboxylic acid,
3,3-dimethyl-7-oxo-6-[[(1-aminocyclohexyl)carbonyl]amino]-
3,3-dimethyl-7-oxo-, dihydrate
+ water Vol 16/17 E363, 364
4-Thia-1-azabicyclo[3,2,0]heptane-2-carboxylic acid,
3,3-dimethyl-7-oxo-6-[(phenylacetyl)amino]
complexed with 2-(diethylamino)-ethyl-4-aminobenzoate (1:1)
+ butanedioic acid, 2,3-dihydroxy-, disodium salt Vol 16/17 E175, 186, 187, 189, 190, 196
+ butanedioic acid, 2,3-dihydroxy-, (multicomponent) Vol 16/17 E175, 186, 187, 190
+ D-gluconic acid (multicomponent) Vol 16/17 E175, 188
+ D-gluconic acid, monosodium salt (multicomponent) Vol 16/17 E175, 188, 190, 197
4-Thia-1-azabicyclo[3,2,0]heptane-2-carboxylic acid,
3,3-dimethyll-7-oxo-6-[(4-methylphenoxyacetyl)amino]-
complexed with N,N´-bis(phenylmethyl)-1,2-ethanediamine (2:1)
+ hydrochloric acid Vol 16/17 E330, 332
+ sodium hydroxide (aqueous) Vol 16/17 E330, 333
+ water Vol 16/17 E330, 331 - 333
4-Thia-1-azabicyclo[3,2,0]heptane-2-carboxylic acid,
3,3-dimethyl-7-oxo-6-[(phenoxyacetyl)amino]-
+ acetic acid, ethyl ester Vol 16/17 E10, 26
+ benzene Vol 16/17 E10, 21
+ benzene, methyl- Vol 16/17 E10, 22
+ benzenemethanol Vol 16/17 E10, 38
+ 1-butanol, 3-methyl- Vol 16/17 E10, 19
+ 1-butanol, 3-methyl-, acetate Vol 16/17 E10, 27
+ 2-butanone Vol 16/17 E10, 29
+ carbon disulfide Vol 16/17 E10, 34
+ cyclohexane Vol 16/17 E10, 20
+ 1,4-dioxane Vol 16/17 E10, 32
+ ethane, dichloro- Vol 16/17 E10, 31
+ ethane, 1,1´-oxybis- Vol 16/17 E10, 30

4-Thia-1-azabicyclo[3,2,0]heptane-2-carboxylic acid,
3,3-dimethyl-7-oxo-6-[(phenoxyacetyl)amino]-
+ 1,2-ethanediol Vol 16/17 E10, 37
+ ethanol Vol 16/17 E10, 17
+ formamide Vol 16/17 E10, 36
+ hydrochloric acid (multicomponent) Vol 16/17 E10, 12
+ hydrochloric acid (quaternary) Vol 16/17 E10, 14
+ methane, tetrachloro- Vol 16/17 E10, 25
+ methane, trichloro- Vol 16/17 E10, 33
+ methanol Vol 16/17 E10, 16
+ pentane, 2,2,4-trimethyl- Vol 16/17 E10, 24
+ petroleum ether Vol 16/17 E10, 23
+ phosphoric acid (multicomponent) Vol 16/17 E10, 15
+ phosphsosric acid, disodium salt (multicomponent) Vol 16/17 E10, 15
+ phosphoric acid, trisodium salt (multicomponent) Vol 16/17 E10, 12
+ 2-propanol Vol 16/17 E10, 18
+ 2-propanone Vol 16/17 E10, 28
+ pyridine Vol 16/17 E10, 35
+ sodium chloride (multicomponent) Vol 16/17 E10, 12
+ sorbitan monooleate, polyoxyethylene derivatives (multicomponent) Vol 16/17 E10, 15
+ sorbitan monooleate, polyoxyethylene derivatives (quaternary) Vol 16/17 E10, 14
+ water Vol 16/17 E10, 11 - 15
4-Thia-1-azabicyclo[3,2,0]heptane-2-carboxylic acid,
3,3-dimethyl-7-oxo-6-[(phenoxyacetyl)amino]-
complexed with N,B´-bis[(1,2,3,4,4a,9,10,10a-octahydro-
1,4a-dimethyl-7-(1-methylethyl)-1-phenanthrenyl)methyl]-
1,2-ethanediamine (2:1)
+ acetic acid, ethyl ester Vol 16/17 E102, 114
+ benzene Vol 16/17 E102, 109
+ benzene, methyl- Vol 16/17 E102, 110
+ benzenemethanol Vol 16/17 E102, 126
+ 1-butanol, 3-methyl- Vol 16/17 E102, 107
+ 1-butanol, 3-methyl-, acetate Vol 16/17 E102, 115, 118
+ 2-butanone Vol 16/17 E102, 117
+ carbon disulfide Vol 16/17 E102, 122
+ cyclohexdane Vol 16/17 E102, 108
+ 1,4-dioxane Vol 16/17 E102, 120
+ ethane, dichloro- Vol 16/17 E102, 119
+ 1,2-ethanediol Vol 16/17 E102, 125
+ ethanol Vol 16/17 E102, 105
+ formamide Vol 16/17 E102, 124
+ methane, tetrachloro- Vol 16/17 E102, 113
+ methane, trichloro- Vol 16/17 E102, 121
+ methanol Vol 16/17 E102, 104
+ pentane, 2,2,4-trimethyl- Vol 16/17 E102, 112
+ petroleum ether Vol 16/17 E102, 111
+ 2-propanol Vol 16/17 E102, 106
+ 2-propanone Vol 16/17 E102, 116
+ pyridine Vol 16/17 E102, 123
+ water Vol 16/17 E102, 103
4-Thia-1-azabicyclo[3,2,0]heptane-2-carboxylic acid,
3,3-dimethyl-7-oxo-6-[(phenoxyacetyl)amino]-
complexed with N,N´-bis(phenylmethyl)-1,2-ethanediamine (2:1)
+ acetic acid, ethyl ester Vol 16/17 E78, 89
+ benzene Vol 16/17 E78, 84
+ benzene, methyl- Vol 16/17 E78, 85
+ benzenemethanol Vol 16/17 E78, 101
+ 1-butanol, 3-methyl- Vol 16/17 E78, 82
+ 1-butanol, 3-methyl-, aceetate Vol 16/17 E78, 90

4-Thia-1-azabicyclo[3,2,0]heptane-2-carboxylic acid,
3,3-dimethyl-7-oxo-6-[(phenoxyacetyl)amino]-
complexed with N,N´-bis(phenylmethyl)-1,2-ethanediamine (2:1)
+ 2-butanone Vol 16/17 E78, 92
+ carbon disulfide Vol 16/17 E78, 97
+ cyclohexane Vol 16/17 E78, 83
+ 1,4-dioxane Vol 16/17 E78, 95
+ ethane, dichloro- Vol 16/17 E78, 94
+ ethane, 1,1´-oxybis- Vol 16/17 E78, 93
+ 1,2-ethanediol Vol 16/17 E78, 100
+ ethanol Vol 16/17 E78, 80
+ formamide Vol 16/17 E78, 99
+ hydrochloric acid Vol 16/17 E72, 75
+ methane, tetrachloro- Vol 16/17 E78, 88
+ methane, trichloro- Vol 16/17 E78, 96
+ methanol Vol 16/17 E78, 79
+ pentane, 2,2,4-trimethyl- Vol 16/17 E78, 87
+ petroleum ether Vol 16/17 E78, 86
+ 2-propanol Vol 16/17 E79, 81
+ 2-propanone Vol 16/17 E78, 91
+ pyridine Vol 16/17 E78, 98
+ sodium hydroxide (aqueous) Vol 16/17 E72, 76
+ water Vol 16/17 E72, 73 - 77
4-Thia-1-azabicyclo[3,2,0]heptane-2-carboxylic acid,
3,3-dimethyl-7-oxo-6-[(phenoxyacetyl)amino]-
complexed with N-(phenylmethyl)benzene ethanamine (1:1)
+ hydrochloric acid Vol 16/17 E68, 70
+ sodium hydroxide (aqueous) Vol 16/17 E68, 71
+ water Vol 16/17 E68, 69 - 71
4-Thia-1-azabicyclo[3,2,0]heptane-2-carboxylic acid,
3,3-dimethyl-7-oxo-6-[(phenoxyacetyl)amino],
calcium salt (2:1)
+ water Vol 16/17 E66, 67
4-Thia-1-azabicyclo[3,2,0]heptane-2-carboxylic acid,
3,3-dimethyl-7-oxo-6-[(phenoxyacetyl)amino],
monopotassium salt
+ acetic acid, ethyl ester Vol 16/17 E39, 53
+ benzene Vol 16/17 E39, 48
+ benzene, methyl- Vol 16/17 E39, 49
+ benzenemethanol Vol 16/17 E39, 65
+ 1-butanol, 3-methyl- Vol 16/17 E39, 46
+ 1-butanol, 3-methyl-, acetate Vol 16/17 E39, 54
+ 2-butanone Vol 16/17 E39, 56
+ carbon disulfide Vol 16/17 E39, 61
+ cyclohexane Vol 16/17 E39, 47
+ 1,4-dioxane Vol 16/17 E39, 59
+ ethane, dichloro- Vol 16/17 E39, 58
+ ethane, 1,1´-oxybis- Vol 16/17 E39, 57
+ ethanol Vol 16/17 E39, 44
+ ethanol, 2-methoxy- Vol 16/17 E39, 64
+ formamide Vol 16/17 E39, 63
+ methane, tetrachloro- Vol 16/17 E39, 52
+ methane, trichloro- Vol 16/17 E39, 60
+ methanol Vol 16/17 E39, 43
+ pentane, 2,2,4-trimethyl- Vol 16/17 E39, 51
+ petroleum ether Vol 16/17 E39, 50
+ phosphoric acid (multicomponent) Vol 16/17 E39, 41
+ phosphoric acid, disodium salt
(multicomponent) Vol 16/17 E39, 41
+ 2-propanol Vol 16/17 E39, 45
+ 2-propanone Vol 16/17 E39, 55
+ pyridine Vol 16/17 E39, 62
+ sorbitan monooleate,
polyoxyethylene derivatives
(multicomponent) Vol 16/17 E39, 41
+ water Vol 16/17 E39, 40 - 42

4-Thia-1-azabicyclo[3,2,0]heptane-2-carboxylic acid,
3,3-dimethyl-7-oxo-6-[(1-oxo-2-phenoxybutyl)amino],
monopotassium salt
+ hydrochloric acid (multicomponent) Vol 16/17 E545, 546, 547
+ polyethylene-23-lauryl ether
(multicomponent) Vol 16/17 E545, 547
+ potassium chloride
(multicomponent) Vol 16/17 E545, 546, 547
+ water Vol 16/17 E545, 546, 547
4-Thia-1-azabicyclo{3,2,0]heptane-2-carboxylic acid,
3,3-dimethyl-7-oxo-6-[(1-oxo-2-phenoxypropyl)amino],
monopotassium salt
+ acetic acid, ethyl ester Vol 16/17 E491, 502
+ benzene Vol 16/17 E491, 498
+ 1-butanol, 3-methyl- Vol 16/17 E491, 517
+ 1-butanol, 3-methyl-, acetate Vol 16/17 E491, 503
+ 2-butanone Vol 16/17 E491, 505
+ carbon disulfide Vol 16/17 E491, 510
+ cyclohexane Vol 16/17 E491, 507
+ 1,4-dioxane Vol 16/17 E491, 508
+ ethane, dichloro- Vol 16/17 E491, 507
+ ethane, 1,1´-oxybis- Vol 16/17 E491, 506
+ 1,2-ethanediol Vol 16/17 E491, 513
+ ethanol Vol 16/17 E491, 496
+ formamide Vol 16/17 E491, 512
+ hydrochloric acid Vol 16/17 E491, 493
+ methane, sulfinylbis- Vol 16/17 E491, 515
+ methane, tetrachloro- Vol 16/17 E491, 501
+ methane, trichloro- Vol 16/17 E491, 509
+ methanol Vol 16/17 E491, 495
+ pentane, 2,2,4-trimethyl- Vol 16/17 E491, 500
+ petroleum ether Vol 16/17 E491, 499
+ 1,2-propanediol Vol 16/17 E491, 514
+ 2-propanol Vol 16/17 E491, 516
+ 2-propanone Vol 16/17 E491, 504
+ pyridine Vol 16/17 E491, 511
+ sodium hydroxide (aqueous) Vol 16/17 E491, 494
+ water Vol 16/17 E491, 492 - 494
4-Thia-1-azabicyclo[3,2,0]heptane-2-carboxylic acid,
3,3-dimethyl-7-oxo-6-(2-phenylacetamaido)-
complexed with 1-(p-chlorobenzyl)-2-(1-pyrrolidinylmethyl)-
benzimidazole (1:1)
+ acetic acid, ethyl ester Vol 16/17 E249, 262
+ benzene Vol 16/17 E249, 258
+ 1-butanol, 3-methyl- Vol 16/17 E249, 256
+ 1-butanol, 3-methyl-, acetate Vol 16/17 E249, 263
+ 2-butanone Vol 16/17 E249, 265
+ carbon disulfide Vol 16/17 E249, 270
+ cyclohexane Vol 16/17 E249, 257
+ 1,4-dioxane Vol 16/17 E249, 268
+ ethane, dichloro- Vol 16/17 E249, 267
+ ethane, 1,1´-oxybis- Vol 16/17 E249, 266
+ 1,2-ethanediol Vol 16/17 E249, 273
+ ethanol Vol 16/17 E249, 254
+ formamide Vol 16/17 E249, 272
+ hydrochloric acid Vol 16/17 E249, 251
+ methane, sulfinylbis- Vol 16/17 E249, 275
+ methane, tetrachloro- Vol 16/17 E249, 261
+ methane, trichloro- Vol 16/17 E249, 269
+ methanol Vol 16/17 E249, 253
+ pentane, 2,2,4-trimethyl- Vol 16/17 E249, 260
+ petroleum ether Vol 16/17 E249, 259
+ 1,2-propanediol Vol 16/17 E249, 274
+ 2-propanol Vol 16/17 E249, 255
+ 2-propanone Vol 16/17 E249, 264
+ pyridine Vol 16/17 E249, 271
+ sodium hydroxide (aqueous) Vol 16/17 E249, 252
+ water Vol 16/17 E249, 250 - 252

4-Thia-1-azabicyclo[3,2,0]heptane-2-carboxylic acid,
3,3-dimethyl-7-oxo-6-(2-phenylacetamido)-
complexed with (-)-2-(methylamino)-1,2-diphenylethanol (1:1)
+ acetic acid, ethyl ester Vol 16/17 E224, 236
+ benzene Vol 16/17 E224, 231
+ benzene, methyl- Vol 16/17 E224, 232
+ benzenemethanol Vol 16/17 E224, 248
+ 1-butanol, 3-methyl- Vol 16/17 E224, 229
+ 1-butanol, 3-methyl-, acetate Vol 16/17 E224, 237
+ 2-butanone Vol 16/17 E224, 239
+ carbon disulfide Vol 16/17 E224, 244
+ cyclohexane Vol 16/17 E224, 230
+ 1,4-dioxane Vol 16/17 E224, 242
+ ethane, dichloro- Vol 16/17 E224, 241
+ ethane, 1,1´-oxybis- Vol 16/17 E224, 240
+ ethanediol Vol 16/17 E224, 247
+ ethanol Vol 16/17 E224, 227
+ formamide Vol 16/17 E224, 246
+ methane, tetrachloro- Vol 16/17 E224, 235
+ methane, trichloro- Vol 16/17 E224, 243
+ methanol Vol 16/17 E224, 226
+ pentane, 2,2,4-trimethyl- Vol 16/17 E224, 234
+ petroleum ether Vol 16/17 E224, 233
+ 2-propanol Vol 16/17 E224, 228
+ 2-propanone Vol 16/17 E224, 238
+ pyridine Vol 16/17 E224, 245
+ water Vol 16/17 E224, 225
4-Thia-1-azabicyclo[3,2,0]heptane-2-carboxylic acid,
3,3-dimethyl-7-oxo-6-[(phenylacetyl)amino]-
complexed with N,N´-bis[(1,2,3,4,4a,9,10,10a-octahydro-
1,4a-dimethyl-7-(1-methylethyl)-1-phenanthrenyl)methyl]-
1,2-ethanediamine (2:1)
+ acetic acid, ethyl ester Vol 16/17 E276, 288
+ benzene Vol 16/17 E276, 283
+ benzene, methyl- Vol 16/17 E276, 284
+ benzenemethanol Vol 16/17 E276, 300
+ 1-butanol, 3-methyl- Vol 16/17 E276, 281
+ 1-butanol, 3-methyl-, acetate Vol 16/17 E276, 289
+ 2-butanone vol 16/17 E276, 291
+ carbon disulfide vol 16/17 E276, 296
+ cyclohexane vol 16/17 E276, 282
+ 1,4-dioxane vol 16/17 E276, 294
+ ethane, dichloro- vol 16/17 E276, 293
+ ethane, 1,1´-oxybis- vol 16/17 E276, 292
+ 1,2-ethanediol vol 16/17 E276, 299
+ ethanol vol 16/17 E276, 279
+ formamide vol 16/17 E276, 298
+ methane, tetrachloro- vol 16/17 E276, 287
+ methane, trichloro- vol 16/17 E276, 295
+ methanol vol 16/17 E276, 278
+ pentane, 2,2,4-trimethyl- vol 16/17 E276, 286
+ petroleum ether vol 16/17 E276, 285
+ 2-propanol vol 16/17 E276, 280
+ 2-propanone vol 16/17 E276, 290
+ pyridine vol 16/17 E276, 297
+ water vol 16/17 E276, 277
4-Thia-1-azabicyclo[3,2,0]heptane-2-carboxylic acid,
3,3-dimethyl-7-oxo-6-[(phenylacetyl)amino]-
complexed with N,N´-bis(phenylmethyl)-1,2-ethanediamine (2:1)
+ acetic acid, ethyl ester vol 16/17 E306, 317
+ benzene vol 16/17 E306, 312
+ benzene, methyl- vol 16/17 E306, 313
+ benzenemethanol vol 16/17 E306, 329

4-Thia-1-azabicyclo[3,2,0]heptane-2-carboxylic acid,
3,3-dimethyl-7-oxo-6-[(phenylacetyl)amino]-
complexed with N,N´-bis(phenylmethyl)-1,2-ethanediamine (2:1)
+ 1-butanol, 3-methyl- vol 16/17 E306, 310
+ 1-butanol, 3-methyl-, acetate vol 16/17 E306, 318
+ 2-butanone vol 16/17 E306, 320
+ carbon disulfide vol 16/17 E306, 325
+ cyclohexane vol 16/17 E306, 311
+ 1,4-dioxane vol 16/17 E306, 323
+ ethane, dichloro- vol 16/17 E306, 322
+ ethane, 1,1´-oxybis- vol 16/17 E306, 321
+ 1,2-ethanediol vol 16/17 E306, 328
+ ethanol vol 16/17 E306, 308
+ formamide vol 16/17 E306, 327
+ hydrochloric acid vol 16/17 E301, 303
+ methane, tetrachloro- vol 16/17 E306, 316
+ methane, trichloro- vol 16/17 E306, 324
+ methanol vol 16/17 E306, 307
+ pentane, 2,2,4-trimethyl- vol 16/17 E306, 315
+ petroleum ether vol 16/17 E306, 314
+ 2-propanol vol 16/17 E306, 309
+ 2-propanone vol 16/17 E306, 319
+ pyridine vol 16/17 E306, 326
+ sodium hydroxide (aqueous) vol 16/17 E301, 304
+ water vol 16/17 E301, 302 - 305
4-Thia-1-azabicyclo[3,2,0]heptane-2-carboxylic acid,
3,3-dimethyl-7-oxo-6-[(phenylacetyl)amino]-
complexed with 2-(diethylamino)ethyl-4-aminobenzoate (1:1)
+ acetic acid, ethyl ester vol 16/17 E200, 211
+ benzene vol 16/17 E200, 206
+ benzene, methyl- vol 16/17 E200, 207
+ benzenemethanol vol 16/17 E200, 223
+ benzoic acid, 4-amino-2-(diethylamino),
ethyl ester, monohydrochloride vol 16/17 E175, 177, 179, 191, 192, 194, 195, 197, 198
+ 1-butanol, 3-methyl- vol 16/17 E200, 204
+ 1-butanol, 3-methyl-, acetate vol 16/17 E200, 212
+ 2-butanone vol 16/17 E200, 214
+ carbon disulfide vol 16/17 E200, 219
+ cyclohexane vol 16/17 E200, 205
+ 1,4-dioxane vol 16/17 E200, 217
+ ethane, dichloro- vol 16/17 E200, 216
+ ethane, 1,1´-oxybis- vol 16/17 E200, 215
+ 1,2-ethanediol vol 16/17 E200, 222
+ ethanol vol 16/17 E200, 202
+ formamide vol 16/17 E200, 221
+ a-D-glycopyranoside, b-D-fructofuranosyl
(multicomponent) vol 16/17 E175, 183, 184, 189, 190, 192 - 196, 198
+ D-glucose vol 16/17 E175, 185
+ D-gluctol (multicomponent) vol 16/17 E175, 182, 184, 198, 190, 193 - 198
+ methane, tetrachloro- vol 16/17 E200, 210
+ methane, trichloro- vol 16/17 E200, 218
+ methanol vol 16/17 E200, 201
+ pentane, 2,2,4-trimethyl- vol 16/17 E200, 209
+ petroleum ether vol 16/17 E200, 208
+ phosphoric acid, disodium salt
(multicomponent) vol 16/17 E175, 178, 179
+ phosphoric acid, monosodium salt
(multicomponent) vol 16/17 E175, 178, 179

4-Thia-1-azabicyclo[3,2,0]heptane-2-carboxylic acid,
3,3-dimethyl-7-oxo-6-[(phenylacetyl)amino]-
complexed with 2-(diethylamino)ethyl-4-aminobenzoate (1:1)
+ 1,2,3-propanetricarboxylic acid,
2-hydroxy- vol 16/17 E175, 180, 189, 190, 193, 195 - 198
+ 1,2,3-propanetricarboxylic acid,
2-hydroxy-, trisodium salt vol 16/17 E175, 180 - 185, 191 - 198
+ 2-propanol vol 16/17 E200, 203
+ 2-propanone vol 16/17 E200, 213
+ pyridine vol 16/17 E200, 220
+ water vol 16/17 E175, 176 - 199
4-Thia-1-azabicyclo[3,2,0]heptane-2-carboxylic acid,
3,3-dimethyl-7-oxo-6-[(phenacetyl)amino]-
complexed with 2-(diethylamino)-ethyl-4-aminobenzoate (1:1)
monohydrate
+ hydrochloric acid vol 16/17 E171, 173
+ sodium hydroxide (aqueous) vol 16/17 E171, 174
+ water vol 16/17 E171, 172 - 174
4-Thia-1-azabicyclo[3,2,0]heptane-2-carboxylic acid,
3,3-dimethyl-7-oxo-6-[(phenylacetyl)amino]-
complexed with N-(phenylmethyl)benzene ethanamine (1:1)
+ hydrochloric acid vol 16/17 E167, 169
+ sodium hydroxide (aqueous) vol 16/17 E167, 170
+ water vol 16/17 E167, 168 - 170
4-Thia-1-azabicyclo[3,2,0]heptane-2-carboxylic acid,
3,3-dimethyl-7-oxo-6-[(phenylacetyl)amino]-,
monoammonium salt
+ 2-propanone (ternary) vol 16/17 E165, 166
+ water vol 16/17 E165, 166
4-Thia-1-azabicyclo[3,2,0]heptane-2-carboxylic acid,
3,3-dimethyl-7-oxo-6-[(phenylacetyl)amino]-,
monopotassium salt
+ acetic acid, butyl ester
(quaternary) vol 16/17 E127, 128, 134
+ acetic acid, butyl ester
(ternary) vol 16/17 E127, 128, 133
+ 1-butanol vol 16/17 E127, 128, 130
+ 1-butanol (quaternary) vol 16/17 E127, 128, 134
+ 1-butanol (ternary) vol 16/17 E127, 128, 131 - 133
+ hydrochloric acid
(multicomponent) vol 16/17 E1, 8, 9
+ 2-propanone (ternary) vol 16/17 E127, 128, 129
+ water vol 16/17 E127, 128, 129, 131, 132, 134
4-Thia-1-azabicyclo[3,2,0]heptane-2-carboxylic acid,
3,3-dimethyl-7-oxo-6-[(phenylacetyl)amino]-,
monosodium salt
+ acetic acid, ethyl ester vol 16/17 E135, 136, 152
+ benzene vol 16/17 E135, 136, 147
+ benzene, methyl- vol 16/17 E135, 136, 148
+ benzenemethanol vol 16/17 E135, 136, 164
+ 1-butanol vol 16/17 E135, 136, 138
+ 1-butanol (ternary) vol 16/17 E135, 136, 139, 140
+ 1-butanol, 3-methyl- vol 16/17 E135, 136, 145
+ 1-butanol, 3-methyl-, acetate vol 16/17 E135, 136, 153
+ 2-butanone vol 16/17 E135, 136, 155
+ carbon disulfide vol 16/17 E135, 136, 160
+ cyclohexane vol 16/17 E135, 136, 146
+ 1,4-dioxane vol 16/17 E135, 136, 158

4-Thia-1-azabicyclo[3,2,0]heptane-2-carboxylic acid, 3,3-dimethyl-7-oxo-6-[(phenylacetyl)amino]-, monosodium salt

+ ethane, dichloro-	vol 16/17	E135, 136, 157
+ ethane, 1,1´oxybis-	vol 16/17	E135, 136, 156
+ 1,2-ethanediol	vol 16/17	E135, 136, 163
+ ethanol	vol 16/17	E135, 136, 143
+ formamide	vol 16/17	E135, 136, 162
+ methane, tetrachloro-	vol 16/17	E135, 136, 151
+ methane, trichloro-	vol 16/17	E135, 136, 159
+ methanol	vol 16/17	E135, 136, 142
+ pentane, 2,2,4-trimethyl-	vol 16/17	E135, 136, 150
+ petroleum ether	vol 16/17	E135, 136, 149
+ 2-propanol	vol 16/17	E135, 136, 144
+ 2-propanone	vol 16/17	E135, 136, 154
+ 2-propanone (ternary)	vol 16/17	E135, 136, 137
+ pyridine	vol 16/17	E135, 136, 161
+ water	vol 16/17	E135, 136, 137, 139 - 141

4-Thia-1-azabicyclo[3,2,0]heptane-2-carboxylic acid, 6-[[(2-ethoxy-1-naphthalenyl)carbonyl]amino]-3,3-dimethyl-7-oxo-, monosodium salt

+ acetic acid, ethyl ester	vol 16/17	E630, 641
+ benzene	vol 16/17	E630, 637
+ 1-butanol, 3-methyl-	vol 16/17	E630, 656
+ 1-butanol, 3-methyl-, acetate	vol 16/17	E630, 642
+ 2-butanone	vol 16/17	E630, 644
+ carbon disulfide	vol 16/17	E630, 649
+ cyclohexane	vol 16/17	E630, 636
+ 1,4-dioxane	vol 16/17	E630, 647
+ ethane, dichloro-	vol 16/17	E630, 646
+ ethane, 1,1´-oxybis-	vol 16/17	E630, 645
+ 1,2-ethanediol	vol 16/17	E630, 652
+ ethanol	vol 16/17	E630, 635
+ formamide	vol 16/17	E630, 651
+ hydrochloric acid	vol 16/17	E630, 632
+ methane, sulfinylbis-	vol 16/17	E630, 654
+ methane, tetrachloro-	vol 16/17	E630, 640
+ methane, trichloro-	vol 16/17	E630, 648
+ methanol	vol 16/17	E630, 634
+ pentane, 2,2,4-trimethyl-	vol 16/17	E630, 637
+ petroleum ether	vol 16/17	E630, 638
+ 1,2-propanediol	vol 16/17	E630, 653
+ 2-propanol	vol 16/17	E630, 655
+ 2-propanone	vol 16/17	E630, 643
+ pyridine	vol 16/17	E630, 650
+ sodium hydroxide (aqueous)	vol 16/17	E630, 633
+ water	vol 16/17	E630, 631 - 633

4-Thia-1-azabicyclo[3,2,0]hept-2-yl-5-tetrazole, 6-[D-2-amino-2-(4-aminophenyl)-acetamido]-3,3-dimethyl-7-oxo-, trihydrate

+ hydrochloric acid (multicomponent)	vol 16/17	E484, 485, 486
+ sodium hydroxide (multicomponent)	vol 16/17	E484, 485, 486
+ water	vol 16/17	E484, 485, 486

5-Thia-1-azabicyclo[4,2,0]oct-2-ene-2-carboxylic acid-3-[(acetyloxy)methyl]-7-[(aminophenylacetyl)amino]-8-oxo-, dihydrate

+ hydrochloric acid (multicomponent)	vol 16/17	E702, 703, 704
+ potassium chloride (multicomponent)	vol 16/17	E702, 703 - 705
+ potassium hydroxide (multicomponent)	vol 16/17	E702, 703, 704
+ water	vol 16/17	E702, 703 - 705

5-Thia-1-azabicyclo[4,2,0]oct-2-ene-2-carboxylic acid-
3-[(acetyloxy)methyl]-8-oxo-7-[(2-thienylacetyl)amino]-,
monosodium salt
+ acetic acid, ethyl ester vol 16/17 E667, 680
+ benzene vol 16/17 E667, 676
+ 1-butanol, 3-methyl- vol 16/17 E667, 674
+ 1-butanol, 3-methyl-, acetate vol 16/17 E667, 681
+ 2-butanone vol 16/17 E667, 683
+ carbon disulfide vol 16/17 E667, 688
+ cyclohexane vol 16/17 E667, 675
+ 1,4-dioxane vol 16/17 E667, 686
+ ethane, dichloro- vol 16/17 E667, 685
+ ethane, 1,1´-oxybis- vol 16/17 E667, 684
+ 1,2-ethanediol vol 16/17 E667, 691
+ ethanol vol 16/17 E667, 672
+ formamide vol 16/17 E667, 690
+ hydrochloric acid vol 16/17 E667, 669
+ methane, sulfinylbis- vol 16/17 E667, 693
+ methane, tetrachloro- vol 16/17 E667, 679
+ methane, trichloro- vol 16/17 E667, 687
+ methanol vol 16/17 E667, 671
+ pentane, 2,2,4-trimethyl- vol 16/17 E667, 678
+ petroleum ether vol 16/17 E667, 677
+ 1,2-propanediol vol 16/17 E667, 692
+ 2-propanol vol 16/17 E667, 673
+ 2-propanone vol 16/17 E667, 682
+ pyridine vol 16/17 E667, 689
+ sodium hydroxide (aqueous) vol 16/17 E667, 670
+ water vol 16/17 E667, 668 - 670
5-Thia-1-azabicyclo[4,2,0]oct-2-ene-2-carboxylic acid,
7-[(amino-1,4-cyclohexadien-1-ylacetyl)amino]-
3-methyl-8-oxo-, monohydrate
+ hydrocloric acid
(multicomponent) vol 16/17 E698, 699, 700
+ potassium chloride
(multicomponent) vol 16/17 E698, 699 - 701
+ potassium hydroxide
(multicomponent) vol 16/17 E698, 699, 700
+ water vol 16/17 E698, 699 - 701
5-Thia-1-azabicyclo[4,2,0]oct-2-ene-2-carboxylic acid,
7-amino-3-(hydroxymethyl)-8-oxo-, acetate ester
+ sodium chloride (aqueous) vol 16/17 E662, 663
+ water vol 16/17 E662, 663
5-Thia-1-azabicyclo[4,2,0]oct-2-ene-2-carboxylic acid,
7-amino-3-methyl-8-oxo-
+ ammonium chloride (aqueous) vol 16/17 E657, 659
+ ammonium chloride
(multicomponent) vol 16/17 E657, 660
+ benzeneacetic acid
(multicomponent) vol 16/17 E657, 660
+ hydrochloric acid
(multicomponent) vol 16/17 E657, 660, 661
+ potassium chloride (aqueous) vol 16/17 E657, 658
+ sodium chloride
(multicomponent) vol 16/17 E657, 660
+ sodium hydroxide
(multicomponent) vol 16/17 E657, 661
+ water vol 16/17 E657, 660, 661
5-Thia-1-azabicyclo[4,2,0]oct-2-ene-2-carboxylic acid-
7-[(aminophenylacetyl)amino]-3-methyl-8-oxo-, monohydrate
+ hydrochloric acid
(multicomponent) vol 16/17 E694, 695, 696
+ potassium chloride
(multicomponent) vol 16/17 E694, 695 - 697
+ potassium hydroxide
(multicomponent) vol 16/17 E694, 695, 696
+ water vol 16/17 E694, 695 - 697

5-Thia-1-azabicyclo[4,2,0]oct-2-ene-carboxlyic acid, 3-methyl-8-oco-7-(2-phenylacetamido)-
+ ammonium chloride (aqueous) vol 16/17 E664, 666
+ potassium chloride (aqueous) vol 16/17 E664, 665
+ water vol 16/17 E664, 665, 666

Thianaphthene
see benzo(b)thiophene

Thiocyanate, ammonium
see Thiocyanic acid, ammonium salt

Thiocyanate, potassium
see Thiocyanic acid, potassium salt

Thiocyanate, sodium
see Thiocyanic acid, sodium salt

Thiocyanic acid
+ sulfur dioxide vol 12 53

Thiocyanic acid, ammonium salt
+ acetamide, N-methyl- vol 11 E294, 295
+ formamide vol 11 E70, 71
+ formamide, N,N-dimethyl- vol 11 E231, 232 - 234
+ formamide, N-methyl- vol 11 E140, 141

Thiocyanic acid, ammonium salt (aqueous)
+ thiocyanic acid, silver salt vol 3 133 - 134, 148 - 149
+ sulfur dioxide vol 12 E37 - E39, 63, 64

Thiocyanic acid, barium salt (aqueous)
+ thiocyanic acid, silver salt vol 3 139 - 140

Thiocyanic acid, calcium salt (aqueous)
+ thiocyanic acid, silver salt vol 3 143 - 144
+ sulfur dioxide vol 12 53

Thiocyanic acid, lithium salt
+ thiocyanic acid, silver salt + ethanol vol 3 217 - 218

Thiocyanic acid, potassium salt
+ acetamide, N,N-dimethyl- vol 11 E331, 332
+ acetamide, N-methyl- vol 11 E282, 283
+ formamide vol 11 E48, 49 - 51
+ formamide, N,N-dimethyl- vol 11 E208, 209 - 211
+ formamide, N-methyl- vol 11 E128, 129
+ thiocyanic acid, silver salt + ethanol vol 3 219
+ thiocyanic acid, silver salt + nitric acid, potassium salt + acetonitrile vol 3 208
+ thiocyanic acid, silver salt + nitric acid, potassium salt + dimethylformamide vol 3 212
+ thiocyanic acid, silver salt + nitric acid, potassium salt + dimethyl sulfoxide vol 3 E213, 214
+ thiocyanic acid, silver salt + nitric acid, potassium salt + formamide vol 3 221
+ thiocyanic acid, silver salt + nitric acid, potassium salt + hexamethylphosphorotriamide vol 3 222
+ thiocyanic acid, silver salt + nitric acid, potassium salt + methanol vol 3 E223, 226
+ thiocyanic acid, silver salt + nitric acid, potassium salt + N-methyl-2-pyrrolidinone vol 3 229
+ thiocyanic acid, silver salt + perchloric acid, potassium salt + 1,2-ethanediol vol 3 220
+ thiocyanic acid, silver salt + perchloric acid, tetraethylammonium salt + tetrahydrothiophene-1,1-dioxide vol 3 232
+ thiocyanic acid, silver salt + 2-propanone vol 3 207

Thiocyanic acid, potassium salt (aqueous)
+ silver cyanide + thiocyanic acid, silver salt + potassium dicyanoargentate vol 3 E46, 65, E102, 122 - 123
+ sulfur dioxide vol 12 E37 - E39, 104, 105
+ thiocyanic acid, silver salt vol 3 E102, E103, E109, 114, 119 - 120, 128 - 129, 130, 131, 135 - 136, 141 - 142
+ thiocyanic acid, silver salt + nitric acid, potassium salt vol 3 E103, 145 - 146, 150 - 151, 152 - 153, 165 - 166, 169 - 170
+ thiocyanic acid, silver salt + nitric acid, potassium salt + dimethylacetamide vol 3 211
+ thiocyanic acid, silver salt + nitric acid, potassium salt + 1,4-dioxane vol 3 E188, 189
+ thiocyanic acid, silver salt + nitric acid, potassium salt + ethanol vol 3 E193, 194 - 195
+ thiocyanic acid, silver salt + nitric acid, potassium salt + methanol vol 3 E200, 201 - 202
+ thiocyanic acid, silver salt + nitric acid, potassium salt + 1,2,3-propanetriol vol 3 199
+ thiocyanic acid, silver salt + nitric acid, potassium salt + 1-propanol vol 3 205
+ thiocyanic acid, silver salt + nitric acid, potassium salt + 2-propanol vol 3 206
+ thiocyanic acid, silver salt + nitric acid, potassium salt + 2-propanone vol 3 E177 - 178 179 - 180
+ thiocyanic acid, silver salt + perchloric acid vol 3 164, 171
+ thiocyanic acid, silver salt + perchloric acid, potassium salt + dimethylformamide vol 3 184
+ thiocyanic acid, silver salt + perchloric acid, potassium salt + dimethyl sulfoxide vol 3 E185, 187
+ thiocyanic acid, silver salt + perchloric acid, potassium salt vol 3 196 - 197, 203 - 204
+ thiocyanic acid, silver salt + perchloric acid, potassium salt + 1,4-dioxane vol 3 E188, 190 - 192
+ thiocyanic acid, silver salt + perchloric acid, potassium salt + ethanol vol 3 E193, 196 - 198
+ thiocyanic acid, silver salt + perchloric acid, potassium salt + methanol vol 3 E200, 203 - 204
+ thiocyanic acid, silver salt + perchloric acid, potassium salt + 2-propanone vol 3 E177 - 178, 181 - 183
+ thiocyanic acid, silver salt + potassium bromide + sulfuric acid vol 3 121
+ thiocyanic acid, silver salt + silver bromide + potassium bromide vol 3 E102, 127
+ thiocyanic acid, silver salt + silver chloride + potassium chloride vol 3 E102, 126

Thiocyanic acid, silver salt
+ ammonia vol 3 233
+ methane, 1,1-sulfinylbis- vol 3 E213
+ methanol vol 3 E223

Thiocyanic acid, silver salt
+ nitric acid, silver salt + perchloric acid, tetraethylammonium salt + dimethyl sulfoxide vol 3 E213, 215
+ sulfur dioxide vol 3 234
+ thiocyanic acid, lithium salt + ethanol vol 3 217 - 218
+ thiocyanic acid, potassium salt + acetone vol 3 207
+ thiocyanic acid, potassium salt + ethanol vol 3 220
+ thiocyanic acid, potassium salt + nitric acid, potassium salt + acetonitrile vol 3 208
+ thiocyanic acid, potassium salt + nitric acid, potassium salt + dimethylacetamide vol 3 211
+ thiocyanic acid, potassium salt + nitric acid, potassium salt + dimethylformamide vol 3 212
+ thiocyanic acid, potassium salt + nitric acid, potassium salt + dimethyl sulfoxide vol 3 E213, 214
+ thiocyanic acid, potassium salt + nitric acid, potassium salt + formamide vol 3 221
+ thiocyanic acid, potassium salt + nitric acid, potassium salt + hexamethylphosphorotriamide vol 3 222
+ thiocyanic acid, potassium salt + nitric acid, potassium salt + methanol vol 3 E223, 226
+ thiocyanic acid, potassium salt + nitric acid, potassium salt + n-methyl-2-pyrrolidinone vol 3 229
+ thiocyanic acid, potassium salt + perchloric acid, potassium salt + 1,2-ethanediol vol 3 220
+ thiocyanic acid, potassium salt + perchloric acid, tetraethylammonium salt + tetrahydrothiophene-1,1-dioxane vol 3 232
+ thiocyanic acid, sodium salt + methanol vol 3 E223, 224 - 225
+ thiocyanic acid, tetraethylammonium salt + perchloric acid, tetraethylammonium salt + 1,2-dimethoxyethane vol 3 209 - 210
+ thiocyanic acid, tetraethylammonium salt + perchloric acid, tetraethylammonium salt + dimethyl sulfoxide vol 3 E213, 216
+ thiocyanic acid, tetraethylammonium salt + nitric acid, tetraethylammonium salt + formamide vol 3 221
+ thiocyanic acid, tetraethylammonium salt + nitric acid, tetraethylammonium salt + methanol vol 3 E223, 226
+ thiocyanic acid, tetraethylammonium salt + perchloric acid, tetraethylammonium salt + 4-methyl-1,3-dioxolan-2-one vol 3 228
+ thiocyanic acid, tetraethylammonium salt + perchloric acid, tetraethylammonium salt + nitromethane vol 3 227
+ thiocyanic acid, tetraethylammonium salt + perchloric acid, tetraethylammonium salt + tetrahydrofuran vol 3 230 - 231
+ water vol 3 e102, e108,
Thiocyanic acid, silver salt (aqueous)
+ acetone vol 3 E177
+ ammonia vol 3 E105, E109 - 110, 116, 132
+ 1,4-dioxane vol 3 E188
+ ethanol vol 3 E193

Thiocyanic acid, silver salt (aqueous)
+ perchloric acid, sodium salt
vol 3 175 - 176
+ perchloric acid, sodium salt + sulfurous acid, sodium salt
vol 3 E103, 173 - 174
+ methane, 1,1-sulfinylbis- vol 3 E185
+ methanol vol 3 E200
+ nitric acid, silver salt vol 3 E104, 115
+ silver bromide + potassium bromide + thiocyanic acid, potassium salt vol 3 E102, 127
+ silver chloride + potassium chloride + thiocyanic acid, potassium salt vol 3 E102, 126
+ silver cyanide + thiocyanic acid, potassium salt + potassium dicyanoargentate vol 3 E46, 65, E102, 122 - 123
+ thiocyanic acid, ammonium salt vol 3 133 - 134, 148 - 149 117 - 118, 124 - 125
+ thiocyanic acid, barium salt vol 3 139, 140
+ thiocyanic acid, calcium salt vol 3 143 - 144
+ thiocyanic acid, potassium salt
vol 3 E102, E103, E109, 114, 119 - 120, 128 - 131, 135 - 136, 141 - 142
+ thiocyanic acid, potassium salt + nitric acid, potassium salt
vol 3 E103, 145 - 146, 150 - 153, 165 - 166, 169 - 170
+ thiocyanic acid, potassium salt + nitric acid, potassium salt + 1,4-dioxane vol 3 E188, 189
+ thiocyanic acid, potassium salt + nitric acid, potassium salt + ethanol vol 3 E193, 194 - 195
+ thiocyanic acid, potassium salt + nitric acid, potassium salt + methanol vol 3 E200, 201 - 202
+ thiocyanic acid, potassium salt + nitric acid, potassium salt + 1,2,3-propanetriol vol 3 199
+ thiocyanic acid, potassium salt + nitric acid, potassium salt + 1-propanol vol 3 205
+ thiocyanic acid, potassium salt + nitric acid, potassium salt + 2-propanol vol 3 206
+ thiocyanic acid, potassium salt + nitric acid, potassium salt + 2-propanone vol 3 E177 - 178, 179 - 180
+ thiocyanic acid, potassium salt + perchloric acid
vol 3 164 - 171
+ thiocyanic acid, potassium salt + perchloric acid, potassium salt vol 3 196 - 197, 203 - 204
+ thiocyanic acid, potassium salt + perchloric acid, potassium salt + dimethylformamide vol 3 184
+ thiocyanic acid, potassium salt + perchloric acid, potassium salt + dimethyl sulfoxide vol 3 E185, 187
+ thiocyanic acid, potassium salt + perchloric acid, potassium salt + 1,4-dioxane vol 3 E188, 190 - 192
+ thiocyanic acid, potassium salt + perchloric acid, potassium salt + ethanol vol 3 E193, 196 - 198
+ thiocyanic acid, potassium salt + perchloric acid, potassium salt + 2-propanone vol 3 E177 - 178, 181 - 183

Thiocyanic acid, silver salt (aqueous)
+ thiocyanic acid, potassium salt + perchloric acid, potassium salt + methanol vol 3 E200, 203 - 204
+ thiocyanic acid, potassium salt + potassium bromide + sulfuric acid vol 3 121
+ thiocyanic acid, sodium salt vol 3 e104, 137 - 138
+ thiocyanic acid, sodium salt + perchloric acid, sodium salt vol 3 e104, 154 - 156, 159 - 163
+ thiocyanic acid, sodium salt + perchloric acid, sodium salt + sulfurous acid, sodium salt vol 3 172
+ thiocyanic acid, tetraethylammonium salt + perchloric acid, tetraethylammonium salt + diemthyl sulfoxide vol 3 e185, 186
+ sulfurous acid, sodium salt vol 3 e103, 168

Thiocyanic acid, sodium salt
+ acetamide, N,N-dimethyl- vol 11 E323, 324
+ acetamide, N-methyl- vol 11 E269, 270
+ formamide vol 11 E26, 27
+ formamide, N,N-dimethyl- vol 11 E181, 182 - 185
+ formamide, N-methyl- vol 11 E111, 112

Thiocyanic acid, sodium salt (aqueous)
+ sulfur dioxide vol 12 E37 - E39, 95 96
+ thiocyanic acid, silver salt vol 3 E104, E109, 137 - 138
+ thiocyanic acid, silver salt + perchloric acid, sodium salt vol 3 E104, 154 - 156, 159 - 163
+ thiocyanic acid, silver salt + perchloric acid, sodium salt + sulfurous acid, sodium salt vol 3 172
+ thiocyanic acid, silver salt + thiocyanic acid, silver sodium salt + perchloric acid, sodium salt vol 3 157 - 158, 162 - 163
+ thiocyanic acid, silver sodium salt + perchloric acid, sodium salt vol 3 E104, 157 - 158

Thiocyanic acid, tetrabutylammonium salt
+ thiocyanic acid, silver salt + perchloric acid, tetrabutylammonium salt + 1,2-dimethoxyethane vol 3 209 - 210
+ thiocyanic acid, silver salt + perchloric acid, tetrabutylammonium salt + tetrahydrofuran vol 3 230 - 231

Thiocyanic acid, tetraethylammonium salt
+ thiocyanic acid, silver salt + perchloric acid, tetraethylammonium salt + dimethyl sulfoxide vol 3 E213, 216
+ thiocyanic acid, silver salt + perchloric acid, tetraethylammonium salt + formamide vol 3 221
+ thiocyanic acid, silver salt + perchloric acid, tetraethylammonium salt + methanol vol 3 E223, 226
+ thiocyanic acid, silver salt + perchloric acid, tetraethylammonium salt + 4-methyl-1,3-dioxolan-2-one vol 3 228
+ thiocyanic acid, silver salt + perchloric acid, tetraethylammonium salt + nitromethane vol 3 227

Thiocyanic acid, tetraethylammonium salt (aqueous)
+ thiocyanic acid, silver salt + perchloric acid, tetraethylammonium salt + dimethyl sulfoxide vol 3 E185, 186

Thionyl choride (ternary)
+ sulfur dioxide vol 12 330, 331
Thionyl chloride
+ sulfur dioxide vol 12 E1, E2
Thiophene, tetrahydro-,1,1-dioxide
+ ethane vol 9 E195 -E199, 228
+ tetraphenylarsonium tetraphenylborate (1-) vol 18 E225, 226, 227
+ silver tetraphenylborate (1-) vol 18 174, 175
+ silver tetraphenylborate + perchloric acid, sodium salt + sodium tetraphenylborate vol 18 176
+ sulfur dioxide vol 12 320, 323
Thiophene, tetrahydro-3-methoxy-, 1,1-dioxide
+ sulfur dioxide vol 12 297
Thiophene, tetrahydro-3-methyl-, 1,1-dioxide
+ sulfur dioxide vol 12 297
Thiosulfuric acid, sodium salt (aqueous)
+ iodic acid, barium salt vol 14 262
+ iodic acid, calcium salt vol 14 E70, E74, 105
+ thiocyanic acid, silver salt + perchloric acid, sodium salt vol 3 172
+ thiocyanic acid, silver salt + thiocyanic acid, sodium salt + perchloric acid, sodium salt vol 3 173 - 174
Thiourea (aqueous)
+ nitric acid, cerium salt vol 13 219
Thiourea
+ water vol 13 219
Thulium nitrate
see Nitric acid, thulium salt
Thymus
see Guinea pig thymus
Thyroid
see Guinea pig thyroids
Tissue
see Guinea pig tissues, Rabbit tissues, Sheep placental tissue
Titanium chloride
+ chlorine vol 12 E354 -E366, 447
Titanium tetrachloride
see Titanium chloride
Toluene
see Benzene, methyl-
Toluidine
see Benzenamine, ar-methyl-
Triacetin
see 1,2,3-propanetriol, triacetate
Tribromomethane
see Methane, tribromo-
Tributylamine
see 1-Butanamine,N,N-dibutyl-
Tributylamine, perfluoro-
see 1-Butanamine,1,1,2,2,3,3,4,4,4-nonafluoro-N,N-bis-(nonafluorobutyl)-
Tributyl phosphate
see Phosphoric acid, tributyl ester
Tributyrin
see Butanoic acid, 1,2,3-propanetriyl ester
1,1,1-Trichloroethane
see Ethane, 1,1,1-trichloro-
1,1,2-Trichloroethane
see Ethane, 1,1,2-trichloro-
1,1,2-Trichloroethene
see Ethene, 1,1,2-trichloro-

1,1,2-Trichloroethylene
see Ethene, 1,1,2-trichloro-
Trichlorofluoromethane
see Methane, trichlorofluoro-
Trichloromethane
see Methane, trichloro-
(Trichloromethyl)benzene
see Benzene, (trichloromethyl)-
2-(Trichloromethyl)pyridine
see Pyridine, 2-(trichloromethyl)-
1,2,3-Trichloropropane
see Propane, 1,2,3-trichloro-
1,1,2-Trichloro-1,2,2-trifluoroethane
see Ethane,1,1,2-trichloro-1,2,2-trifluoro-
1,1,1-Trichloro-2-hydroxyethane
see Ethanol, 2,2,2-trichloro-
3,4,5-Trichloro-2-(dichloromethyl)pyridine
see Pyridine, 3,4,5-trichloro-2-(dichloromethyl)-
3,4,5-Trichloro-2-(trichloromethyl)pyridine
see Pyridine, 3,4,5-trichloro-2-(trichloromethyl)-
1,1,1,2,4,4,5,7,7,8,10,10,11,13,13,14,16,16,17,17,18,18,18-Tricosa fluoro-5,8,11,14-tetrakis(trifluoromethyl)-3,6,9,12,15-pentaoxaoctadecane
see 3,6,9,12,15-pentaoxaoctadecane, 1,1,1,2,4,4,5,7,7,8,10,10, 11,13,13,14,16,16,17,17,18,18,18-tricosafluoro-5,8,11,14-tetrakis(trifluoromethyl)-
Tricresyl phosphate
see Phosphoric acid, tris(methyl phenyl) ester
Tricyclo[3,3,1,3,7]decane, tetradecafluorobis(trifluoromethyl)-
+ oxygen vol 7 341
Tridecane
+ argon vol 4 136
+ helium vol 1 56
+ hydrogen vol 5/6 142
+ krypton vol 2 49
+ neon vol 1 201
+ nitrogen vol 10 E119, 142
+ nitrous oxide vol 8 E160, 175
+ oxygen vol 7 E214, E215, 234 239
+ sulfur dioxide vol 12 E116, E117, 133
+ xenon vol 2 161
Tridecanoic acid
+ radon-222 vol 2 309
Triethanolamine
see Ethanol, 2,2´,2´´-nitrilotris-
Triethylamine
see Ethanamine, N,N-diethyl-
Triethylamine nitrate
see Ethanamine, N,N-diethyl-, nitrate
Triethylenetetramine
see 1,2-ethanediamine, N,N-bis(2-aminoethyl)-
Triethyl phosphate
see Phosphoric acid, triethyl ester
(Trifluoromethyl)benzene
see Benzene, (trifluoromethyl)-
Triglyceride oil
+ oxygen vol 7 305 - 453
Trihexanoin
see Hexanoic acid, 1,2,3-propanetriyl ester
Trimethylammonium tetraphenylborate
+ water vol 18 115
Trimethylborane
+ nitrogen vol 10 497
1,3,5-Trimethylbenzene
see Benzene, 1,3,5-trimethyl-

2,2,4-Trimethylpentane
see Pentane, 2,2,4-trimethyl-
2,3,3-Trimethyl-2-butanol
see 2-Butanol, 2,3,3-trimethyl-
2,3,3-Trimethyl-3-pentanol
see 3-Pentanol, 2,3,3-trimethyl-
1,3,5-Trimethyl benzene
see Benzene, 1,3,5-trimethyl-
N,N,N-Trimetylmethanamine hydroxide
see Methanamine, N,N,N-trimethyl-, hydroxide
Trioctanoin
see Octanoic acid, 1,2,3-propanetriyl ester
1,3,5-Trioxane, 2,4,6-trimethyl-
+ oxygen vol 7 305, 453
Triphenyl methane
see Benzene, 1,1´,1´´-methylidynetris-
Tripotassium phosphate
see Phosphoric acid, potassium salt
Tripropylphosphate
see Phosphoric acid, tripropyl ester
Triisobutyl phosphate
see Phosphoric acid, tris(2-methyl propyl) ester
Trisodium phosphate
see Phosphoric acid, sodium salt
Tris(methylphenyl)phosphate
see Phosphoric acid, tris(methylphenyl)ester
Tris(o-phenanthroline)ruthenium (11) tetraphenylborate (1-)
+ acetic acid, butyl ester vol 18 120
+ acetic acid, ethyl ester vol 18 127
+ acetic acid, 2-methylpropyl ester vol 18 133
+ acetic acid, propyl ester vol 18 137
+ benzene, chloro- vol 18 121
+ benzene, methanol vol 18 117
+ butane, epoxy- vol 18 138
+ 1-butanol, 3-methyl- vol 18 131
+ 1-butanol, 3-methyl-, acetate vol 18 132
+ 2-butanol vol 18 119
+ 2-butanone vol 18 118
+ 2-butanone, 3,3-dimethyl- vol 18 125
+ ethane, 1,2-dichloro- vol 18 124
+ ethane, 1,1´-oxybis(2-chloro)- vol 18 122
+ ethanol vol 18 126
+ methane, trichloro- vol 18 123
+ methanol vol 18 130
+ 3-pentanone, 2,2,4,4-tetramethyl- vol 18 139
+ propane, nitro- vol 18 134
+ propane, 2,2´-oxybis- vol 18 129
+ 2-propanol vol 18 128
+ 2-propanone vol 18 135, 136
Tris (hydroxymethyl) aminomethane
see 1,3-propanediol, 2-amino-2-(hydroxymethyl)-
Tritolyl phosphate
see Phosphoric acid, tris(methyl phenyl) ester
Trout blood
see Blood, trout
Turpentine
+ sulfur dioxide vol 12 E147, 169
+ radon-222 vol 2 334
Tyrosine, 3,5-diiodo-monosodium salt (aqueous)
+ iodic acid, calcium salt vol 14 E76, E77, 129
Tyrosine, monosodium salt (aqueous)
+ iodic acid, calcium salt vol 14 E76, E77, 131

Ultrasene		
+ krypton	vol 2	69
+ xenon	vol 2	191
Umbilical cord		
see human umbilical cord		
Undecafluoro (trifluoromethyl) cyclohexane		
see cyclohexane, undecafluoro (trifluoromethyl)-		
Undecane		
+ argon	vol 4	132
+ helium	vol 1	54
+ hydrogen	vol 5/6	140
+ krypton	vol 2	43
+ neon	vol 1	199
+ nitrogen	vol 10	E119, 141
+ nitrous oxide	vol 8	E160, 173
+ oxygen	vol 7	E214, E215, 232, 239
+ sulfur dioxide	vol 12	E116, E117, 130,
+ xenon	vol 2	159
Undecanoic acid		
+ radon-222	vol 2	307
1-Undecanol		
+ hydrogen	vol 5/6	E189, 209
+ nitrogen	vol 10	E177, E178, 215
+ nitrous oxide	vol 8	205
+ oxygen	vol 7	E267, 288
+ water	vol 15	412
Unguentolan		
+ radon-222	vol 2	321
Uranium dioxosulfato- (aqueous)		
+ helium	vol 1	262
+ hydrogen	vol 5/6	E30, 56, 57
+ krypton	vol 2	E12 - E13, 17, 18
+ oxygen	vol 7	E64, E65, 109 - 116
+ xenon	vol 2	150
Uranium, dioxosulfato- (Multicomponent)		
+ oxygen	vol 7	E64, E65, 117, 118
Uranium, difluorodioxo- (aqueous)		
+ oxygen	vol 7	E64, E65, 107, 108
Uranyl fluoride		
see Uranium, difluorodioxo-		
Uranyl sulfate		
see Uranium, dioxosulfato-		
Urea + water	vol 13	35, 36, 216, 217, 334, 380, 403, 428, 433, 444, 453, 460, 464
Urea (aqueous)		
+ ethane	vol 9	E64, 71 - 73
+ hydrogen	vol 5/6	90
+ iodic acid, barium salt	vol 14	311
+ iodic acid, calcium salt	vol 14	175
+ nitric acid, cerium salt	vol 13	216, 217
+ nitric acid, erbium salt	vol 13	444, 445
+ nitric acid, gadolinium salt	vol 13	403
+ nitric acid, holmium salt	vol 13	433
+ nitric acid, lutetium salt	vol 13	464
+ nitric acid, neodymium salt	vol 13	334
+ nitric acid, praseodymium salt	vol 13	276
+ nitric acid, thulium salt	vol 13	453
+ nitric acid, ytterbium salt	vol 13	460

Urea (aqueous)
+ nitrogen vol 10 304 - 306
+ nitrous oxide vol 8 135 - 137
+ neon vol 1 183
+ oxygen vol 7 397 - 399
+ potassium tetraphenylborate vol 18 30
+ radon vol 2 258
Urea mononitrate
+ water vol 13 218, 404
Urea mononitrate (aqueous)
+ nitric acid, cerium salt vol 13 218
+ nitric acid, gadolinium salt vol 13 404
+ nitric acid, praseodymium nitrate vol 13 277
Uterine muscle
see Human uterine muscle
Urine, humam
+ nitrogen vol 10 299

Valeric acid,
see Pentanoic acid
Vaseline
+ radon-222 vol 2 331
Vaseline, oil
+ radon-222 vol 2 334
Vegetable oil
see Oil, vegetable
Vitreous body
see Rabbit vitreous body

Water : Aqueous solid-liquid and liquid-liquid systems are indexed under organic components not under water, only gas-liquid systems are indexed below.

Water		
+ air	vol 10	510 - 516
+ argon	vol 4	E1 - 7, 8 - 19, 25 - 26, E33 - 36, 37 - 68, 77 - 78, E80 - 84, 85 - 106, 245 - 247, E256, 257 - 260
+ chlorine	vol 12	E333, E334, 335 - 343
+ chlorine dioxide	vol 12	E454, 455, 456
+ chlorine monoxide	vol 12	449 - 453
+ deuterium	vol 5/6	279, 280
+ ethane	vol 9	E1, 2, 3 - 15, E16, 17 - 22
+ ethane + ethene	vol 9	E23, 24 - 26
+ helium	vol 1	E1, 5, E257, 258
+ helium-3	vol 1	15
+ hydrogen	vol 5/6	E1 - E3, 4 - 16
+ hydrogen deuteride	vol 5/6	E299
+ krypton	vol 2	E1 - 3, 4 - 8
+ krypton-85	vol 2	7, 121, 124
+ neon	vol 1	E124, 127
+ nitric oxide	vol 8	E260, E261, 262 - 264
+ nitrogen	vol 10	E1 - E4, 5- 29, E45 - E56, 57 - 82, E83, 84 - 103, 286, 294, 304 - 314, E333, 334 - 352
+ nitrous oxide	vol 8	E1, E2, 3 - 22, 241
+ oxygen		E1 - E5, 6 - 40, 305, 456 - 464
+ oxygen (18)	vol 7	39, 40
+ ozone	vol 7	481 - 483
+ radon-219	vol 2	E339, 340
+ radon-220	vol 2	E343, 344
+ radon-222	vol 2	E227 - 229, 230 - 234
+ silver azide	vol 3	E1
+ silver cyanide	vol 3	E45
+ sulfur dioxide	vol 12	E3 - E5, 6 - 33
+ xenon	vol 2	E134 -E136, 137 - 145, 220
+ xenon-133	vol 2	207, 214, 215, 217, 218, 225
Water (ternary)		
+ hydrogen	vol 5/6	36 - 99, 102 - 117, 313 - 320

Water (ternary)		
+ nitric oxide	vol 8	E265 -E267, 268 - 305, 311, 314 - 318, 328 - 335
+ nitrous oxide	vol 8	E27 - E40, 41 - 159, 248 - 255
Water (quaternary)		
+ hydrogen	vol 5/6	100, 101
Water-D2		
+ argon	vol 4	E20 - 21, 22 - 26, E33 - 36, 69 - 76
+ argentate(1-), bis(cyano-C)-potassium	vol 3	89
+ ethane	vol 9	E195 - 199, 229
+ helium	vol 1	14
+ hydrogen	vol 5/6	314
+ neon	vol 1	137
+ nitrogen	vol 10	30
+ radon-222	vol 2	235
+ xenon	vol 2	146
Wax ester oil		
+ oxygen	vol 7	443
Wine		
+ oxygen	vol 7	426 - 429

Xanthan			
+ oxygen	vol 7		424
Xenon			
+ helium	vol 1	E255,	353
Xylene			
see Benzene, dimethyl-			
o-xylene			
see Benzene, 1,2-dimethyl-			
p-xylene			
see Benzene, 1,3-dimethyl-			
m-xylene			
see Benzene, 1,4-dimethyl-			
Xylidine			
see benzenamine,ar,ar-dimethyl-			
Xylitol			
+ hydrogen	vol 5/6	317,	318
Xylose (aqueous)			
+ hydrogen	vol 5/6	315,	316
Yeast extract			
+ oxygen	vol 7		421
Yttrium hydroxide (aqueous)			
+ nitric acid, yttrium salt	vol 13	26,	27
Yttrium nitrate			
see Nitric acid, yttrium salt			
Zinc acetate			
see acetic acid, zinc salt			
Zinc bisulfite			
see Sulfurous acid zinc salt			
Zinc nitrate			
see Nitric acid, zinc salt			
Zinc sulfate			
see Sulfuric acid, zinc salt			

REGISTRY NUMBER INDEX

Page numbers preceded by E refer to evaluation text whereas those not preceded by E refer to compiled tables.

50-01-1	VOL 9	E29,E30,41
50-04-4	VOL 7	380,381
50-59-9	VOL 16/17	E706,707-732
50-70-4	VOL 5/6	319,320
50-70-5	VOL 16/17	E175,182,184,189,190,193-198
50-99-7	VOL 5/6	109,110
	VOL 7	397,412,413
	VOL 10	312
	VOL 16/17	E175,185
51-83-2	VOL 7	397
52-90-4	VOL 7	400
54-21-7	VOL 14	E75,E76,124
54-35-3	VOL 16/17	E175,176-199,E200,201-223
55-21-0	VOL 13	108
56-18-8	VOL 5/6	E483,505
56-23-5	VOL 2	88,176
	VOL 4	213,E214,E253-255,309
	VOL 5/6	E238,245,246,E281,290
	VOL 7	304,306-308,452,486-488,492
	VOL 8	218-220,E336,347,348
	VOL 9	E138-E141,152,153
	VOL 10	236,252,258
	VOL 12	E1,E2,298,300,303,304,316,317,E395-E397, 398-409,411
	VOL 16/17	E10,25,E39,52,E78,88,E102,113,E135,E136,151, E200,210,E224,235,E249,261,E276,287,E306, 316,E334,345,E367,379,E410,423,E457,464, E491,501,E518,528,E548,559,E576,586,E603, 613,E630,640,E667,679,E706,718,E740,751
56-40-6	VOL 5/6	96
	VOL 7	400
	VOL 10	307
	VOL 14	E76,E77,137,141,158,177,235,E252,281-284
56-41-7	VOL 7	400
56-65-5	VOL 7	404
56-81-5	VOL 2	81,286,287
	VOL 3	199
	VOL 4	E80-E84,98-100
	VOL 5/6	104,105
	VOL 7	207,305,409
	VOL 8	123-125,185
	VOL 10	E84,100-102
	VOL 12	E190,231,232,234
	VOL 13	119
	VOL 14	E74,169,E251,304

56-87-1	VOL 7	401
57-00-1	VOL 7	402
57-09-0	VOL 4	260
	VOL 7	360
57-13-5	VOL 18	30
57-13-6	VOL 1	183-184
	VOL 2	258
	VOL 5/6	90
	VOL 7	397-399
	VOL 8	135-137
	VOL 9	E64,71-73
	VOL 10	304-306
	VOL 13	4,5,35,36,216,217,276,334,380,403,428,433, 444,445,453,460,464
	VOL 14	E74,175,E251,311
57-14-7	VOL 1	101,103
	VOL 4	234,236
	VOL 10	272,273
57-48-7	VOL 10	311
57-50-1	VOL 2	257
	VOL 5/6	111,112
	VOL 7	414-416,417,426-428
	VOL 10	314
	VOL 14	E77,103,159
	VOL 16/17	E175,183,184,189,190,192-196,198
57-55-6	VOL 16/17	E249,274,E367,392,E410,438,E457,477,E491, 514,E518,541,E548,572,E576,599,E603,626, E630,653,E667,692,E706,731
57-59-1	VOL 9	E64,68
	VOL 4	249
	VOL 7	393,396
	VOL 10	300
58-64-0	VOL 7	404
58-71-9	VOL 16/17	E667,668-693
59-46-1	VOL 16/17	E175,177,179,191,192,194,195,197,198
59-56-3	VOL 7	407
60-01-5	VOL 2	315
60-29-5	VOL 2	288
60-29-7	VOL 4	E253-E255,294
	VOL 5/6	E213,216,217
	VOL 7	298,299,304,453
	VOL 8	196,E336,344
	VOL 9	E232,E233,234
	VOL 10	E105,109,110,227,228,521
	VOL 13	37,132,133,E221,222,223,E278,281-283,E335, 336,337,448
	VOL 16/17	E10,30,E39,57,E78,93,E102,118,135,E136,151, E200,215,E224,240,E249,254,E276,292,E306, 321,E334,350,E367,384,E410,429,E457,469, E491,506,E518,533,E548,564,E576,591,E603, 618,E630,645,E667,684,E706,723,E740,756

60-31-1	VOL 7	403
60-33-3	VOL 2	310
60-34-4	VOL 1	100
	VOL 4	233
	VOL 10	271
60-35-5	VOL 5/6	95
	VOL 10	E84,93
	VOL 11	E245,246,E247,248,E249,250,E251,E252,253, 254
	VOL 13	447
62-53-3	VOL 2	188,189,320
	VOL 5/6	265
	VOL 7	305,354,454
	VOL 8	214,215
	VOL 10	265,266
	VOL 12	E1,E2,287
	VOL 13	146
62-56-6	VOL 13	219
62-75-9	VOL 7	350
63-42-3	VOL 7	417
63-53-3	VOL 9	E195-E199,212
64-17-5	VOL 1	35-36,E82,83-84,181-182,E227,228-229
	VOL 2	E73,74,256,E273,274-277,342
	VOL 3	82,E193,194-198,217-219
	VOL 4	E34-E35,E80-E84,86-91,E169-E170,E176,177-179,E253-E255,283
	VOL 5/6	97-99,E187,194-199
	VOL 7	E190,E191,195-202,E262,270-276,303,304, 426-428,452
	VOL 8	118,185,186,188-191,294,306,307-310,312,313, 318,320,321,325-327,E336,343
	VOL 9	E64,69,70,E166,E167,172-175
	VOL 10	E83,88-91,E104,E105,106-114,E174,E177,E178, 185-190,447
	VOL 12	111,112,176,E181,182-188
	VOL 13	111
	VOL 14	E74,166,E210,224,E251,299-301
	VOL 16/17	E1,6,E10,17,E39,44,E78,80,E102,105,E135, 136,143,E200,202,E224,227,E249,254,E276, 279,E306,334,337,E367,372,E410,413,414, E457,459,E491,496,E518,523,E548,554,E576, 581,E603,608,E630,635,E667,672,E706,711, E740,743,765,766
64-18-6	VOL 2	290
	VOL 4	E83
	VOL 8	128,322
	VOL 12	198,200
		453
64-19-7	VOL 2	84,171,E291,292,293
	VOL 4	E80-E84,103-104,E169-E170,204
	VOL 5/6	94,E227,228,229
	VOL 7	453,492
	VOL 8	132,206-208,323
	VOL 9	E195-E199,208
	VOL 10	E45-E47,E54,77,221,357-360

64-19-7	VOL 12	198-201,E354-E366,386,387
	VOL 16/17	E1,4,5
	VOL 18	19,47,60,78,146,179
64-20-0	VOL 4	E33-E34,39-40
	VOL 9	E29,E30,37
64-56-1	VOL 2	E71,72,272
67-17-5	VOL 18	25,26,69,85,126,163,208
67-27-5	VOL 18	87
67-30-0	VOL 16/17	E276,280
67-48-1	VOL 7	403
67-56-1	VOL 1	E20,34,81,179-180,226,E255
	VOL 3	28,E200,201-204,E223,224-226
	VOL 4	E34-E35,79,E80-E84,85,97,99,E196-E171,172-175,E253-E255,282
	VOL 5/6	E186,191-193,E435,436-439,440,567,568
	VOL 7	188,189,E190-E192,193,194,E262,E263,268,269,304,419,452
	VOL 8	117,183-187,319,E336,342
	VOL 9	E166,E167,168-171,E232,E233,247,248
	VOL 10	E83,86,87,E174,E177,E178,179-183,446,495
	VOL 12	108,109,E174,175-180
	VOL 13	110
	VOL 14	E74,165,E251,298
	VOL 16/17	E10,16,E39,43,E78,79,E102,104,E135,E136,142,E200,201,E224,226,E249,253,E276,278,E306,307,E334,336,E367,371,E410,411,412,E457,458,E491,495,E518,522,E548,553,E576,580,E603,607,E630,E630,634,E667,671,E706,710,E733,735,E740,742
	VOL 18	27-29,38,61-63,71,88,101,130,E166,167-169,E217,218,219,220
67-63-0	VOL 2	279
	VOL 3	206
	VOL 4	E80-E84,93
	VOL 5/6	E188,212,E441,447-449
	VOL 7	E191,E192,205,206,E263,E264,278,304
	VOL 8	121,186
	VOL 10	E88,98,99,E175,195,196
	VOL 12	192
	VOL 13	117
	VOL 15	2
	VOL 16/17	E10,18,E39,45,E78,81,E102,106,E135,E136,144,E200,203,E224,228,E249,255,E306,309,E334,338,E367,373,E410,415,416,E491,516,E518,543,E548,574,E576,601,E603,628,E630,655,E667,673,E706,712,E740,744
	VOL 18	86,128,147,148
67-64-1	VOL 1	89,234
	VOL 2	83,289
	VOL 3	E177,179-183,207
	VOL 4	E34-E35,E82-E83,E169-E170,203,E253-E255,297-298
	VOL 5/6	153,168,212,E221,223,224
	VOL 7	290-293,303,304,453
	VOL 8	193-196,310,324
	VOL 9	E195-E199,206,E232,E233,246
	VOL 10	E105,106,114,E216,217-219
	VOL 12	178,E189,207-212,216
	VOL 13	56,137

67-64-1	VOL 14	E74,172,E251,305
	VOL 16/17	E10,28,E39,55,E78,91,E102,116,E127,E128,129, E135,E136,137,154,E165,166,E200,213,E224,238, E249,264,E276,290,E306,319,E334,348,E367,382, E410,426,427,E457,467,E491,504,E518,531,E548, 562,E576,589,E603,616,E630,643,E667,682,E706, 721,E733,736,E740,754
	VOL 18	2,4,15,16,E21,22,23,E40-41,42,48,51,79,135, 136,173,224
67-66-3	VOL 2	88,175,318
	VOL 4	212
	VOL 5/6	E238,247
	VOL 7	452,484,492
	VOL 8	220-222
	VOL 10	232,233
	VOL 12	E1,E2,E298,299-302,E354-E366,394
	VOL 16/17	E10,33,E39,60,E78,96,E102,121,E135,E136,159, E200,218,E224,243,E249,269,E276,295,E306,324, E334,353,E367,387,E410,433,E457,472,E491,509, E518,536,E548,567,E576,594,E603,621,E630,648, E667,687,E706,726,E733,738,E740,759
	VOL 18	123
67-68-5	VOL 1	99,244
	VOL 2	101,184
	VOL 3	E23,18,24-25,84,E185,186-187,E213,214-216
	VOL 4	26
	VOL 5/6	100,101,E259,260,261,E281,292,293
	VOL 7	346-348
	VOL 9	E64,67,E195-E199,221,222
	VOL 10	E253,258
	VOL 12	113,E318,319-322
	VOL 16/17	E249,275,E367,393,E410,439,440,E457,478,E491, E491,515,E518,542,E548,573,E576,600,E603,627, E630,654,E667,693,E706,732
	VOL 18	24,158-162,181,E203,204-207
68-04-2	VOL 16/17	E175,180-185,189,191-198
68-12-2	VOL 3	22,184,212
	VOL 4	E34-E35,E83
	VOL 9	E195-E199,220
	VOL 11	E148,149,150,E151,152-155,E156,157,158,E159, 160,161,E162,163-167,E168,169,170,E171,172- 175,E176,177,178,E179,180,E181,182-185,E186, 187-193,E194,195-197,E198,199-203,E204,205, E206,207,E208,209-211,E212,213-215,E216,217- 219,E220,221,E222,223-225,E226,227,228,E229, 230,E231,232-234,E235,236,237,E238,239,E240, 241,242,E243,244
	VOL 12	E261,E262,269-275,297,E354-E366,390
	VOL 14	E74,176,207,E251,310
	VOL 18	E154,155-157,180,199-202
68-41-7	VOL 16/17	E740,741-766
69-52-3	VOL 16/17	E367,368-393
69-53-4	VOL 16/17	E394,E395,396-409,E410,411-441
69-57-8	VOL 16/17	E135,E136,137-164
69-65-8	VOL 5/6	314
69-79-4	VOL 7	417

70-54-2	VOL 7	401
71-23-6	VOL 3	205
71-23-8	VOL 2	75,170,278
	VOL 4	E80-E84,92,E169-E170,E180-E181,182-184,E253-E255,284
	VOL 5/6	E187,200,201,E441,446
	VOL 7	E191,E192,203,204,263,277
	VOL 8	119,120,186,192
	VOL 9	E64,65,E166,E167,176-178
	VOL 10	E104,115,116,E174,E175,E177,E178,191-194
	VOL 12	114,115
	VOL 13	116
	VOL 14	E74,167,E251,302
	VOL 15	1
	VOL 16/17	E410,417
	VOL 18	6,105,111,114
71-32-8	VOL 10	E84,96,97
71-36-3	VOL 2	76,280
	VOL 4	E81,E169-E170,E185-E186,187-190,E253-E255,285
	VOL 5/6	E188,202,203,E441,452-455
	VOL 7	E264,279,280,304,453
	VOL 8	197
	VOL 9	E166,E167,179-181
	VOL 10	E175,E177,E178,197-200
	VOL 13	120
	VOL 15	E32-E38,39-91
	VOL 16/17	E127,E128,130-134,E135,E136,138-140
71-41-0	VOL 2	78,284,285,342
	VOL 4	E169-E170,193
	VOL 5/6	E189,204,205,E441,464
	VOL 7	E265,282
	VOL 8	200
	VOL 9	E61,62,63,E166,E167,182-184
	VOL 10	E176-E178,202-205
	VOL 13	124
	VOL 15	E162-E167,168-195,453
71-43-2	VOL 1	E68-E69,70-72,E214,215-217,E255
	VOL 2	62,166,264,341
	VOL 4	E158,159-161,211,248-250,E253-E255,279
	VOL 5/6	154,E159,E160,162-167,253,E281,287,E406,E407,408-414,554,555
	VOL 7	E250,E251,252-256,305,394-396,453
	VOL 8	180-182,E336,339
	VOL 9	E138-E141,155-157,163-165,E232,E233,238-244
	VOL 10	E104,111,112,E162,163-168,252,300-302,443,444
	VOL 12	E145-E147,148-153,E354-E366,375-379
	VOL 16/17	E10,21,E39,48,E78,84,E102,109,E135,E136,147,E200,206,E224,231,E249,258,E276,283,E306,312,E334,341,E367,376,E410,420,E457,461,E491,498,E518,525,E548,556,E576,583,E603,610,E630,637,E667,676,E706,715,E740,747
71-48-7	VOL 10	E45-E47,E54,77
71-55-6	VOL 12	E1,E2
71-91-0	VOL 1	E20,25
	VOL 2	E12-E13,15
	VOL 4	E33-E34,39-40

75-15-0	VOL 7	345
	VOL 8	259
	VOL 9	E138-E141,161,162
	VOL 10	E253,254-257
	VOL 12	328,E354-E366,392,411
	VOL 16/17	E10,34,E39,61,E78,97,E102,122,E135,E136,160, E200,219,E224,E249,270,E276,298,E306,325, E334,354,E367,388,E410,434,457,473,E491, 510,E518,537,E548,568,E576,595,E603,622, E630,649,E667,688,E706,727,E740,760
75-21-8	VOL 9	E195-E199,213,214
	VOL 10	231,454
75-25-2	VOL 2	88
75-28-5	VOL 5/6	E356,375
	VOL 9	E110-E112,117,118
	VOL 10	414-418
75-29-6	VOL 12	E1,E2
75-31-0	VOL 5/6	E482,494
75-34-3	VOL 12	E1,E2
	VOL 18	67,99,103,109,112,192
75-35-4	VOL 12	E1,E2
75-43-4	VOL 12	106,107
75-45-6	VOL 4	E253-E255,308
	VOL 7	489
	VOL 10	455,456
75-50-3	VOL 12	297
75-52-5	VOL 1	104,245
	VOL 2	103,187
	VOL 3	85,227
	VOL 4	E80-E84,106,227
	VOL 7	349
	VOL 10	260
	VOL 12	264,265
	VOL 18	171,222
75-58-1	VOL 1	E20,25
	VOL 2	E12-E13,15
75-59-2	VOL 5/6	100,101
75-65-0	VOL 1	37-38
	VOL 4	E34-E35,E80-E84,94-95,E169-E170,190
	VOL 5/6	106-108
	VOL 7	E192,208-210
	VOL 12	192
	VOL 13	123
	VOL 15	29-31
75-69-4	VOL 7	485,489
75-71-8	VOL 1	E254,280-281
	VOL 2	87,173
	VOL 10	235,457
75-72-9	VOL 5/6	515,516
	VOL 7	489

78-93-3	VOL 7	294,453
	VOL 12	210
	VOL 16/17	E10,29,E39,56,E78,92,E102,117,E135,E136,155, E200,214,E224,239,E249,265,E276,291,E306,320, E334,349,E367,383,E410,428,E457,468,E491,505, E518,532,E548,563,E576,590,E603,617,E630,644, E667,683,E706,721,E740,755
	VOL 18	118,119
78-96-6	VOL 8	140
79-00-5	VOL 12	E1,E2
79-01-6	VOL 7	452
	VOL 12	E1,E2,306
79-09-4	VOL 2	E294,295,296
	VOL 4	E169-E170
	VOL 5/6	102
	VOL 7	453,492,493
	VOL 8	133,134
	VOL 10	E84,94,95,
	VOL 12	213
79-10-7	VOL 2	297
79-11-8	VOL 5/6	91
79-16-3	VOL 1	105
	VOL 4	229
	VOL 9	E195-E199,215
	VOL 10	261
	VOL 11	E255,256,257,E258,E259,260,261,E262,263-265, E266,267,268,E269,270,E271,272,273,E274,E275, E275,276,277,E278,E279,280,281,E282,283,E284, E285,286,287,E288,E289,290,E292,E294,295, E296,297,E298,299,E300,301,E302,303,304,E305, 306,E307,308
79-20-4	VOL 8	209
79-20-9	VOL 5/6	E230,232
	VOL 7	302,304,453,492
	VOL 9	E195-E199,204,E232,E233,245
	VOL 10	223
	VOL 12	178,E189,214-216,218
	VOL 13	140
79-31-2	VOL 2	298
	VOL 10	E84,103,222
79-34-5	VOL 1	94,239
	VOL 5/6	E240,249
	VOL 12	E1,E2,E354-E366,414
81-25-4	VOL 16/17	E394,E395,408,E442,E443,454
84-66-2	VOL 12	E354-E366,389
84-74-2	VOL 2	85
	VOL 12	E354-E366,389
84-76-4	VOL 12	E354-366,389
87-08-1	VOL 16/17	E10,11-38
87-68-3	VOL 12	E354-E366,415

87-69-4	VOL 16/17	E175,186,190
87-99-0	VOL 5/6	317,318
88-72-2	VOL 12	289,E354-366,415
88-73-3	VOL 12	E354-E366,439
89-61-2	VOL 12	E354-E366,436
90-05-1	VOL 12	236
90-12-0	VOL 5/6	427-430
	VOL 10	172
90-80-2	VOL 16/17	E175,188
91-17-8	VOL 5/6	155,393
	VOL 9	E138-E141,151
	VOL 12	129,140,143,144
91-22-5	VOL 5/6	510
	VOL 12	278
91-66-7	VOL 7	354
	VOL 12	295,296
92-51-3	VOL 5/6	158,394
	VOL 9	E138-E141,150
	VOL 10	161
93-58-3	VOL 12	245
93-89-0	VOL 8	310
95-15-8	VOL 5/6	517,518
95-47-6	VOL 1	76,221
	VOL 2	64
	VOL 4	165
	VOL 7	258
	VOL 12	E354-E366,382
95-49-8	VOL 12	E354-E366,427
95-50-1	VOL 3	83
	VOL 12	309,E354-366,417
95-53-4	VOL 7	354
	VOL 13	145
95-57-8	VOL 12	313
96-18-4	VOL 12	E1,E2
96-22-0	VOL 4	E169-E170,E253-E255,301
	VOL 7	453
96-48-0	VOL 7	185
97-95-0	VOL 15	E237,238,239
97-99-4	VOL 12	233
98-00-0	VOL 12	233

98-01-1	VOL 9	E195-E199,219
98-07-7	VOL 12	E354-E366,423
98-08-8	VOL 12	E354-E366,423,430,431
98-82-8	VOL 5/6	182,183,425
	VOL 7	260
98-83-9	VOL 5/6	E178,179-181,183
98-86-2	VOL 12	239
	VOL 13	13
98-87-3	VOL 12	E354-E366,425
98-94-2	VOL 5/6	E483,506
98-95-3	VOL 1	106,246
	VOL 2	104,190,321
	VOL 4	228
	VOL 5/6	266
	VOL 7	353,454
	VOL 9	E195-E199,210,211
	VOL 10	264,265
	VOL 12	E261,E262,264,282-286,297,E354-E366,391
98-95-3	VOL 8	E336,350
99-54-7	VOL 12	E354-E366,435
100-00-5	VOL 12	E345-366,441
100-41-4	VOL 7	258,453
	VOL 12	E354-E366,385
100-44-7	VOL 12	E354-E366,426
100-46-9	VOL 12	293
100-47-0	VOL 12	267,288,297
100-51-6	VOL 4	E253-E255,293
	VOL 9	E166,E167,194
	VOL 12	196,244
	VOL 13	131
	VOL 16/17	E10,38,E39,65,E78,101,E102,126,E135,E136,164, E200,223,E224,248,E276,300,E306,329,E334,358, E740,764
	VOL 18	117
100-52-7	VOL 2	342
	VOL 4	E253-E255,296
	VOL 8	212,213
100-54-3	VOL 16/17	E733,739
100-61-8	VOL 7	354
	VOL 12	290
100-66-3	VOL 12	248
100-81-5	VOL 9	E195-E199,203
	VOL 5/6	560-561
	VOL 10	173
102-21-1	VOL 14	E74.168,E251,303

102-71-6	VOL 8	143
102-76-1	VOL 2	314
	VOL 4	E253-E255,306
	VOL 5/6	480
	VOL 10	452
102-82-9	VOL 5/6	E483,509
	VOL 12	277
103-50-4	VOL 4	E253-255,295
	VOL 7	453
103-65-1	VOL 7	260
103-69-5	VOL 7	354
	VOL 12	294
103-73-1	VOL 12	249
103-82-2	VOL 16/17	E1,8,E657,660
104-51-8	VOL 7	260
104-76-7	VOL 5/6	E442,470
	VOL 15	E359,360,361
105-30-6	VOL 15	E240,241,242
105-55-3	VOL 8	310
106-33-2	VOL 12	259
106-37-6	VOL 2	94
	VOL 4	220
106-42-3	VOL 1	80,225
	VOL 2	66,67,92,94,96,168
	VOL 4	167,E168,218,220,222
	VOL 7	258,259,305,453
	VOL 12	E354-366,384
106-43-4	VOL 12	E354-E366,429
106-46-7	VOL 2	92
	VOL 4	E80-E84,96-97,100,E169-E170,199-200
106-93-4	VOL 8	220,223,224
	VOL 12	E354-E366,412
106-97-8	VOL 5/6	E299,E346,347-353
	VOL 9	E110-E112,116,137
	VOL 10	385,386,E409,410-415,418
106-99-0	VOL 10	418
107-06-2	VOL 5/6	E239,248
	VOL 8	220
	VOL 12	E1,E2,306,307
	VOL 18	124
107-10-8	VOL 5/6	E482,491-493
	VOL 7	453
107-15-3	VOL 5/6	E482,489,490
	VOL 8	139

107-15-3	VOL 13	143
107-18-6	VOL 13	118
107-21-1	VOL 3	220
	VOL 5/6	E441,442
	VOL 7	305,453
	VOL 8	122
	VOL 9	E166,E167,190,192
	VOL 10	184
	VOL 12	E189,203-206
	VOL 13	112
	VOL 16/17	E10,37,E78,100,E102,125,E135,E136,163,E200, 222,E224,247,E249,273,E276,299,E306,328,E334, 357,E367,391,E410,437,E457,476,E491,513,E518, 540,E548,571,E576,598,E603,625,E630,652,E667, 691,E706,730
107-31-3	VOL 7	453
	VOL 12	202
107-66-2	VOL 18	50,68,100,104,110,113,193,231
107-87-9	VOL 4	E196-E170,E253-E255,300
	VOL 7	295,453
	VOL 13	11
107-92-6	VOL 2	299,300
	VOL 4	E80-E84,105,E169-E170,205
108-10-1	VOL 12	210
108-11-2	VOL 15	E257,E258,259-262
108-20-3	VOL 12	234
	VOL 13	9,284
	VOL 18	129
108-24-7	VOL 7	492
	VOL 12	E189,E190,220-223,E354-E366,388
108-32-7	VOL 3	228
	VOL 4	E253-E255,305
	VOL 5/6	477,478
	VOL 7	183,184,186,187
	VOL 9	E195-E199,223,224
	VOL 10	448
	VOL 12	E190,224,225
	VOL 18	172,223
108-38-3	VOL 1	E77,78-79,E222,223-224
	VOL 2	65
	VOL 4	166
	VOL 5/6	176,E419,420-422,558,559
	VOL 7	258
	VOL 8	E336,341
	VOL 12	E146,E147,158-160,E354-E366,383
108-39-4	VOL 5/6	475,476
108-41-8	VOL 12	E354-E366,428
108-44-1	VOL 7	354
108-67-8	VOL 2	69,168
	VOL 5/6	424,563
	VOL 12	E146,E147,161-163

108-82-7	VOL 15	E387,388,389
108-86-1	VOL 1	96,241
	VOL 2	93,179
	VOL 4	219
	VOL 7	311,453
	VOL 12	E261,E262,311
108-87-2	VOL 1	62,208
	VOL 2	55,165
	VOL 4	E149,150-152,E253-E255,277
	VOL 5/6	391,392
	VOL 7	244
	VOL 10	155,355,356,441,442
108-88-3	VOL 1	E73,74-75,E218,219-220
	VOL 2	63,167,168,191,E265,266-269,341
	VOL 4	E162,163-164,E253-E255,280
	VOL 5/6	161,E169,170-175,E281,288,E415,416-418,569, 570
	VOL 7	256,257,453
	VOL 8	E336,340
	VOL 9	E138-E141,154
	VOL 10	E162,169,170,445
	VOL 12	E145-E147,151,154-157,E354-E366,380,381
	VOL 16/17	E10,22,E39,49,E78,85,E102,110,E135,E136,148, E200,207,E224,232,E276,284,E306,313,E334,342, E740,748
	VOL 18	86,147,148
108-90-7	VOL 1	95,240
	VOL 2	91,178
	VOL 4	217
	VOL 5/6	E240,251
	VOL 7	311,312,453
	VOL 8	225
	VOL 9	E128-E141,158,159
	VOL 10	240
	VOL 12	E298,308-310,E354-E366,419-422
	VOL 18	121
108-91-8	VOL 5/6	263
	VOL 9	E195-E199,202
108-93-0	VOL 1	88,233
	VOL 2	82
	VOL 4	E169-E170,E201,E253-E255,292
	VOL 5/6	E189,210,211
	VOL 7	E267,289
	VOL 8	185,202,E336,345
	VOL 9	E166,E167,189
	VOL 10	207,357,358
	VOL 12	196,197
	VOL 13	130
	VOL 15	E210-E213,214-224
108-94-1	VOL 4	E253-E255,304
	VOL 5/6	225
	VOL 7	297
	VOL 8	201
	VOL 10	220,357-360,449,450
	VOL 12	E190,238-240
	VOL 13	138
108-95-2	VOL 5/6	226
	VOL 9	E166,E167,191

108-95-2	VOL 12	235,236
109-52-4	VOL 2	301
109-55-7	VOL 5/6	E482,501,502
109-60-4	VOL 7	453
	VOL 18	137
109-66-0	VOL 1	40,185
	VOL 2	25,155
	VOL 4	107,271
	VOL 5/6	E119,125,E354,358
	VOL 7	216,452
	VOL 8	163,164
	VOL 9	E77,E78,79,E110-E112,119,120,137
	VOL 10	123,E419,420
	VOL 12	E116,E117,118
109-69-3	VOL 12	E1,E2
109-73-9	VOL 7	453
	VOL 12	277
109-76-2	VOL 5/6	E482,497
109-86-4	VOL 5/6	E442,450
	VOL 12	217
	VOL 13	113,E278,279
	VOL 16/17	E39,64,E740,763
109-87-5	VOL 7	305
	VOL 12	234
109-94-4	VOL 7	453
	VOL 12	202
	VOL 13	139
109-99-9	VOL 3	230-231
	VOL 5/6	E215,219
	VOL 7	301
	VOL 10	229
	VOL 12	226,227
	VOL 14	E74,173,206,E251,306
	VOL 16/17	E410,441
110-19-0	VOL 5/6	E231,235
	VOL 10	225
	VOL 18	133
110-34-3	VOL 10	E119,E120,124-128
110-54-3	VOL 1	41,186,E255
	VOL 2	E26,27-79,156,262
	VOL 4	E108,109-110,E253-E255,273
	VOL 5/6	E119,126-129,E354,359-363,554,555
	VOL 7	217,218,219,452
	VOL 8	165,166,E336,337
	VOL 9	E77,E78,E79-E84,109,E110-E112,121-124
	VOL 10	E419,422-424
	VOL 12	E116,E117,119-121
110-56-5	VOL 12	E1,E2
110-62-3	VOL 4	E169-E170
110-63-4	VOL 5/6	E442,462

110-71-4	VOL 3	209-210
	VOL 5/6	E442,460
	VOL 12	234
110-80-5	VOL 5/6	E442,461,462
	VOL 9	E195-E199,205
	VOL 13	114,E278,280
110-82-7	VOL 1	E59,60-61,E204,205-207
	VOL 2	E52,53,54,E162,163,164,263
	VOL 4	E145,146-148,E253-E255,277
	VOL 5/6	148-154,E380,381-390,554,555
	VOL 7	241,242,243
	VOL 8	179,E336,337
	VOL 9	E138-E141,148,149
	VOL 10	E104,117,118,E148,149-154,353,354,439,440
	VOL 12	140-142,E354-E366,374
	VOL 16/17	E10,20,E39,47,E78,83,E102,108,E135,E136,146, E200,205,E224,230,E249,257,E276,282,E306, 311,E334,340,E367,375,E410,419,E457,460, E491,497,E518,524,E548,555,E576,582,E603, 609,E630,636,E667,675,E706,714,E740,746
110-83-8	VOL 5/6	149
	VOL 7	241
	VOL 10	154
110-86-1	VOL 3	10,29
	VOL 5/6	265
	VOL 7	352
	VOL 8	216,217,310
	VOL 10	263
	VOL 12	278-280
	VOL 16/17	E10,35,E39,62,E78,98,E102,123,E135,E136,161, E200,220,E224,245,E249,271,E276,297,E306,E326, E334,355,E367,389,E410,435,E457,474,E491,511 E518,538,E548,569,E576,596,E603,623,E630,650, E667,689,E706,728,E740,761
110-89-4	VOL 5/6	265,E482,500
	VOL 7	352
	VOL 10	263
110-91-8	VOL 13	144
110-97-4	VOL 8	142
110-98-5	VOL 9	E166,E167,193
111-01-3	VOL 5/6	E123,147
	VOL 9	E77,E78,108
111-13-7	VOL 4	E253-E255,303
	VOL 12	250
111-14-8	VOL 2	303
111-15-9	VOL 12	241
111-26-2	VOL 13	264,265
111-27-3	VOL 4	E169-E170,194,E253-E255,289
	VOL 5/6	205,E442,466
	VOL 7	E265,282
	VOL 8	200
	VOL 9	E166,E167,185,186
	VOL 10	E177,E178,205,206

111-27-3	VOL 12	193
	VOL 13	128,129
	VOL 15	E263-E268,269-290
111-29-5	VOL 8	127
111-42-2	VOL 8	141
	VOL 9	74,76
111-43-3	VOL 5/6	E214,218
	VOL 7	300
	VOL 10	230
111-44-4	VOL 9	E195-E199,209
	VOL 18	122
111-46-6	VOL 4	E80-E84,101
	VOL 8	126
	VOL 12	204
	VOL 13	134
111-55-7	VOL 12	241
111-65-9	VOL 1	E43,44-45,E188,189-190
	VOL 2	E35,36-38,159
	VOL 4	E114-E115,116-119,E253-E255,274
	VOL 5/6	E121,130,134,135,E281,285,E300,E355,365-367
	VOL 7	222-224,238,452
	VOL 8	169
	VOL 9	E77,E78,89-93,E110-E112,127,128
	VOL 10	E119,E120,131-134,E419,431,432,517
	VOL 12	E116,E117,123,124,E354-E366,370
111-66-0	VOL 5/6	E400,405
111-70-6	VOL 4	E169-E170,195
	VOL 5/6	E189,206,207
	VOL 7	E265,283,284,453
	VOL 8	203
	VOL 9	E166,E167,187
	VOL 10	E177,E178,208-210,520
	VOL 12	E194
	VOL 13	7
	VOL 15	E329-E333,334-351
111-76-2	VOL 5/6	E442,467
	VOL 12	217
111-77-3	VOL 5/6	E443,465
111-83-1	VOL 12	315
111-84-2	VOL 1	50,195
	VOL 2	39,159
	VOL 4	E125,126-127
	VOL 5/6	E122,137,138,562
	VOL 7	228,229,452
	VOL 8	171
	VOL 9	E77,E78,91-93
	VOL 10	E119,E120,137,138
	VOL 12	E116,E117,127
111-85-3	VOL 12	314
111-87-2	VOL 4	E125,126-127

111-87-5	VOL 1	86,231
	VOL 2	79
	VOL 4	E169-E170,196-197,E253-E255,290
	VOL 5/6	E189,206,207
	VOL 7	E266,283,284,285,453
	VOL 8	203
	VOL 9	E166,E167,186
	VOL 10	E176-E178,209-212,520
	VOL 12	E189,193,195
	VOL 15	E364-E366,367-381
111-92-2	VOL 5/6	E483,507
111-96-6	VOL 12	252
112-05-0	VOL 2	305
112-08-8	VOL 10	E176-E178,213,214,324
112-24-3	VOL 12	297
112-27-6	VOL 5/6	E483,468
112-30-1	VOL 1	87,232
	VOL 2	80
	VOL 4	E169-E170,198
	VOL 5/6	E189,208
	VOL 7	E266,E267,286,287,453
	VOL 8	204
	VOL 10	E176-E178,213,214,324
	VOL 15	E402,E403,404-411
112-32-3	VOL 7	453
112-35-6	VOL 5/6	E189,469
112-36-7	VOL 7	453
112-37-8	VOL 2	307
112-40-3	VOL 1	55,200
	VOL 2	44-46,160
	VOL 4	E133,134-135
	VOL 5/6	E122,141,356
	VOL 7	233
	VOL 8	174
	VOL 9	E77,E78,95,96,E110-E112,132-135
	VOL 10	E119,142,437,519
	VOL 12	E116,E117,131,132
112-42-5	VOL 5/6	E189,209
	VOL 7	E267,288
	VOL 8	205
	VOL 10	E177,E178,215
	VOL 15	412
112-53-8	VOL 5/6	E189,209
	VOL 7	E267,288
	VOL 8	205
	VOL 10	177,178,215,324,
	VOL 15	E413,414-417
112-72-1	VOL 15	E418,419,420
112-73-2	VOL 7	453
112-80-1	VOL 2	311

112-80-1	VOL 10	280
112-92-5	VOL 15	427
112-95-8	VOL 9	E77,E78,102,103,E110-E112,136
113-24-6	VOL 14	E75,E76,111
113-98-4	VOL 16/17	E1,9,E127,E128,129-134
115-07-1	VOL 4	E253-E255,276
	VOL 5/6	E400,401,402,540,547-551,553
	VOL 9	E232,E233,235
	VOL 10	400,401
115-11-7	VOL 12	E1,E2
115-20-8	VOL 12	305
117-84-0	VOL 12	E354-E366,389
119-36-8	VOL 12	246,247
119-64-2	VOL 2	61
	VOL 5/6	185,426,556-561
	VOL 10	280
	VOL 12	129,140
120-51-4	VOL 12	253
120-80-9	VOL 12	236
120-82-1	VOL 12	E354-E366,418
120-94-5	VOL 5/6	E482,498
121-44-8	VOL 5/6	E482,503
	VOL 13	95
121-69-7	VOL 7	354
	VOL 12	E261,E262,291-293,297
121-73-3	VOL 12	E354-E366,440
122-99-6	VOL 12	217
123-19-3	VOL 4	E169-E170
123-38-6	VOL 4	E169-E170
123-39-7	VOL 11	E88,89-92,E93,94,95,E96,97-100,E101,102-105, E106,107-110,E111,112,E113,114-117,E118, 119-112,E123,124-127,E128,129,E130,131-133, E134,135-137,E138,139,E140,141,E142,143,E144, 145,E146,147
123-51-3	VOL 2	283
	VOL 8	185,198,199
	VOL 12	193
	VOL 13	6,126
	VOL 15	E140-E142,143-158
	VOL 16/17	E10,19,E39,46,E78,82,E102,107,E135,E136,145, E200,204,E224,229,E249,256,E276,281,E306,310, E334,339,E367,E374,E410,418,E491,517,E518, 544,E548,575,E576,602,E603,629,E630,656,E667, 674,E706,713,E740,745
	VOL 18	131

123-63-7	VOL 7	305,453
123-72-8	VOL 4	E169-E170
123-75-1	VOL 5/6	266,E482,499
	VOL 7	352
	VOL 10	263
123-84-4	VOL 16/17	E127,E128,133,134
123-86-4	VOL 2	85
	VOL 7	453
	VOL 12	418
	VOL 18	120
123-91-1	VOL 3	E188,189-192
	VOL 4	E80-E84,102,E169-E170,202
	VOL 5/6	E215,220,221
	VOL 7	301,305,453
	VOL 9	E64,66,E195-E199,207
	VOL 10	229
	VOL 12	226-228-229
	VOL 13	135,136,E221,224,285
	VOL 14	E74,170,171,E251,307,308
	VOL 16/17	E10,32,E39,59,E78,95,E102,120,E135,E136,158, E200,217,E224,242,E249,268,E276,294,E306,323, E334,352,E367,386,E410,431,432,E457,471,E491, 508,E518,535,E548,566,E576,593,E603,620,E630, 647,E667,686,E706,725,E740,758
123-92-2	VOL 12	218
	VOL 16/17	E10,27,E39,54,E78,90,E102,115,E135,E136,153, E200,212,E224,237,E249,263,E276,289,E306,318, E334,347,E367,381,E410,425,E457,466,E491,503, E518,530,E548,562,E576,588,E603,615,E630,642, E667,681,E706,720,E740,753
	VOL 18	132
123-96-6	VOL 13	8
123-96-9	VOL 15	E382,383-385
124-07-2	VOL 2	304
124-09-4	VOL 5/6	E483,504
124-18-5	VOL 1	E51,52-53,E196,197-198
	VOL 2	E35,40-42,159
	VOL 4	E128,129-131
	VOL 5/6	E122,130,139,E355,368-370
	VOL 7	230,231,238
	VOL 8	172
	VOL 9	E77,E78,93,94,E110-E112,129-131
	VOL 10	E199-E121,139,140,E419,435,436,517
	VOL 12	E166,E117,128,129,E354-E366,372
124-38-9	VOL 1	E298,299-301
	VOL 4	E253-E255,320-321
	VOL 5/6	E299,E610,611-618
	VOL 7	434,435,468-472
	VOL 10	76,466,488,489-495,498-502
124-47-0	VOL 13	218,277,404
125-04-2	VOL 7	380,381

151-21-3	VOL 9	E32,54-56,E61,62,63
	VOL 16/17	E394,E395,407,E442,E443,453,E479,482,E487, 490
151-50-8	VOL 3	E45,68-69,81,72-73,77-78
	VOL 11	E46,47,E204,205
151-73-5	VOL 7	380,381
156-59-2	VOL 12	E1,E2
156-60-5	VOL 12	E1,E2
292-64-8	VOL 1	63,209
	VOL 2	56
	VOL 4	157
	VOL 7	249
	VOL 10	156
298-14-6	VOL 8	E38,112
302-01-2	VOL 1	103,113
	VOL 4	232,236
	VOL 10	270,273
302-17-0	VOL 5/6	92,93
	VOL 7	211
	VOL 8	144
	VOL 10	E84,92,237
302-72-7	VOL 14	E76,E77,142,178,E252,285,286
306-94-5	VOL 7	336
	VOL 10	527
307-34-6	VOL 7	326
	VOL 10	526
311-89-7	VOL 1	107-108
	VOL 2	97,185
	VOL 4	E230-E231
	VOL 5/6	267,E281,295
	VOL 7	356-359,386
	VOL 9	E138-E141,142,143
	VOL 10	267-269,532
330-20-0	VOL 11	E48,49-51,E128,129,E208,209-211,E282,283, E331,332
333-20-0	VOL 3	E46,65,E102-E103,E109,114,119-123,126-131, 135-136,141-142,145-146,150-153,164-166, 169-171,E177-E178,179-184,E185,187,E188, 189-192,E193,194-199,E200,201-208,211-212, E213,214,219-222,E223,226,232
	VOL 12	E37-E39,104,105
334-48-5	VOL 2	306
335-02-2	VOL 10	244
335-36-4	VOL 7	327,328,385
	VOL 10	247
335-57-9	VOL 1	90
	VOL 4	206
	VOL 5/6	E237,243,244,E281,289
	VOL 7	321

141-78-6	VOL 18	127
141-97-9	VOL 12	242
142-62-1	VOL 2	302
	VOL 12	213,243
142-68-7	VOL 5/6	E215,219
	VOL 7	301
	VOL 10	229
142-71-2	VOL 10	E45-E47,E54,77
142-82-3	VOL 5/6	E120,126,130-133,E281,283,284,E300,E354,364
142-82-5	VOL 2	30,157
	VOL 4	E111,112-113
	VOL 7	220,221,238
	VOL 8	167,168
	VOL 9	E77,E78,85-89,E110-E112,125,126
	VOL 10	E119,E120,128-131,355,356,E425,426-430,517
	VOL 12	E116,E117,122,E354-E366,367-389,452
142-83-5	VOL 1	42,187
142-96-1	VOL 12	251
	VOL 13	10
143-07-7	VOL 2	308
143-08-8	VOL 5/6	E189,207
	VOL 7	E266,287,453
	VOL 8	204
	VOL 10	E177,E178,213
	VOL 15	E392,E393,394-397
143-24-8	VOL 10	453
	VOL 12	258
143-33-9	VOL 3	79-80
	VOL 11	E176,177,178
143-66-8	VOL 18	E3,4-6,141-145,150-153,155,156,160,161,163,164,167-173,176
143-82-5	VOL 7	219
144-55-8	VOL 7	160
	VOL 8	E35,93
	VOL 10	E45-E47,E53,71
144-62-7	VOL 8	129-131
147-48-8	VOL 16/17	E66,67
149-65-5	VOL 14	E76,E77,121
150-78-7	VOL 12	252
150-83-4	VOL 14	E75,E76,119
151-10-0	VOL 12	252
151-21-3	VOL 4	259
	VOL 7	212

126-33-0	VOL 3	232
	VOL 9	E195-E199,228
	VOL 12	320,323
	VOL 18	E174,175,176,E225,226,227
126-71-6	VOL 9	E195-E199,230
	VOL 12	325
126-73-8	VOL 4	E253-E255,310
	VOL 5/6	519
	VOL 9	E195-E199,230
	VOL 12	324,326,327
	VOL 13	73,74,77,78,147,E221,225,240
127-09-3	VOL 7	162
	VOL 10	E45-E47,E54,77
	VOL 16/17	E1,5
127-18-4	VOL 7	304,452
	VOL 12	E1,E2,E354-E366,413
127-19-5	VOL 3	21,211
	VOL 11	E309,E310,311,312,E313,314,E315,316,E317,318, E319,320,E321,322,E323,324,E325,326,E327,328, E329,330,E331,332,E333,334,E335,336,E337,338, E339,340,E341,342,E343,344
	VOL 12	276,297
	VOL 18	E194,194-197
131-11-3	VOL 12	237,E354-E366,389
132-92-3	VOL 16/17	E518,519-543
132-93-4	VOL 16/17	E491,492-517
132-98-9	VOL 16/17	E39,40-65
133-37-9	VOL 16/17	E175,187
133-42-6	VOL 7	411
135-98-8	VOL 7	260
137-32-6	VOL 4	E253-E255,288
	VOL 15	E126,E127,128-130
140-11-4	VOL 12	253
140-29-4	VOL 12	267
141-43-5	VOL 8	138
	VOL 9	E74,75,E195-E199,226
	VOL 12	277
	VOL 13	115
141-78-6	VOL 2	312
	VOL 5/6	E230,233,234
	VOL 7	304,453
	VOL 8	310,E336,346
	VOL 9	E195-E199,201
	VOL 10	E105,107,108,224
	VOL 12	218,230
	VOL 13	141
	VOL 14	E74,174,E251,309
	VOL 16/17	E10,26,E39,53,E78,89,E102,114,E135,E136,152, E200,211,E224,236,E249,262,E276,288,E306,317, E334,346,E367,380,E410,424,E457,465,E491,502, E518,529,E548,560,E576,587,E603,614,E630,641, E667,680,E706,719,E733,737,E740,752

335-57-9	VOL 9	E138-E141,144,145
	VOL 10	245
	VOL 12	E354-E366,416
338-69-2	VOL 5/6	103
	VOL 10	308
355-02-2	VOL 1	91,235
	VOL 2	86,172
	VOL 4	E207,208-209,E253-E255,307
	VOL 7	320
373-02-4	VOL 10	E45-E47,E54,77
375-96-2	VOL 7	332
392-56-3	VOL 1	92,236
	VOL 2	90
	VOL 4	216
	VOL 7	314,315
	VOL 10	242
420-05-3	VOL 3	E38
423-55-2	VOL 7	325
	VOL 10	525
462-06-6	VOL 1	93,237
	VOL 2	89,177
	VOL 4	215
	VOL 7	311
463-56-9	VOL 12	53
463-82-1	VOL 4	E253-E255,272
464-07-3	VOL 15	E234,235,236
464-49-3	VOL 12	E190,254-257
493-01-6	VOL 5/6	156
493-02-7	VOL 5/6	157
495-19-8	VOL 10	E45-E47,E53,71
	VOL 5/6	E33,76
	VOL 7	139,163
	VOL 8	E35,93
504-63-2	VOL 5/6	E442,451
505-22-6	VOL 13	338
506-61-6	VOL 3	E45,63-65,74-76,79-80,83,89,E102,122-123
506-64-9	VOL 3	E45-E46,59-73,75-81,84-88
506-93-4	VOL 13	92,259
507-20-0	VOL 12	E1,E2
512-56-1	VOL 9	E195-E199,230
513-08-6	VOL 9	E195-E199,230
	VOL 12	325
513-36-0	VOL 12	E1,E2

527-07-1	VOL 16/17	E175,188,190,197
532-94-5	VOL 14	E76,E77,130
537-03-1	VOL 13	67,68
538-23-8	VOL 2	317
540-54-5	VOL 12	E1,E2
540-67-0	VOL 7	304,453
540-72-7	VOL 3	E104,E109,137-138,154-163,172
	VOL 11	E26,27,E111,112,E181,182-185,E269,270,E323, 324
	VOL 12	37-39,95,96
540-73-8	VOL 1	102
	VOL 4	235
	VOL 10	274
540-84-1	VOL 1	49,194
	VOL 2	34,158
	VOL 4	E123,124,E253-E255,275
	VOL 5/6	E122,136,E281,286,E356,376-379
	VOL 7	213,225-227,238,304
	VOL 8	170
	VOL 9	106
	VOL 10	E104,113,115-118,E121,135,136,433,434,517, 518
	VOL 12	125,126,E354-E366,371
	VOL 16/17	E10,24,E39,51,E78,87,E102,112,E135,E136,150, E200,209,E224,234,E249,260,E276,286,E306, 315,E334,344,E367,378,E410,422,E457,463, E491,500,E518,527,E548,558,E576,585,E630, 639,E667,678,E706,717,E740,750
542-15-4	VOL 13	104,105,215,271,424
543-49-7	VOL 15	352,353
543-53-3	VOL 13	102,103,213,214,269,270,331,397,423,441
543-59-9	VOL 12	E1,E2
544-10-5	VOL 5/6	E240,250
	VOL 7	313
	VOL 10	241
	VOL 12	E1,E2
544-76-3	VOL 1	58,203
	VOL 2	49,161
	VOL 4	141
	VOL 5/6	E123,145,146,373,374
	VOL 7	237
	VOL 8	178
	VOL 9	E77,E78.97-100,109
	VOL 10	E119,E121,146,147,E419,438
	VOL 12	E116,E117,136-138
548-94-1	VOL 1	47,192
551-16-6	VOL 16/17	E1,2-9
554-13-2	VOL 7	139

556-50-3	VOL 14	E76,E77,143,288
556-67-2	VOL 1	109,248
	VOL 2	107
	VOL 4	239
	VOL 5/6	268,520
	VOL 9	E195-E199,227
	VOL 10	275
557-34-6	VOL 10	E45-E47,E54,77
563-45-1	VOL 12	E1,E2
565-60-6	VOL 15	E250,251,252
565-67-3	VOL 15	E247,248,249
575-52-0	VOL 16/17	E334,335-358
583-60-8	VOL 12	240
584-02-1	VOL 13	125
	VOL 15	E203,E204,205-209
584-08-7	VOL 5/6	E35,89
	VOL 7	139
	VOL 8	E38,112
	VOL 10	E45-E47,E55,81,82
584-94-1	VOL 2	32
	VOL 4	121
589-38-8	VOL 4	E169-E170,E253-E255,302
589-43-5	VOL 1	48,193
	VOL 2	33
	VOL 4	122
589-55-9	VOL 15	357
589-81-1	VOL 1	46,191
	VOL 2	31
	VOL 4	120
589-82-2	VOL 15	E354,355,356
590-01-2	VOL 13	14
590-28-3	VOL 3	E38-E39,43-44
590-28-5	VOL 11	E206,207
590-36-3	VOL 15	E243,244-246
591-07-1	VOL 13	220,370,446
591-23-1	VOL 12	238
591-50-4	VOL 1	97,242
	VOL 4	221
	VOL 7	311,453
	VOL 12	312
591-78-6	VOL 7	296
	VOL 13	12
592-27-8	VOL 16/17	E603,612

592-41-6	VOL 5/6	E400,403
592-76-7	VOL 5/6	E400,404
593-45-3	VOL 9	E77,E78,99,101
593-90-8	VOL 10	497
594-42-3	VOL 12	E354-E366,410,411
594-60-5	VOL 15	E231,232,233
594-82-2	VOL 15	299
595-41-5	VOL 15	E307,308,309
597-49-9	VOL 15	E318,319,320
597-96-6	VOL 15	E325,326,327
598-06-1	VOL 15	363
598-75-4	VOL 15	E159,160,161
600-36-2	VOL 15	E314,315,316
616-25-1	VOL 15	120
617-29-8	VOL 15	324
621-70-5	VOL 2	316
623-37-0	VOL 15	E295,296-298
623-93-8	VOL 15	401
624-29-3	VOL 1	67,213
	VOL 2	60
	VOL 4	156
	VOL 7	248
	VOL 10	159
624-38-4	VOL 2	96
	VOL 4	222
624-48-6	VOL 12	237
624-51-1	VOL 15	399
625-06-9	VOL 15	E311,312,313
625-23-0	VOL 15	E321,322,323
625-25-2	VOL 15	362
626-89-1	VOL 15	256
626-93-7	VOL 15	E291,292-294
627-13-4	VOL 7	351
	VOL 10	262,531
627-59-8	VOL 15	328
628-63-7	VOL 2	313
	VOL 5/6	E230,236

628-63-7	VOL 8	208,210,211
	VOL 9	E195-E199,200
	VOL 10	226
628-97-7	VOL 10	280
628-99-9	VOL 15	398
629-04-9	VOL 5/6	E242,254
	VOL 7	317
629-06-1	VOL 12	E1,E2
629-50-5	VOL 1	56,201
	VOL 2	47,161
	VOL 4	136
	VOL 5/6	E122,142
	VOL 7	234
	VOL 8	175
	VOL 10	E119,143
	VOL 12	E116,E117,133
629-59-4	VOL 1	57,202
	VOL 2	48,161
	VOL 4	E137,138-139
	VOL 5/6	E122,143,E356,372
	VOL 7	235
	VOL 8	176
	VOL 10	E119,144
	VOL 12	E116,E117,134,E354-E366,373
629-62-9	VOL 1	58,203
	VOL 2	49,161
	VOL 5/6	E122,144
	VOL 7	236
	VOL 8	177
	VOL 10	E119,145
	VOL 12	E116,E117,135
629-76-5	VOL 15	421
629-78-7	VOL 9	E77,E78,98
	VOL 12	E166,E117,138,139,316
629-97-0	VOL 9	E77,E78,104
630-08-0	VOL 1	E254,295-297
	VOL 4	E253-E255,319
	VOL 5/6	E299,471-474,528,529,E601,602-608,631-633
638-04-0	VOL 1	66,212
	VOL 2	59
	VOL 4	155
	VOL 7	246
	VOL 10	160
638-38-0	VOL 10	E45-E47,E54,77
638-53-9	VOL 2	309
661-95-0	VOL 7	309
680-31-9	VOL 3	27,222
	VOL 9	E195-E199,218
	VOL 18	165,E213,214-216
696-29-7	VOL 5/6	564-566

751-84-8	VOL 16/17	E167,168-170
776-35-2	VOL 5/6	432,433
814-80-2	VOL 14	160
815-24-7	VOL 18	139
821-09-0	VOL 15	122
868-18-8	VOL 16/17	E175,186,187,189,190,196
872-50-4	VOL 3	30,229
	VOL 5/6	511
	VOL 9	E195-E199,216,217
	VOL 10	462
	VOL 12	281
	VOL 18	5,39,80,102,170,221
872-93-5	VOL 12	297
917-61-3	VOL 3	E38-39,42
917-63-1	VOL 11	E179,180
917-70-4	VOL 13	66
929-06-6	VOL 9	E195-E199,225
957-68-6	VOL 16/17	E662,663
985-16-0	VOL 16/17	E630,631-656
993-20-4	VOL 3	18,20,E23,24-26
1003-03-8	VOL 5/6	262,E281,295
1016-09-7	VOL 12	248
1068-52-6	VOL 14	263
1070-36-9	VOL 14	E75,E76,114
1070-67-3	VOL 14	E76,E77,120
1119-40-0	VOL 5/6	479
	VOL 10	451
1120-21-4	VOL 1	54,199
	VOL 2	43,159
	VOL 4	132
	VOL 5/6	E122,140
	VOL 7	232
	VOL 8	173
	VOL 10	E119,141
	VOL 12	E116,E117,130
1128-16-1	VOL 12	E354-E366,434
1173-88-2	VOL 16/17	E603,604-629
1185-33-7	VOL 15	E228,229,230
1186-50-1	VOL 13	316
1197-03-1	VOL 12	E354-E366,438

1201-30-5	VOL 12	E354-E366,432
1245-44-9	VOL 16/17	E545,546,547
1300-21-6	VOL 7	452,492
	VOL 16/17	E10,31,E39,58,E78,94,E102,119,E135,E136,157, E200,216,E224,241,E249,267,E276,293,E306,322, E334,351,E367,385,E410,430,E457,470,E491,507, E518,534,E548,565,E576,592,E603,619,E630,646, E667,685,E706,724,E740,757
1300-73-8	VOL 12	287,297
1303-33-9	VOL 8	145,148
1305-62-0	VOL 14	E69,102,154
1309-33-7	VOL 8	145-147
1310-58-3	VOL 1	E20,28,E141,167
	VOL 4	E33-E34,60
	VOL 5/6	E33,E34,77-82
	VOL 7	131,162,165-169,188,189
	VOL 8	E35,94
	VOL 14	E68,E69,132
	VOL 16/17	E359,360,E394,E395,402,403,E442,E443,448,449, E479,480,481,E487,488,E694,695,696,E698,699, 700,E702,703-704
1310-65-2	VOL 7	131
1310-73-2	VOL 3	75-76,79-80
	VOL 5/6	E31,64,65
	VOL 7	131,154-161
	VOL 10	E45-E47,E49,E50,65,66
	VOL 14	E68,99,E191,E195,202,E252,282,286,301
	VOL 16/17	E68,71,E72,76,E167,170,E171,174,E249,252, E301,304,E330,333,E367,370,E394,E395,404, E442,E443,451,E484,485,486,E491,494,E518, 521,E548,552,E576,579,E603,606,E630,633, E657,661,E667,670,E706,709
	VOL 18	27-29,61,62,84,98
1310-82-3	VOL 7	131
1314-64-3	VOL 1	262
	VOL 2	E12-E13,17-18,150
	VOL 5/6	E30,56,57
	VOL 7	109-118
1321-94-4	VOL 9	E138-E141,160
1330-20-7	VOL 2	168,270,271
	VOL 4	E253-E255,281
	VOL 5/6	176
	VOL 10	171
	VOL 12	151
1330-78-5	VOL 2	106
	VOL 12	324
1333-41-1	VOL 12	278
1333-74-0	VOL 1	E307,308-309,E310,311-322
1333-84-2	VOL 5/6	E30,53

1333-84-2	VOL 7	101
	VOL 10	E45-48,60
1335-47-3	VOL 12	E146,E147,165,166
1336-21-6	VOL 5/6	E28,42
	VOL 7	97
	VOL 14	E247,294
1342-98-2	VOL 8	150,151
1454-85-9	VOL 15	426
1495-28-9	VOL 7	489
1538-09-6	VOL 16/17	E301,302-305,E306,307-329
1569-50-2	VOL 15	121
1600-27-7	VOL 10	E45-E48,E54,77
1643-19-2	VOL 1	E20,26
	VOL 4	E33-E34,39-40
	VOL 9	E29,E30,40
1701-93-5	VOL 3	E46,65,E102-E113,114-156,159-176,E177-E178, 179-184,E185,186-187,E188,189-192,E193,194-199,E200,201-212,E213,214-222,E223,224-234
1715-33-9	VOL 7	380,381
1762-95-4	VOL 3	133-134,148-149
	VOL 11	E70,71,E140,141,E231,232-234,E294,295
	VOL 12	E37-E39,63,64
1923-70-2	VOL 3	209-210,230-231
1929-82-4	VOL 12	E354-E366,437
1941-24-8	VOL 13	323
1941-26-0	VOL 3	20,E23,24,26,28,221,E223 226
1941-30-6	VOL 4	E33-E34,39-40
	VOL 9	E29,E30,39
2092-16-2	VOL 3	143-144
	VOL 12	76
2092-17-3	VOL 3	139-140
2140-69-3	VOL 3	57,75-76,82
2207-01-4	VOL 1	64,210
	VOL 2	57
	VOL 4	153
	VOL 7	245
	VOL 10	157
2207-03-6	VOL 1	66,212
	VOL 4	155
	VOL 7	246
	VOL 10	160

2207-04-7	VOL 1	67,213
	VOL 4	156
	VOL 7	248
	VOL 10	159
2207-23-6	VOL 2	59
2207-34-7	VOL 2	60
2370-12-9	VOL 15	300
2375-03-3	VOL 7	380,381
2392-39-4	VOL 7	380,381
2430-22-0	VOL 15	391
2551-62-4	VOL 10	259
2567-83-1	VOL 3	18,E23,25,84-85,E213,215-216,227-228,232
2696-92-6	VOL 8	E351
2836-32-0	VOL 14	E75,E76,106-108
3087-82-9	VOL 18	E52-53,54-63,E64,65-71
3121-79-7	VOL 15	317
3244-41-5	VOL 18	E7-10,11-20,E21,22-30,E31-32,33-39,E40-41,42
3251-23-8	VOL 8	E29,E30,51
	VOL 13	196-198
3315-16-0	VOL 3	E38-E39,40-44
3330-16-3	VOL 7	338
3344-16-9	VOL 16/17	E276,277-300
3384-87-0	VOL 3	E31,32-33
3425-46-5	VOL 3	E90,94-97
3474-12-2	VOL 7	452
3485-14-1	VOL 16/17	E359,360-362,E363,364
3648-20-2	VOL 12	E354-E366,389
3674-54-2	VOL 3	209-210,230-231
3811-04-9	VOL 14	51,52,57,58,E253,267
3970-62-5	VOL 15	E301,302,303
4328-04-5	VOL 9	E29,E30,42
4371-52-2	VOL 7	400
4377-37-1	VOL 12	E354-E366,442
4390-04-9	VOL 9	107

4587-19-3	VOL 3	E185,186,E213,216,221,E223,226-228
4712-38-3	VOL 15	92,93
4712-39-4	VOL 15	118,119
4798-44-1	VOL 15	226
4798-45-2	VOL 15	225
4798-58-7	VOL 15	227
4911-70-0	VOL 15	E304,305,306
5075-92-3	VOL 4	E253-E255,311
	VOL 5/6	512
	VOL 10	463
5169-33-5	VOL 3	E90-E91,92-101
5328-37-0	VOL 10	310
5928-84-7	VOL 16/17	E72,73-77,E78,79-101
5932-79-6	VOL 15	400
5971-93-7	VOL 18	E43,44-51
6000-44-8	VOL 14	E75,E76,109,110
6011-39-8	VOL 16/17	E249,250-275
6032-29-7	VOL 15	E196,E197,198-202
6091-45-8	VOL 13	100,101,211,212,268,329,401
6130-64-9	VOL 16/17	E171,172-174
6143-53-9	VOL 13	327,396,422
6143-55-1	VOL 13	395
6305-71-1	VOL 15	310
6484-52-2	VOL 2	E239,243
	VOL 5/6	E29,46
	VOL 8	E29,50,51
	VOL 12	E37-E39,62
	VOL 13	32-34,85,185,186,203,204,390
6591-72-6	VOL 16/17	E102,103-126
6727-90-8	VOL 18	185,191
6876-23-9	VOL 1	65,211
	VOL 2	58
	VOL 4	154
	VOL 7	247
	VOL 10	158
6899-06-5	VOL 7	397,401
6912-67-0	VOL 7	397,400

6928-94-5	VOL 18	96
6941-45-3	VOL 13	266,267,328
7041-22-7	VOL 12	E354-E366,433
7041-25-0	VOL 12	E354-E366,438
7081-44-9	VOL 16/17	E576,577-602
7177-43-7	VOL 16/17	224-248
7177-48-2	VOL 16/17	E442,E443,444-456,E457,458-478
7294-05-5	VOL 15	358
7439-90-9	VOL 1	E254,323,E369,370-372
	VOL 2	E1-E3,4-8,E9,10,11,E12-E13,14-25,E26,27-34, E35,36-51,E52,53-70,E71,72,E73,74-107,E108-E190,110-111,E114-E115,116,118-119,122-123, 126,128,132
	VOL 4	240
7440-01-9	VOL 1	124-252,357-385
	VOL 5/6	E581,582-588
7440-37-1	VOL 1	110,249,283,284-294,E359,360-368
	VOL 5/6	E299,E571,572-580
	VOL 10	37,38,482
7440-59-1	VOL 1	1-14,16-18,20-107,109-123,254-301,304-306, 310-356
7440-63-3	VOL 2	E134-E136,137,148,E149,150-161,E162,163-192, E193-E194,195,196,E199-E201,202-204,206-211, 213,219,220,222,224
7446-09-5	VOL 1	E255
	VOL 3	88,234
	VOL 5/6	E299,621
	VOL 10	315,475
	VOL 12	1-332
7446-70-0	VOL 4	E33,42
	VOL 5/6	E30,54
	VOL 7	99
7447-39-4	VOL 8	E266,314,315,318-320,322-324,326,327
7447-40-7	VOL 1	E20,29,E141,168,177-178
	VOL 2	E12-E13,21-E240,255,340
	VOL 3	E45,63,E102,126
	VOL 4	E33-E36,47,49,61
	VOL 5/6	E34,44,83-85
	VOL 7	133,170-175
	VOL 8	E35,E36,E39,96-100
	VOL 9	E32,57
	VOL 10	E45-E47,E54,E55,78,79
	VOL 11	E28,29,30-34,E133,114-117,E186,187-193,E247, 248,E271,272,273,E325,326
	VOL 12	E37-E39,97-99,353
	VOL 14	57,58,E70,E72,100,101,133-137,230,E247,264-266,E252,283
	VOL 16/17	E359,360,E394,E395,400-403,410,412,414,416, 417,427,432,440,441,E442,E443,448,449,E479, 480,481,483,E487,488,E545,546,547,E657,658, E664,665,E694,695-697,E698,699-701,E702,703-705

7447-41-8	VOL 1	E20,29,E141,161,173,175,178
	VOL 2	E12-E13,21
	VOL 4	E33-E36,44-48,69
	VOL 5/6	E31,44,61-63
	VOL 7	129,130,133,183
	VOL 8	E32,E39,76-78
	VOL 9	E30,E31,44,45
	VOL 10	E45-E47,E49,64
	VOL 11	E1,2,3,E88,89-92,E151,152-155,E255,257,E309, E310,311,312
	VOL 14	E70,E71,91,92
	VOL 18	25-29,51,61,62,69,85,87
7487-88-9	VOL 1	E141,148
	VOL 5/6	E31,58
	VOL 7	122,123
	VOL 8	E30,65-67
	VOL 10	E45-E48,345
	VOL 14	E70,E74,152,153
7487-94-7	VOL 2	E239,247
7488-54-2	VOL 7	137
7493-90-5	VOL 10	309
7550-35-8	VOL 7	134
	VOL 11	E4,5,6,E93,94,95,E156,157,158,E313,314
7550-45-0	VOL 12	E54-E366,447
7558-02-3	VOL 11	E274,E275,276,277,E327,328
7558-79-4	VOL 8	E35,92
	VOL 16/17	E10,15,E39,41,E175,178,179,E733,734
7558-80-7	VOL 16/17	E175,178,179
7601-54-9	VOL 2	E12-E13,23,E149,154
	VOL 7	164
	VOL 8	E35,92
	VOL 16/17	E10,12,E359,362,E365,366,E442,E443,455,E487, 489
7601-89-0	VOL 3	E1,15-16,E104,154-163,172-176
	VOL 7	136
	VOL 18	151,156,160,163,170-173,176
7601-90-3	VOL 1	E20,24
	VOL 3	79-80,164,171-172
	VOL 4	E33,37
	VOL 14	195,203
7631-86-9	VOL 8	149
7631-99-4	VOL 1	E20,33,E141,166
	VOL 3	21-22,26-28,30
	VOL 4	E33,59
	VOL 5/6	E33,44,74,75
	VOL 7	138,153,160
	VOL 8	E35,90,91
	VOL 13	57,173
	VOL 14	301
	VOL 18	143,150,155,157,164,167

7632-00-0	VOL 5/6	E33,73
7642-01-0	VOL 3	71
7647-01-0	VOL 1	E20,23,E141,144
	VOL 2	E12-E13,16
	VOL 5/6	E27,36
	VOL 7	77
	VOL 8	E28,E40,287,290,291,314
	VOL 12	40,41,107,E344,345-348
	VOL 14	E68,98,E246,295
	VOL 16/17	E1,3,8,9,E10,12,14,E68,70,E72,75,E167,169, E171,173,E249,251,E301,303,E330,332,E359, 360,362,E365,366,E367,369,E394,E395,402,403, 405,406,E442,E443,448-450,452,455,E479,480-481,E484,485,486,E487,488,489,E491,493,E518, 520,E545,546,547,E548,551,E576,578,E603,605, E630,632,E657,660,661,E667,669,E694,695,696, E698,699,700,E702,703,704,E706,708
7647-14-5	VOL 1	E20,29-31,E141,163,173,176,187,261
	VOL 2	E12-E13,20,21,126,128,132,133,E149,151-153, E240,251,252
	VOL 4	E33-E36,47-53,70
	VOL 5/6	E32,44,66-70
	VOL 7	133,140-146,397
	VOL 8	E33,80-84,E266,332,334,335
	VOL 9	E31,E32,46-50
	VOL 10	E45-E47,E50-E52,67-71,384-351,513,514
	VOL 11	E9,10-13,E96,97-100,E162,163-167,E258,E259, 260,201,E317,318
	VOL 12	E37-E39,88,352
	VOL 14	E70,E72,94-103,E191,E195,200,201,E210,215
	VOL 16/17	E1,8,9,E10,12,E359,362,E365,366,E442,E443, 455,E487,489,E657,660,E662,663
7647-15-6	VOL 1	E20,30,E141,164
	VOL 4	E33-E36,54
	VOL 7	134,147
	VOL 8	E33,85
	VOL 9	E32,51
	VOL 11	E14,E15,16-20,E101,102-105,E168,169,170,E262, 263-265,E319,320
	VOL 12	E37-E39,89
	VOL 18	168,169
7647-17-8	VOL 1	E141,171
	VOL 4	E33-E36,66,74
	VOL 7	133,182
	VOL 8	E38,E39,114-115
	VOL 9	E33,59
	VOL 11	E54,55,E212,213-215
7664-38-2	VOL 8	E29,E40,44,45,E351,352
	VOL 16/17	E10,15,E39,41
7664-39-3	VOL 3	87
	VOL 7	430
	VOL 12	E1,E2,448
7664-41-7	VOL 1	E255,282
	VOL 2	105
	VOL 3	E1,E7-E8,17,E38,44,E47,E54-E55,59,66-67
	VOL 4	E314,315-318
	VOL 5/6	297,E299,E622,623-629,634
	VOL 10	E476,477-483
	VOL 14	E69,87,88,E247,293

7664-93-9	VOL 2	E12-E13,17-18,340
	VOL 3	E86,121
	VOL 5/6	E27,E33,37-40,72
	VOL 7	78-83,117,118,149,162
	VOL 8	E28,E34,E40,42,89,289,311,317
	VOL 9	E29,E32,34,53
	VOL 10	E45-48,E53,57,58,74,76,515,516
	VOL 12	42-52,350
7681-11-0	VOL 1	E20,30,E141,170,173,177
	VOL 2	E12-E13,21,E149,151
	VOL 3	81
	VOL 4	E33-E36,48,64-65,73
	VOL 5/6	E34,44,86
	VOL 7	135,177
	VOL 8	E37,103,104
	VOL 9	E32,E33,58
	VOL 10	E45-48,E55,80
	VOL 11	E39,E40,41-45,E123,124-127,E198,199-203,E251, E252,253,254,E278,E279,280,E329,330
	VOL 12	E37-E39,101
	VOL 14	E70,139
7681-17-8	VOL 9	E32,E33
7681-38-1	VOL 12	92-94
7681-49-4	VOL 7	132
	VOL 11	E7,8,E159,160,161
7681-55-2	VOL 14	31,32,E69,164
7681-82-5	VOL 1	E20,34,E141,165,176,178
	VOL 4	E33-E36,55-58,71,79
	VOL 7	135
	VOL 9	E32,52
	VOL 11	E21,22-25,E106,107-110,E171,172-175,E245,246, E266,267,268,E321,322
	VOL 18	144,145
7697-37-2	VOL 1	E20,23
	VOL 2	E12-E13,16
	VOL 3	33,E86
	VOL 5/6	E28,41
	VOL 7	84-96,487,488
	VOL 8	E28,E40,43
	VOL 10	59
	VOL 13	3,22,23,25,28,29,34,53-55,149,151-154,167-170,229-233,235,238,239,252-254,289,292, 294,296,299,313-315,342,345,347,349,351, 352,364-367,371-375,379,389,390,405-408, 412,413,420,431,432,439,440,451,452,459,463
	VOL 14	E246,297
7705-08-0	VOL 1	E141,146
	VOL 8	E266,312-313
7719-09-7	VOL 12	E1,E2,330-331
7720-78-7	VOL 2	E239,249
	VOL 8	E29,E30,58,E265,E266,270,279,281,282,284,288, 293,299-303,305,331
7722-64-7	VOL 2	E240,254

7727-37-9	VOL 1	E328,329-345,E373,374-379
	VOL 4	312-313
	VOL 5/6	E299,312,530,531,552,E589,E590,591-600,631-634
	VOL 7	434,435
	VOL 10	E1-E4,5-30,E31-E33,34-44,E45-E56,57-82,E83-E85,86-103,E104,E105,106-118,E119-E122,123-147,E148,149-161,E162,163-173,E174-E178,179-215,E216,217-252,E253,254-278,E279,280-331,E333,334-360,E361,E362,363-390,E391,392-401,E402,403-408,E409,410-418,E419,420-424,E425,426-475,E476,477-487,E488,489-580
7732-02-2	VOL 2	E239,246
7732-18-5	VOL 1	E1-E4,5-13,15-33,35-39,120-123,E124-E126,127-136,138-184,E257,258-262
	VOL 2	E1-E3,4-8,E12-E13,14-24,121-126,128,132-133,E134-E136,137-145,E149,150-154,209-211,216-220,222,224,E227-E229,230-234,E238-E241,242-258,340,E343,344
	VOL 4	E1-E7,8-19,25-26,E33-E36,77-78,E80-E84,85-106 245-247,E256,257-260
	VOL 5/6	E1,E2,3-16,E17,E18,19-24,E25-E35,36-118,278,279,E299,E303,E304,305-312,314-320
	VOL 7	E1-E40,E41-E43,44-55,E56-E76,77-183,188,189,E190-E192,193-212,305,355,360,377-379,386,387,389-392,397-417,420-425,451,456-464,E474-E480,481-483,493
	VOL 8	E1,E2,3-22,E27-40,41-159,241,248-255,E260,E261,262-264,E265,E267,268-305,311,314-318,328-335
	VOL 9	E1,E2,3-15,E16,17-22,E23,24-26,E27-33,34-60,E61,62,63,E64,65-73,E74
	VOL 10	E1-E4,5-29,E45-E46,57-82,E83-E85,86-103,237,281,286,293,294,297,303-314,E333,334-352,510-516
	VOL 12	E3-E5,6-33,E37-39,40-115,199,E333,E334,335-343,E344,345-348,E349,350-353,449-453,E454,455-456
	VOL 13	E1,2-14,E15-E19,20-36,E38-E47,48-108,110-114,116-131,134,136-143,145,146,E148,149-155,E156-E162,163-220,E226-E227,229-240,E241-E246,247-277,E286-E288,289-300,E301-E308,309-334,E339-E341,342-353,E354-E359,360-375,E376,E377,378-380,E381-E385,386-408,E409,410-413,E414-E417,418-428,E429,430-433,E434-E436,437-447,E449,450-453,E454-E456,457-460,E461,462-464
	VOL 14	E1-E3,4,5,E6,E7,8-10,E11-E18,19-43,E44-E47,48-58,E59-E82,83-178,E179-E181,182-185,E186-E188,189,190,E191-E197,198-207,E208-E212,213-224,E225-E227,228-241,E242-E255,256-311
	VOL 16/17	E1,2-5,7-9,E10,11-15,E39,40-42,E66,67,E68,69-71,E72,73-77,E102,103,E127,E128,129,131,132,134,E135,E136,137,139-141,E165,166,E167,168-170,E171,172-174,E175,176-199,E224,225,E249,250-252,E276,277,E301,302-305,E330,331-333,E334,335,E359,360-362,E363,364,E365,366,E367,368-370,E394,E395,396-409,E410,412,414,416,417,427,432,440,441,E442,E443,444-456,E484,485,486,E487,488-490,E491,492-494,E518,519-521,E545,546,547,E548,549-552,E576,577-579,E603,604-606,E630,631-633,E657,658-661,E662,663,E664,665,666,E667,668-670,E694,695-697,E698,699-701,E706,707-709,E733,734,E740,741,766

7732-18-5	VOL 18	1,E3,4,E7-10,11-20,E21,22,24-30,E43,44-48,E52, E53,54-63,E72-73,74-79,81-86,89-98,E106,107, 108,115,E140,141-148,E177,178,179,E182,183-186,E228
7733-02-0	VOL 5/6	E29,47
	VOL 8	52
	VOL 12	E37-E39,82-87
7740-37-1	VOL 4	1-321
7740-63-3	VOL 1	E255,353-355
7757-79-1	VOL 2	E12-E13,24
	VOL 3	9,E31,33,81,E90,96,97,E103,145,146,150-153, 165-166,169,170,E177,E178,179,180,E188,189, E193,194,195,199,E200,201,202,205,206,208, 211,212,E213,214,221,222,223,226
	VOL 5/6	E35,44,87,88
	VOL 7	138,180
	VOL 8	E37,E38,107-110
	VOL 12	E37-E39,103
	VOL 13	58,59,174,175
	VOL 14	E70,E74,146,232,E248-E250,271-275
7757-82-6	VOL 1	E20,32
	VOL 2	E12-E13,22
	VOL 5/6	E32,E33,71,72
	VOL 7	137,148-151,161
	VOL 8	E33-E35,86-89,E266,278,332
	VOL 9	E32,53
	VOL 10	E45-E47,E52,E53,73-76,347,515,516
	VOL 12	E37-E39,90,91
	VOL 14	E74,104,E210,216
	VOL 18	17,18,55,56,76,77
7757-83-7	VOL 3	E103,168
	VOL 7	152
7758-01-2	VOL 14	231,269
7758-02-3	VOL 1	E141,169
	VOL 2	E12-E13,21
	VOL 3	E45,64,E102,121,127
	VOL 4	E33-E36,62-63,72
	VOL 7	134,176
	VOL 8	E36,101,102
	VOL 11	E35,36-38,E118,119-122,E194,195-197,E249, 250
	VOL 12	E37-39,100
	VOL 14	E70,138
7758-05-6	VOL 14	33,34,E69,140-143,E195,203,270,E252,284,287
7758-94-3	VOL 8	E265,E266,268,269,271,280,283,285,287,291, 304,307,309,310
7758-98-7	VOL 2	E12-E13,17,18,E239,244
	VOL 5/6	E29,48
	VOL 7	102,117,118
7761-88-8	VOL 2	E239,245
	VOL 3	E47,60,66,E90-E91,98-99,E104,115,147,E213, 215
	VOL 18	185,191
7764-93-9	VOL 7	490,491

7772-98-7	VOL 14	E74,105,262
7773-01-5	VOL 8	E266,330,331
7775-14-6	VOL 7	161
7778-53-2	VOL 7	164
7778-74-7	VOL 3	E177-E178,181-184,187,E188,190-192,E193, 196-198,E200,203-204,220
	VOL 7	136
7778-80-5	VOL 7	137,178,179
	VOL 8	E37,105,106
	VOL 12	E37-E39,102
	VOL 14	E70,144,145
7779-88-6	VOL 8	E29,E30,53
	VOL 13	82-84,256,318,368,369
7782-18-5	VOL 14	E246,296
7782-39-0	VOL 1	E254,302-305
	VOL 5/6	278,279,E280-282,283-296,313,E354,E355,575, E581,582,597
7782-41-4	VOL 1	E254,306
	VOL 12	448
7782-44-7	VOL 1	E348,349-352,E380,381-385
	VOL 10	37,38,76,353,354,466
7782-50-5	VOL 7	465
	VOL 12	333-447
7782-68-5	VOL 14	21-24,204,205
7782-79-8	VOL 3	E5
7783-06-4	VOL 5/6	E299,562-568,619,620
	VOL 9	E232,E233,244
	VOL 10	468-470,495
7783-08-6	VOL 8	295
7783-20-2	VOL 5/6	E28,45
	VOL 7	98
	VOL 8	29,49
	VOL 12	E37-E39,56
7783-41-7	VOL 10	467
7783-54-2	VOL 10	486
7783-86-0	VOL 8	265,266,275,308
7783-90-6	VOL 3	E45,63,71,E102,126
7783-93-9	VOL 3	E90-E91,100-101
7783-96-2	VOL 3	81
7785-23-1	VOL 3	E45,64,E102,121,127
7785-45-9	VOL 8	E266,316,321,325
7785-87-7	VOL 5/6	E30,52
	VOL 7	106
	VOL 8	E29,E30,60

7786-30-3	VOL 1	E141,147
	VOL 4	E33,43
	VOL 7	119-121
	VOL 8	E30,64
	VOL 11	E72,73,E142,143,E235,236,237,E333,334
	VOL 14	E70,E73,148-150,276
7786-81-4	VOL 5/6	E29,49
	VOL 7	103
	VOL 8	E29,E30,56,E266,328,331
7787-34-0	VOL 14	E242-E255
7787-69-1	VOL 7	134
	VOL 11	E56,57,E216,217-219
7789-00-6	VOL 13	24
7789-17-5	VOL 7	135
	VOL 11	E58,59-60,E220,221
7789-18-6	VOL 1	E141,172
	VOL 7	138
	VOL 13	61,178,179
7789-20-0	VOL 1	14,137
	VOL 2	146,235
	VOL 3	89
	VOL 4	E20-E21,22-26,E33-E36,69-76
	VOL 5/6	313
	VOL 9	E195-E199,229
	VOL 10	30
	VOL 15	92,93,118,119
7789-23-3	VOL 1	E141,174
	VOL 7	132
	VOL 8	E39,95
7789-24-4	VOL 7	132
	VOL 11	E148,149,150
7789-36-8	VOL 14	E6,E7,8,9
7789-39-1	VOL 7	134
	VOL 11	E52,53
7789-41-5	VOL 1	E141,152
	VOL 11	E78,79,E146,147,E243,244,E298,299,E339,340
7789-42-6	VOL 12	E37-E39,79
7789-46-0	VOL 8	E265,E266,274,292,306
7789-80-2	VOL 14	40,E59-E82,83-178
7790-21-8	VOL 8	E38,111
7790-29-6	VOL 7	135
7790-32-1	VOL 14	E11-E18,19-43,E69,151
7790-36-5	VOL 14	E191-E197,198,200,201
7790-69-4	VOL 1	E141,162
	VOL 7	138
	VOL 13	171,172
	VOL 14	93,E191,E195,202,261

7790-80-9	VOL 12	E37-E39,80
7790-86-5	VOL 13	192
7790-91-2	VOL 10	471
7791-03-9	VOL 7	136,185-187
	VOL 14	93,191
7791-10-8	VOL 14	E179-E181,182-185
7791-11-9	VOL 1	E141,172
	VOL 4	E33-E36,68,76
	VOL 7	133,181
	VOL 8	E38,113
7791-19-7	VOL 14	E1-E3
7791-21-1	VOL 12	449-453
7791-25-5	VOL 12	E1,E2,329,E354-E366,445
7803-63-6	VOL 12	E37-E39,60,61
9005-65-6	VOL 16/17	E10,14,15,E39,41
9048-46-8	VOL 2	126
10022-31-8	VOL 1	E141,160
	VOL 2	E239,250
	VOL 3	81
	VOL 13	65,191
	VOL 14	E210,219-222,234,241,E250,279,280,292
10022-50-1	VOL 10	485
10022-52-3	VOL 14	E186-E188
10024-97-2	VOL 1	E255,346-347
	VOL 2	102,186
	VOL 8	1-259
	VOL 10	316,484
10024-97-3	VOL 7	466,467
10026-04-7	VOL 10	252
	VOL 12	E354-E366,446
10028-15-6	VOL 7	E474-480,481-493
10028-22-5	VOL 8	E29,E30,59,E266,311
10031-32-0	VOL 14	E59-E61,E81,E82
10031-33-1	VOL 14	E59-E61,E81,E82
10031-51-3	VOL 13	E434-E436
10031-53-5	VOL 13	E376,E377
10035-00-4	VOL 13	E454-E456
10035-05-9	VOL 14	E44-E47
10035-10-6	VOL 8	292

10036-47-2	VOL 10	487
10038-98-9	VOL 10	252
10042-76-9	VOL 13	189,190
10043-01-3	VOL 7	100
	VOL 8	E29,E30,62
10043-02-4	VOL 8	E265,E266,278
10043-27-3	VOL 13	E409,410-413
10043-52-4	VOL 1	E141,15
	VOL 4	E33,43,E74,75-77,E144,145,E240,241,242,E296, 297,E337,338
	VOL 5/6	E31,59
	VOL 7	125,126
	VOL 8	31,70-72
	VOL 9	E30,43
	VOL 12	E37-E39,78
	VOL 14	53-58,E70,E73,155-158,277
10043-53-4	VOL 10	E45-E49,346
10045-89-3	VOL 8	E265,266,272,273,277,289,290,297,298
10045-95-1	VOL 13	71-74,255,279,280,282,284,E301-E308,309-334, E335,336-353
10049-04-4	VOL 12	E454,455,456
10099-58-8	VOL 5/6	E30,55
	VOL 13	69
10099-59-9	VOL 13	28,E38-E47,48-108,E109,110-147
10099-67-9	VOL 13	E461,462-464
10099-74-8	VOL 2	E239,248
10101-53-8	VOL 8	E29,E30,61
10102-43-9	VOL 5/6	E299,609
	VOL 8	259-351
10102-44-0	VOL 7	85
	VOL 8	258
10102-68-8	VOL 1	E141,153
	VOL 4	E341,342
10108-73-3	VOL 13	E156-E162,163-225
10124-36-4	VOL 12	E37-E39,81
10124-37-5	VOL 1	E141,154
	VOL 7	124
	VOL 8	E31,73
	VOL 13	187,188
	VOL 14	278
10124-43-3	VOL 5/6	E29,50
	VOL 7	104
	VOL 8	E29,E30,57,E266,329,331
10137-74-3	VOL 14	E44-E47,48-58
10138-01-9	VOL 13	E376,E377,378-380

10141-05-6	VOL 13	30,80
10143-38-1	VOL 13	E414-E417,418-428
10168-80-6	VOL 13	E434-E436,437-448
10168-81-7	VOL 13	29,E381-E385,386-404
10168-82-8	VOL 13	E429,430-433
10192-30-0	VOL 12	57-59
10257-55-3	VOL 12	E37-E39,67,69
10277-43-7	VOL 13	E38-E47
10294-38-9	VOL 14	E208-E212,213,214
10294-41-4	VOL 13	E156-E162
10294-54-9	VOL 7	137
10308-78-8	VOL 13	330,402
10325-94-7	VOL 13	199,200
10326-21-3	VOL 14	E1-E3,4,5
10326-26-8	VOL 14	E225-E227
10361-37-2	VOL 1	E20,27,E141,157
	VOL 2	E12-E13,19
	VOL 4	E33,43,E84,85,E305,306
	VOL 5/6	E31,60
	VOL 7	127,128
	VOL 8	E32,75
	VOL 10	E45-E47,E49,62,63
	VOL 12	351
	VOL 14	E210,217,218,E227,236,237
10361-80-5	VOL 13	70,E241-E243,247-277,E278-E285
10361-83-8	VOL 13	75-78,317,E354-E359,360-370
10361-93-0	VOL 13	E15-E19,20-37
10377-48-7	VOL 7	137
	VOL 8	E30,E31,68,69
10377-51-2	VOL 1	E20,29,E141,176,178
	VOL 4	E315,316
	VOL 7	135
10377-58-9	VOL 4	E238,239,E335,336
10377-60-3	VOL 1	E141,149
	VOL 13	62-64,182-186,346
	VOL 14	10,37-39,233
10377-66-9	VOL 13	79,193,194
10476-81-0	VOL 1	E141,156
	VOL 4	E82,83,E302,303,304
	VOL 14	185
10476-85-4	VOL 1	E141,155
	VOL 4	E33,43,E80,81,E330,301,E343,344
	VOL 8	E31,E32,74

10544-72-6	VOL 1	111
	VOL 4	237-238
	VOL 7	96,432,433
	VOL 10	317,318
10553-31-8	VOL 1	E141,158
	VOL 4	E86,87,E307,308
	VOL 14	E210,221,222,240,291
10567-69-8	VOL 14	41,E242-E255,256-311
10579-83-6	VOL 3	172-174
	VOL 10	E45-E47,E52,E53,72,75
12027-06-4	VOL 4	E67,68,69,E138,139,E229,230,E292,293
12099-10-4	103	104,105
12101-96-1	VOL 5/6	569,570
12124-97-9	VOL 4	E33-E34,39-40,E64,65,66,E134,135-137, E226,227,228,E288,E289,290,201
	VOL 8	E29,48
	VOL 9	E29,E30,36
	VOL 12	E37-E39,55
12125-02-9	VOL 1	E20,25,E141,145
	VOL 2	E12-E13,14,239,242
	VOL 4	E33,38,E61,62,63,E130,131-133,E222,223-225, E284,E285,286,287
	VOL 5/6	E28,43,44
	VOL 8	E29,46,47
	VOL 9	E29,E30,35
	VOL 12	E37-E39,54
	VOL 16/17	E1,7,8,E657,659,660,E664,666
12259-21-1	VOL 5/6	E29,51
	VOL 7	105
	VOL 10	E45-E48,61
13083-04-0	VOL 13	E226,234,235
13126-12-0	VOL 7	138
	VOL 13	25,60,176,177,254,315
13138-45-9	VOL 13	81,183,184,195
13400-13-0	VOL 7	132
13412-64-1	VOL 16/17	E548,549-575
13435-20-6	VOL 3	84-85
13444-71-8	VOL 8	E40
13446-09-8	VOL 14	25,26
13446-17-8	VOL 14	E11-E18
13446-74-7	VOL 7	132
13446-76-9	VOL 14	35,36
13451-19-9	VOL 13	E409
13464-97-6	VOL 13	86,87,201,202,320,321
13464-98-7	VOL 13	88,89,257,258

13465-60-6	VOL 13	E1,2-14
13470-01-4	VOL 14	E191-E197,198-207
13470-40-1	VOL 13	E15-E19
13473-90-0	VOL 8	E29,E30,63
	VOL 13	31,319
13476-05-6	VOL 13	E434-E436
13477-00-4	VOL 14	E208-E212,E213-E224,E227,238,239,289,290
13494-98-9	VOL 13	E15-E19
13510-42-4	VOL 7	136
13536-84-0	VOL 7	107,108
13550-46-4	VOL 13	E162,E226,E227
13566-21-7	VOL 13	E148,155
13568-66-6	VOL 13	342-346
13597-99-4	VOL 13	180,181
13637-63-3	VOL 10	472,473
13718-50-8	VOL 1	E141,159
13759-83-6	VOL 13	E354-E359
13765-03-2	VOL 14	27-30,E69,161-163
13768-67-7	VOL 13	E454-E456,457-460
13773-30-3	VOL 13	E281-E385
13773-54-1	VOL 13	233
13773-69-8	VOL 13	E15-E19
13774-25-9	VOL 12	73-75
13780-03-5	VOL 12	69,70,73-75
13826-42-1	VOL 13	E148,149,150
13839-85-5	VOL 13	E454-E456,458,459
13847-66-0	VOL 3	E2,12-13
13863-88-2	VOL 3	E1-E8,9-30
13943-58-2	VOL 2	E240,253
13943-58-3	VOL 14	E70,E74,147
13967-90-3	VOL 14	E225-E227,228-241
13983-20-5	VOL 5/6	E299
13983-27-2	VOL 2	7-8,17-18,98,99,112,113,117,120,121,124,125, 127,129-131,133
13984-18-4	VOL 13	E45-E47

14013-86-6	VOL 8	E265,E266,286
14024-11-4	VOL 12	330,331
14202-62-1	VOL 15	386
14338-74-5	VOL 3	133
14483-17-1	VOL 13	E241-E246
14483-18-2	VOL 13	E429
14485-20-2	VOL 18	1,2
14517-29-4	VOL 13	E301-E308
14519-17-6	VOL 14	E6,E7,8,9,10
14519-18-7	VOL 14	E186-E188,189,190
14520-95-7	VOL 13	E45-E47
14637-31-1	VOL 18	E177,178-181
14637-34-4	VOL 18	E72-73,74-80
14637-35-5	VOL 18	E140,141-150,152-165,E166,167-170,172,173, E174,175,176
14762-55-1	VOL 1	15,19,108,E254,302-303,307-309
14835-02-0	VOL 2	E339,340-342
14859-67-7	VOL 2	E227-E229,230-237,E238-E241,242-258,E259-E261,262-264,E265,266-272,E273,274-290,E291,292-293,E294,295-321,E322,323-327,E328,329-338
14932-42-4	VOL 2	181,182,197,198,205,207,208,212-215,217,218, 221,223,225,226
14977-61-8	VOL 12	E354-E366,443,444
14985-19-4	VOL 13	E449,450-453
14995-63-1	VOL 2	207,208,221
15123-75-8	VOL 14	42,43
15276-91-2	VOL 13	228,229
15276-92-3	VOL 13	E226,E227
15318-60-2	VOL 13	E226,234,235
15362-24-0	VOL 4	E33-E34,41
15457-98-4	VOL 12	84-87
15522-59-5	VOL 18	98-102
15525-13-0	VOL 18	E106,107-111
15525-15-2	VOL 18	E228,229-231
15556-39-5	112	113,114

15627-12-0	VOL 18	E182,183-185,E187,188-193,E194,195-197,E198,199-202,E203,204-208,E209,210-212,E213,214-216,E217,218-224,E225,226,227
15627-12-5	VOL 18	186
15857-43-9	VOL 8	E265,E266,294-296
15878-72-5	VOL 13	E38-E47
15878-75-8	VOL 13	E38-E47,E156-E162
15878-76-9	VOL 13	E156-E162
15878-77-0	VOL 13	E241-E246
15878-79-2	VOL 13	E241-E246
16454-60-7	VOL 13	E301-E308
16469-22-0	VOL 13	26,27
16655-52-0	VOL 14	E76,E77,131
16742-92-0	VOL 18	84-88
16774-21-3	VOL 13	236-240
16999-44-3	VOL 13	E1
16999-46-5	VOL 13	E1,5
17032-40-5	VOL 3	137
17194-00-2	VOL 14	E210,223
17203-49-5	VOL 13	E339-E341
17242-52-3	VOL 5/6	296
17611-22-2	VOL 7	184
17647-86-8	VOL 9	60
18851-79-1	VOL 13	E45-E47
18851-80-4	VOL 13	E45-E47
18939-61-2	VOL 8	E266,317
19287-45-7	VOL 5/6	630
	VOL 10	496
19478-66-1	VOL 13	E286-E288
19495-49-9	VOL 14	E191-E197,198
19549-79-2	VOL 15	390
19598-90-4	VOL 13	E381-E385
19944-52-6	VOL 14	E75,E76,125-UIM
20340-68-3	VOL 13	E162,E226,E227
20394-14-3	VOL 13	E226,E227

20627-66-1	VOL 12	297
20748-72-5	VOL 13	391
20762-60-1	VOL 3	14
20829-66-7	VOL 13	98,99,209,210,262,263,394,421
21351-79-1	VOL 7	131
21640-15-3	VOL 13	106,107,272,332,333,398,425,442
22113-87-7	VOL 13	205,206
22202-75-1	VOL 16/17	E702,703-705
22252-43-3	VOL 16/17	E657,658-661
22465-27-6	VOL 13	152
22481-88-7	VOL 2	E343,344-345
23228-90-2	VOL 7	337
	VOL 10	250
23325-78-2	VOL 16/17	E694,695-697
23338-69-4	VOL 14	E76,E77,115
24581-35-9	VOL 13	E354-E359
25238-43-1	VOL 13	90,91,322
25321-09-9	VOL 12	164
25322-01-4	VOL 18	134
25322-20-7	VOL 7	492
25322-68-3	VOL 12	260
25512-65-6	VOL 5/6	E214,218
	VOL 10	229
2556-92-1	VOL 12	E146,E147,167,168
25822-29-1	VOL 13	153
25990-60-7	VOL 5/6	315,316
26140-60-3	VOL 2	69
26249-20-7	VOL 18	138
26545-61-9	VOL 7	322
26628-22-8	VOL 3	E1,E6,3-11,15-16,21-22,26-28,30
26635-06-3	VOL 13	E301-E308
26738-51-2	VOL 7	342
26915-12-8	VOL 12	287
27096-29-3	VOL 13	393
27096-30-6	VOL 13	93,94,208,260,324
27096-31-7	VOL 13	96,97,261,325

27099-40-7	VOL 13	E46,E47
27255-72-7	VOL 16/17	E657,661,E664
28876-81-5	VOL 13	E286-E288
28876-82-6	VOL 13	E339-E341
28896-83-7	VOL 13	E354-E359
29063-28-3	VOL 4	E253-E255,201
29611-84-5	VOL 12	297
29990-16-7	VOL 13	E162
30781-73-8	VOL 13	207
31176-55-3	VOL 13	352,353
31828-50-9	VOL 16/17	E698,699-701
32074-07-0	VOL 13	E286-E288,289-291
32074-08-1	VOL 13	371
32221-81-1	VOL 14	E76,E77,122
33412-01-0	VOL 13	E46,E47
33412-02-1	VOL 13	E46,E47
34216-91-6	VOL 13	E286-E288
34342-98-8	VOL 13	E286-E288
34992-36-4	VOL 14	E59-E61,E81,E82
35725-30-5	VOL 13	E414-E417
35725-31-6	VOL 13	E429
35725-33-8	VOL 13	E449
35725-34-9	VOL 13	E454-E456
36153-28-3	VOL 13	E286-E288
36354-70-8	VOL 13	E46,E47
36355-97-2	VOL 14	E1-E3
36548-87-5	VOL 13	E449
36549-50-5	VOL 13	E461,463
36653-82-4	VOL 15	E422,423-425
37131-73-0	VOL 13	E354-E359
37131-74-1	VOL 13	E376,E377
37131-76-3	VOL 13	E409
37131-78-5	VOL 13	E429
37131-79-6	VOL 13	E434-E436

37131-80-9	VOL 13	E449
37340-18-4	VOL 7	333
	VOL 10	249
37486-69-4	VOL 7	343
	VOL 10	528
40464-54-8	VOL 7	324
	VOL 10	248
41240-13-5	VOL 5/6	E483,508
41521-39-5	VOL 14	E76,E77,116
41719-16-8	VOL 7	333
	VOL 10	249
50279-29-3	VOL 7	318
	VOL 10	523
50328-28-4	VOL 18	97
50402-70-5	VOL 14	E75,E76,113
51016-92-3	VOL 18	115
51294-16-7	VOL 7	340
51537-72-5	VOL 13	E46,E47
52152-93-9	VOL 16/17	E733,734-739
52181-51-8	VOL 12	E354-E366,430
52788-53-1	VOL 13	E381-E385
53368-21-1	VOL 13	151
53694-97-6	VOL 18	90
56077-96-4	VOL 10	474
56852-63-2	VOL 16/17	E484,485-486
57432-67-4	VOL 13	392
57584-28-8	VOL 13	E15-E19
58214-38-3	VOL 14	E75,E76,117,118
58979-41-2	VOL 7	344
59141-08-1	VOL 4	E33-E36,67,75
60337-02-2	VOL 18	94
60491-92-1	VOL 13	274,275,399,426
60491-93-2	VOL 13	400,427
61192-73-2	VOL 13	E45-E47
61336-70-7	VOL 16/17	E479,480-483
61391-42-2	VOL 13	E45-E47,E148

62928-04-5	VOL 3	119,135
62928-05-6	VOL 3	E104,157-158
63026-01-7	VOL 13	E376,378,379
63243-37-8	VOL 7	319
	VOL 10	524
63243-38-9	VOL 7	331
63267-58-3	VOL 7	341
65907-05-3	VOL 13	E148
66418-41-5	VOL 3	E104,137,157-158,162-163
67728-22-7	VOL 7	310
67728-31-8	VOL 7	329
67728-32-9	VOL 7	334
67728-33-0	VOL 7	315,335
67728-34-1	VOL 7	323
67728-35-2	VOL 7	330
67728-37-4	VOL 7	339
68028-01-3	VOL 13	E301,E305,314
69502-97-2	VOL 18	83
69502-98-3	VOL 18	89
71973-92-7	VOL 13	E434,440
75333-20-9	VOL 16/17	E165,166
76082-02-5	VOL 16/17	E68,69-71
76082-03-6	VOL 16/17	E330,331-333
76629-97-5	VOL 14	E11-E18
76637-28-0	VOL 13	E286-E288,294,295
76841-98-0	VOL 14	E76,E77,129
77076-82-5	VOL 13	E45-E47
80573-05-3	VOL 13	E38-E47
80573-06-4	VOL 13	E38-E47
80573-07-5	VOL 13	E45-E47
80573-09-7	VOL 13	E46,E47
80573-10-0	VOL 13	E46,E47
81201-28-7	VOL 13	E156,E158-E161
81201-31-2	VOL 13	E241,252

81201-33-4	VOL 13	E241,E243,249,252
81201-38-9	VOL 13	E354,E356,E359,366
81201-40-3	VOL 13	E381,E384,389
81201-55-0	VOL 13	E449,452
82150-36-5	VOL 14	E6,E7,8,9
82150-37-6	VOL 14	E179-E181
82150-38-7	VOL 14	E1-E3
82150-39-8	VOL 14	E11-E18
82808-59-1	VOL 14	E44-E47
82808-60-4	VOL 14	E44-E47
82808-61-5	VOL 14	E76,E77,128
82808-62-6	VOL 14	E76,E77,123
84682-49-5	VOL 13	373
84682-50-8	VOL 13	151
84682-51-9	VOL 13	152
84682-52-0	VOL 13	153
84682-53-1	VOL 13	154
84682-54-2	VOL 13	230
84682-55-3	VOL 13	E226,230
84682-56-4	VOL 13	231
84682-57-5	VOL 13	232
84682-58-6	VOL 13	E286-E288,292,293
84682-59-7	VOL 13	E286,296,297
84682-60-0	VOL 13	E286,298
84682-61-1	VOL 13	E286-288,298
84682-62-2	VOL 13	E286-E288,299,300
84682-63-3	VOL 13	E399-E401,347,348
84682-64-4	VOL 13	E399-E401,347,348
84682-65-5	VOL 13	E399-E401,349,350
84682-66-6	VOL 13	E399,351
84682-67-7	VOL 13	E399,351
84682-68-8	VOL 13	372

84682-69-9	VOL 13	372
84682-70-2	VOL 13	373
84682-71-3	VOL 13	374
84682-72-4	VOL 13	374
84682-73-5	VOL 13	375
84682-74-6	VOL 13	375
84682-75-7	VOL 13	405
84682-76-8	VOL 13	405
84682-77-9	VOL 13	406
84682-78-0	VOL 13	406
84682-79-1	VOL 13	407
84682-80-4	VOL 13	407
84682-81-5	VOL 13	408
84682-82-6	VOL 13	408
84697-20-1	VOL 13	E399-E401,349,350

AUTHOR INDEX

Page numbers preceded by E refer to evaluation texts, those not preceded by E refer to compiled tables.

ABATE, K.
VOL 3 E213, 215
ABDULLAYEV, Y.A.
VOL 5/6 E610,E611
VOL 10 E488
ABDULSATTAR, A.H.
VOL 12 E34
ABEGG,R.
VOL 3 E45-E56,E102-E113, 116
ABRAHAM,M.H.
VOL 18 6, 50, 67, 68, 99, 100, 103, 104, 105, 109, 110, 111, 112-114, 192, 193, 230, 231
ABRAMENKOV,A.
VOL 8 E39, 95, 115
ABROSIMOV,V.K.
VOL 1 E1-E4, 14,E124-E126, 137
VOL 4 E1-E7,E20,E21, 24, 26,E33-E36, 46, 53, 58, 63, 66-76
ABYKEEV.K.
VOL 13 E156-E162, 219,E414-E417,E429
ACKLEY,R.D.
VOL 2 181,182
ACKMAN,R.G.
VOL 7 361-363, 371, 443
ADAMS,F.W.
VOL 12 E333,E344
ADDISON C.C.
VOL 13 E109
ADENEY,W.E.
VOL 7 E1-E5, 12
VOL 10 E1-E4, 11
ADITYA,S.
VOL 3 E102-E113, 148, 149
ADLIVANKINA,M.A.
VOL 10 234
AFANAS EV.YU.A.
VOL 13 E15-E19, 23,E381-E385, 389,E409, 412, 413,E414-E417, 420,E429, 431, 432, E434-E436, 439, 440,E449, 451, 452,E454-E456, 459,E461, 463
AHANGAR, A.M.
VOL 5/6 523-525
AHO,L.
VOL 7 365
AKERLOF,G.
VOL 1 E1-E4,7, E20-E22, 24, 29, 33
VOL 4 E1-E7,E33-E36, 37, 42, 43, 47, 59
AKERS,W.W.
VOL 5/6 E589,E590, 595,E601, 631-633
VOL 10 E409, 411, E425, 427
ALBERT,A.
VOL 3 E1-E8
ALBRIGHT,L.F.
VOL 12 E194, 250, 258, 259,E261, 262, 272, 276, 283
ALDER, B.J.
VOL 5/6 E120-E122, 131, 132, 134, 136,E159,E160, 164,E169, 171,E237,E238, 243, 246,E255,258,E280-E282, 283-291
ALDRICH,E.W.
VOL 5/6 E17,E18
ALEKSANDROU,YU.A.
VOL 7 492,E493
ALEKSEENKO,L.A.
VOL 13 E381-E385
ALEKSEEVA,L.S.
VOL 1 E141-E143, 146, 147, 151-153, 155-159, 161, 163-165, 169-171
ALESKOVSKII,V.B.
VOL 7 E250,E251, 256, 258, 260

ALEXANDER D.M.
VOL 1 E124-E126
VOL 2 E1-E3,E134-E136
VOL 4 E1-E7
ALEXANDER,R.
VOL 3 20-22,E23, 24, 26-28, 208, 211, 212,E213, 214, 221,222, E223, 226
VOL 11 E52,E54, 55,E56, 57,E58, 59,E171,E186, 192,E194, 197,E198, 202,E212, 215,E216, 219,E220, 221
VOL 18 E31,E32, 35-37,E52,E53, 54,E64, 65, 70, 71,E140, 143,E149, 150, 151, 153,E154, 155, 156,E158, 159, 160, 163-165,E166, 167, 170-173,E174, 176,E182, 183, 184,E187, 188-190,E194, 195-197,E198, 199-201,E203, 204-206, 208,E209, 210, 211, 212,E213, 214-216,E217, 218-224,E225, 226, 227
ALIEV,Z.M.
VOL 12 349
ALI,J.K.
VOL 5/6 E178
ALLEN,J.A.
VOL 7 E1-E5
AL´YANOV,M.I.
VOL 4 E33-E36
AMBROSE,D.
VOL 5/6 E356,E415,E419
AMBROZHII,M.N.
VOL 13 E381-E385,E414-E417
AMICK,A.H.
VOL 5/6 E346, 352, 353
AMLIE,R.F.
VOL 5/6 E1-E3, 10,E25-E27,E33,E35, 40, 80
AMSTER,A.B.
VOL 7 351
VOL 10 262, 531
ANDERSON,A.M.
VOL 2 E134-E136, 139,E149,E193,E194, 197,E199-E201, 205, 212, 215, 217
ANDERSON,C.J.
VOL 2 E1-E2, 7, 8,E12,E13, 17, 18,
VOL 7 E65,E76, 117, 118
ANDO,N.
VOL 8 E27-E39,51, 64, 72, 104, 110, 333-335
ANDREW,M.L.
VOL 16/17 E10, 11, 16-38,E39, 40, 43-65,E72, 73,E78, 79-101,E102, 103-126,E135,E136, 141164,E175, 199,E200, 201-223,E224, 225-248,E249,E276, 277-300,E301,305,E306, 307-329,E334, 335-358,E367,E410,E457,E491,E518,E548,E576,E603,E630, E667,E707,E740
ANDROSOV,D.I.
VOL 5/6 E442, 471-474
ANGELOV,I.I.
VOL 13 E156-E162, 167-169,E226,E227, 235, 237-239
ANSCHULTZ.A.
VOL 14 E208,E209,E211,E212, 213, 214,E225-E227, 228,E242-E246, 256
ANTHONY,R.G.
VOL 9 E16, 20,E23, 24-26
ANTROPOFF,A.
VOL 2 E1-E3,E134-E136
VOL 4 E1-E7
AOKI,I.
VOL 16/17 E733, 734-739
APEL´BAUM,L.A.
VOL 5/6 E33,E35, 78
VOL 7 E71,E72,E76, 166
APPLEBY,M.P.
VOL 12 65,66

ARAI,Y,
VOL 5/6 E321, 328,E332, 338,E395, 399, 537-539, 543-551
VOL 4 E253-E254, 308, 320, 321
VOL 10 456,E488, 493, 498, 499
ARKADIEV,V.
VOL 12 E333,E334
ARKHIPOV.S.M.
VOL 14 E11-E18, 29, 30, 40,E59-E69, 163
ARMISHEVA.L.N.
VOL 13 E38-E47, 58, 59
ARMITAGE,D.A.
VOL 9 E138-E140, 147, 155, 164, 165,E232
ARMOR,J.N.
VOL 8 E260,E261, 264
AROYAN,H.J.
VOL 5/6 E346, 349, 350
ARRIAGA,J.L.
VOL 9 E77,E78, 94, 95
ARTEMOVICH,V.S.
VOL 5/6 E406, 411
AR,A.
VOL 10 293
ASAI,S.
VOL 7 416
VOL 8 E1,E2, 17,E27-E39, 93
ASHTON,J.T.
VOL 4 E1-E7, 15
ASSALI,N.S.
VOL 8 244
ATROSHCHENKO,A.V.
VOL 7 E62,E76, 86, 87,91, 92
ATWELL.L.L.
VOL 10 E409, 411
AUBRY,M.
VOL 13 E1,E2
AUDYKOWSKI,T.
VOL 7 E190-E192,E250,E251, 254,E261-E267, 273, 304, 305
AUGOOD,D.R.
VOL 5/6 E299,E321, 325,E346, 351,E571, 572,E589,E590, 594,E601, 604, 609,E610,E611,612, 619, 621,E622, 625
AUSTIN,K.W.B.
VOL 16/17 E394, 396,E442,E443, 444
AUTUNBETOV K.
VOL 13 E434-E436, 444-447
AVDEEVA,O.I.
VOL 1 E20-E22, 31
VOL 4 E33-E36, 51
VOL 7 E68,E69,E76, 143
VOL 9 E27-E33, 48
VOL 10 E50,E51,E56, 70
AYSCOUGH,P.B
VOL 12 E1,E2
AZARNOOSH.A.
VOL 10 E419, 435, 436
AZAROVA,L.A.
VOL 14 E11-E18, 23, 24, 27, 28,E59-E69,E78, 161, 162,E191-E197, 204, 205
AZHIPA.L.T.
VOL 13 E15-E19, 23,E381-E385, 389,E409, 412, 413,E414-E417, 420,E429, 431, 432,E434-E436, 439, 440,E449, 451, 452,E454-E456, 459,E461, 463
AZUMA,Y.
VOL 7 416
AZZAM,A.M.
VOL 3 E45-E56, 77

BABIEVSKAYA.I.Z.
VOL 13 E15-E19, 28, 29, 34,E38-E47,E381-E385, 390
BADELIN,V.G.
VOL 1 E124-E126
VOL 4 E33-E36, 46, 53, 58, 63, 65, 66-76
BAER,S.
VOL 4 E1-E7, 11,E80-E84, 87
BAES C.F.
VOL 13 E1
BAHAL,C.K.
VOL 16/17 E359, 361,E363, 364,E395, 409,E442,E443, 456
BAHE,L.W.
VOL 11 E88, 91, 92,E93,E96, 99, 100,E101, 104, 105,E106, 109, 110,E113, 116, 117,E118, 121, 122,E123, 126, 127,E130, 132, 133,E134, 136, 137,E138
BAILEY,S.M.
VOL 3 E39,E45-E56
BAIRD,W.R.
VOL 7 185, 186
BAKER J.L.
VOL 13 E38-E47,E354-E359, 363,E381-E385, 387,E409, 411
BAKHTIAROVA,G.A.
VOL 13 E38-E47, 66
BALDWIN,R.R.
VOL 5/6 270
VOL 7 E214,E215, 226
VOL 10 E119-E122, 136, 319, 327, 510, 518, 537, 541
BALEJ,J.
VOL 12 E261,E262, 292
BALOG-MEGYERY,K.
VOL 7 E261-E267, 269, 276-280, 282, 284, 287, 288
VOL 10 SEE MEGYERY-BALOG,K.
BANCROFT,W.D.
VOL 12 E37-E39
BARANOVICH,Z.N.
VOL 4 E253-E255, 273
VOL 10 E419, 424, 432
BARANOV,A.V.
VOL 7 E1-E5,E61,E62,E76, 88-90, 93-95
BARCLAY,I.M.
VOL 12 E181, 185
BARDIN,J.C.
VOL 3 85, 227
BARGER,A.C.
VOL 2 E114,E115
BARINOV,N.S.
VOL 5/6 569, 570
BARNARD,A.J.
VOL18 E3
BARON,J.D.
VOL 10 E402
BARRICK,P.L.
VOL 1 E298, 300, 301,E324, 325, 326,E348
VOL 5/6 E581, 583, 584,E610, 614, 615
BARSUK.S.D.
VOL 10 E409, 413
BARTON,J.R.
VOL 9 E195-E199, 208
BAR-ELI,K.
VOL 5/6 296
BASSETT,H.
VOL 3 E45-E55, 69
BASSETT,J.
VOL 10 E333
BATES,R.G.
VOL 3 E102,E113
BATISTE,M.
VOL 7 E214,E215, 239

BATTERSBY,A.C.
VOL 2 E114,E115
BATTINO R.
VOL 1 41, 42,E43,44-50,E51, 52, 53, 55, 57,E59, 61, 63-67,E68,E69, 71,E73, 74, 76,E77, 79,80, 85-87, 92, 93, 95-97, 106, 119,E124-E126, 187,E188, 189-195,E196, 197, 198, 200, 202,E204, 206, 209-213,E214, 216,E218, 219, 221,E222, 224, 225, 230-232, 236, 237, 240-242, 246, 248
VOL 2 E26, 27, 30-34,E35, 36, 38, 39, 40, 42, 44, 48,E52, 53, 56-60, 62-65, 67, 77, 79, 80, 89-91, 93, 95, 104, 107, 167
VOL 4 E1-E7,E108, 109,E111, 112,E114,E115, 116, 119-122,E123, 124,E125, 126,E128, 129, 131,E133, 134,E137, 138,E145, 147,E149, 151-157,E158, 160, 161,E162,163, 165, E169,E170, 191, 197, 198,E207, 215-217, 219, 221, 228, 239,E241, 244, 248-250
VOL 5/6 E169, 172
VOL 7 E1-E5, 35,E214,E215,E250,E251,E261-E267
VOL 8 E2,E260,E261
VOL 9 E1,E2, 3, 4
VOL 10 E1-E4, 19, 28,E119-E122, 134, 140,E148, 152, 155-160, E162, 167, 170, E174-E178, 201, 212, 214, 242, 275, E279, 282, 300-302
VOL 12 E354-E366
BATYUK,A.G.
VOL 13 E434-E436, 444, 445
BAUP,S.
VOL 3 E45-E55, 57, 82
BECKER B.
VOL 11 E48, 50
BECKER,H.G.
VOL 7 E1-E5, 12
VOL 10 E1-E4, 11
BECKLAKE,M.R.
VOL 8 238, 239
BEDEL,C.
VOL 3 E102-E113
BEENNAKKER,J.J.M.
VOL 5/6 E581
BEERBOWER,A.
VOL 9 E195-E198
BEHNKE,A.R.
VOL 1 E1-E4, 8, 118
VOL 4 E241, 242
VOL 10 E1-E4
BEKAREK,V.
VOL 12 E174,177-179,E189,E190,209,215,216
BELL,G.
VOL 2 E114,E115
BELL,P.R.F.
VOL 2 E114,E115
BELL,R.P.
VOL 14 E59-E70,E77-E81, 132, 144
BENCH,R.
VOL 3 E45-E56
BENEDICKS,C.
VOL 13 E381-E385
BENHAM.A.L.
VOL 5/6 E321, 324,E339, 540-542, 553
BENOIT,R.L.
VOL 12 142,E145-E147, 153,E189-E191, 225, 227, 230, E261,E262, 265, 267, 275, 280, 286, 307,E318, 322, 323
BENSON,B.B.
VOL 1 E1-E4, 13,E124-E126, 134
VOL 2 E1-E3, 5,E134-E136, 142
VOL 4 E1-E7, 10
VOL 7 E1-E5, 22, 23, 37, 39, 40,E41-E43
VOL 10 E1-E4, 21, 22,E31-E33, 37, 38

BEN-NAIM,A.
VOL 4 E1-E7, 11, 13,E20,E21, 22, 23, 25,E33-E36, 38, 44, 52, 56, 61, 62,E80-E84, 85, 87, 96, 102,E169,E171, 175, 200, 202
VOL 9 E1,E2, 6-8,E27-E33, 35, 45, 50-52, 57,E64, 59, 65-71,E138-E140, 148,E166,E167, 169, 173, 177, 180, 183, 186,E195-E199, 207, 229,E232,E233
BEN´YAMINOVIC,O.A.
VOL 1 E254-E256, 276, 277
VOL 10 E361,E362,E409, 413
BERARDELLI,M.L.
VOL 11 E1, 3,E4, 6,E9, 11,E14,E15, 17,E21, 23,E28,E29, 31,E35, 37,E39,E40, 42,E72, 73,E74, 75,E78, 79,E88, 90,E93, 95, E96, 98,E101, 103,E106, 108,E113, 115,E118, 120,E123, 125,E142, 143,E144, 145,E146, 147
BERENGARTEN,M.G.
VOL 8 166, 179, 182
VOL 10 E119-E121, 127,E148, 153,E162, 168
VOL 12 E116,E117, 121, 141, E145-E147, 152
BERGER,E.V.
VOL 1 E1-E4, 5
BERGER,J.E.
VOL 11 E255, 256, 257,E259, 261,E262, 264, 265,E266, 268,E269, 270,E271, 273,E275, 277,E279, 281,E282, 283,E285, 287, E289, 291,E292, 293,E294, 295,E296, 297,E298, 299,E302, 304
BERNE,D.H.
VOL 18 E31,E32, 34,E43, 46, 49, 51,E52,E53, 58, 61,E64, 66, 69,E228, 229
BERNINGER,E.
VOL 7 E1-E5
VOL 8 E1,E2
BERTHELOT, M.
VOL 12 344
BERTONI,A.
VOL 14 E208-E212, 215-217
BERTY,T.E.
VOL 5/6 E380, 388, 389
BESLEY,L.
VOL 4 E27-E29
VOL 10 E31-E33
BESSERER,G.J.
VOL 9 E111, 117, 118
VOL 10 E419, 420, 421, 468, 469
BEUSCHLEIN, W.L.
VOL 12 E3-E5, 21-22,E37-E39, 71, 72
BEVAN,D.J.M.
VOL 13 E241-E246
BEZYULEV,V.V.
VOL 5/6 E622, 629
BHATNAGAR,O.N.
VOL 11 E259,E275,E279,E285,E289,E300,E305,E307
BIERI,R.H.
VOL 4 E27-E29
VOL 5/6 E18
BIKOV,M.M.
VOL 7 171,E172
VOL 10 E54-E56, 78, 79
BILJ, A.
VOL 1 E1-E4
BILLETT,F.
VOL 5/6 E1-E3, 8, 55, 62, 68, 86
VOL 7 E1-E5, 16
VOL 9 E1,E2, 11,E27-E33, 44, 47, 58
VOL 10 E1-E4, 17,E49-E52, 55, 56, 64, 69, 80
BIL´SHAU,K.V.
VOL 12 E145-E147, 164-168
BINEAU,A.
VOL 12 E189-E191, 254

BIRCHER,L.J.
VOL 4 E169-E171, 173,E176, 178, 183,E185,E186, 188, 193-196
VOL 9 E166,E167, 168, 172, 176, 179, 182, 185, 187, 188
E232,E233
VOL 10 E174-E178, 181, 188, 192, 198, 204, 206, 208, 211
BIRCKENBACH,L.
VOL 3 34-37,E39, 40, 47,E90,E91, 92, 93
BISCERDY,B
VOL 3 E45-E46,E102-E113
BISSON,C.S.
VOL 14 E59-E68,E70,E72,E75, 135, 137, 177, 178,E225-E227, 230,
235,E242,E243,E246-E248,E252, 266, 281, 283, 285
BITTRICH,H.J.
VOL 5/6 E1-E3, 12,E32,E35, 43, 63, 69, 85, 477
BJERRUM,J.
VOL 3 E1-E8,E45-E56,E102-E113
BLACK,C.A.
VOL 5/6 E622, 634
BLAGOI,YU,P
VOL 2 50, 51
VOL 4 E253-E255, 276
VOL 10 400
BLAIR, M.C. JR.
VOL 12 E395-E397, 403
BLANC,C.
VOL 7 E214,E215, 239
BLAZNINA,V.A.
VOL 12 E349
BLOOMER,O.T.
VOL 10 E361,E362, 363, 364,E391, 392
BLYTAS, G.C.
VOL 3 233
BOARDMAN,M.
VOL 7 E41-E43, 52, 53
BOBRINEV,YU.M
VOL 4 E33-E36
BODLANDER,G.
VOL 3 E45-E56,E102-E113
BOER,H.
VOL 7 484
BOEVA,N.K.
VOL 13 E301-E308, 320-333
BOGARDUS,J.B.
VOL 16/17 E484, 485, 486
BOGDANOVA L.P.
VOL 4 E253-E256, 273
BOGDANOVSKAYA,R.L.
VOL 13 E38-E47, 57-59,E156-E162, 171-179, 187-191
BOGEATZES,A.S.
VOL 12 297
BOHR,C.
VOL 7 E1-E5, 10,E59,E60,E75, 80
VOL 10 E1-E4, 9,E47,E48,E53,E56, 58
BOL´SHCHIKOVA,T.P.
VOL 3 E31, 33
BONDAREVA,M.M.
VOL 4 E1-E7,E256
BONNELL,W.S.
VOL 5/6 E346, 348
BONNER, O.D.
VOL 11 E259
BON´SHAKOV,P.E.
VOL 10 E476
BOOMER.E.H.
VOL 10 E425, 426
BORCSIK,M
VOL 7 E41-E43, 52, 53
BORGSTEDT,H.H.
VOL 8 13

BORINA,A.F.
VOL 1 E124-E126, 132, 133, 135, 136,E141-E143, 144-149, 151-161, 163-165, 167-171, 174-178, 183, 184
BORN,M.
VOL 14 E71,E251
BORODKIN,V.F.
VOL 4 E33-E36
BOTTGER,W.
VOL 3 E45-E55, 61, 62,E102-E113 117, 118, 124, 125
BOTTOMLEY,G.A.
VOL 4 E27-E29
VOL 10 E31-E33
BOUSQUET,J.
VOL 14 E59-E68,E77-E81, 85, 86,E191-E197, 200, 201,E242-E243, E253,E254, 258
BOYER,F.L.
VOL 4 E169-E171, 173,E176, 178,E180, 181, 183,E185,E186, 188, 193-196
VOL 9 E166,E167, 168, 172, 176, 179, 182, 185, 187, 188, E232,E233
VOL 10 E174-E178, 181, 188, 192, 198, 204, 206, 208, 211
BOYLE,R.W.
VOL 2 E227-E229, 232, 236,E260,E261,E265, 268,E273, 275, 285,E343
BOZHOVAKAYA,T.K.
VOL 1 E20-E22, 31
VOL 4 E33-E36, 51
VOL 7 E68,E69,E76, 143
VOL 9 E27-E33, 48
VOL 10 E50,E51,E56, 70
BRADSTREET,E.D.
VOL 1 E1-E4,E257
VOL 4 E1-E7,E27-E29
VOL 7 E41-E43, 44
VOL 10 24,E31-E33,E83-E85, 352
BRAINARD,A.J.
VOL 5/6 E380, 390,E406, 414, 554,
BRANDE,H.
VOL 9 E232,E233, 249-251
VOL 10 505, 506, 546, 547
BRANDT,L.W.
VOL 10 E361-E362
BRANION,R.M.R.
VOL 7 418
BRASTED,R.C.
VOL 10 E1-E4,E54,E56, 77
BRAUN,A.
VOL 16/17 E10, 12,E359, 362,E442,E443, 455,E487, 489,E767, 768
BRAUN,L.
VOL 5/6 E1-E3,E31,E32,E35, 60, 67, 102
VOL 10 E1-E4,E49-E51,E56, 62, 63, 67, 68,E83-E85, 94, 95, 304, 305
BREANT,M.
VOL 3 220
BRECE, D.B.
VOL 12 E37-E39, 44
BRESTKIN,A.P.
VOL 5/6 E406, 411
BRIDGEMAN,O.C.
VOL 5/6 E17,E18
BRINER,E.
VOL 7 E474-E480, 486
BRINKLEY,S.R.
VOL 3 E102-E113
BRISCOE,W.A.
VOL 2 E114,E115, 121
BRITTON,H.T.S.
VOL 3 E45-E56
BRODEN,A.
VOL 7 E68,E69,E76, 160

BRODSKAYA,L.F.
VOL 5/6 E571, 577, 578
BROMLEY,L.A.
VOL 12 E34
BROOKER,R.E.
VOL 11 E7, 8,E9, 13,E14,E15, 19,E21, 25,E28,E29, 33, E39, 44, E46, 47,E48, 51,E74, 77
BROOMELL,H.T.
VOL 8 232, 242
BROTZ,W.
VOL 18 E7-E10
BROUTY,M.-L.
VOL 3 E1-E8, 12, 13
BROUWER,J.P.
VOL 5/6 E581
BROWN,M.A.
VOL 13 E38-E47,E354-E359, 363,E381-E385, 387,E409, 411
BROWN,S.F.
VOL 14 E59-E69, 84, 164
BRUHN,G.
VOL 5/6 E25-E30,E35, 39, 42, 45, 48-50, 65, 76
VOL 7 E57,E58,E60,E62-E64,E67,E68,E76, 81, 97, 98
BRUMMER,S.B.
VOL 12 331
BRUNISHOLZ,G.
VOL 13 E38-E47, 49, 70-72, 75, 76, 83,E156-E162, 164, E241-E246, 248, 255, 256,E301-E308, 309, 317, 318,E354-E359, 360, 368, 369, E376, E377
BRUNNER,E.
VOL 5/6 E441-E444, 445, 446, 448-451, 454, 455-458-470, E481-E483, 488, 490, 493, 494, 497-509
VOL 16/17 E68, 69-71,E72, 74-76,E167, 168-170,E171, 172-174,E301, 302-304,E330, 331-33
BRUNNER.G.
VOL 10 355, 356,E425,429, 442
BRUNS.B.P.
VOL 16/17 E1, 2-6,E740, 765, 766
BRYANT,J.M.
VOL 13 E109, 147
BUB,G.K.
VOL 7 293-296
BUCKWATER,F.H.
VOL 16/17 E175, 176, 177
BUCKLEY,P.S.
VOL 3 E223, 224, 225
BUCKLEY,W.D.
VOL 9 E111,E112
BUDNEVICH,S.S.
VOL 1 E328
BUECHL,H.
VOL 18 E3
BUI VAN TUAN
VOL 13 E38-E47, 73, 74,E301-E308
BUKINA,V.K.
VOL 12 389
BUKOWIECKI,A.
VOL 7 E190-E192,E250,E251, 254,E261-E267, 273, 304, 305
BULTER,J.N.
VOL 3 E45-E56
BULYCHEVA,M.S.
VOL 16/17 E662, 663
BUNCEL,E.
VOL 5/6 278,E280-E282, 293
BUNSEN,R.
VOL 5/6 E1-E3
VOL 7 E1-E5
VOL 9 E1,E2
VOL 10 E1-E4
VOL 12 E3-E5, 6, 333,E334, 335, 336

BURCH,R.J.
VOL 1 E328, 337,E373, 374
BURFIELD,D.W.
VOL 1 E298, 299
BURISS,W.L.
VOL 5/6 E339, 341
BUTT,J.B.
VOL 12 419
BUTT,M.A.
VOL.8. E1,E2,20,E116, 139-143
BYERLEY,J.J
VOL 12 E3-E5, 33, 110
BYRNE,J.E.
VOL 1 76,E77, 79, 80, 221,E222, 224, 225
VOL 2 64, 65, 67
VOL 4 E158, 161, 165, 166, 168, 248-250
VOL 7 E250,E251, 255
VOL 10 E162, 167, 300-302

C

CADY,G.H.
VOL 12 451, 452
CADY,H.P.
VOL 1 E1-E4, 5
CAILLET,A.
VOL 3 209, 210, 230. 231
CALADO,J.C.G.
VOL 5/6 E321, 329-331,E571, 579, 580,E589,E590, 599, 600
CAMPBELL,D.
VOL 8 E160-E162, 163, 185, 196, 208, 220
CAMPOS CARLES,A.
VOL 1 121
VOL 5/6 277
VOL 8 247
CANDER,L.
VOL 1 120
VOL 8 243
CANNON P.
VOL 5/6 520
VOL 7 441
CANNON,W.A.
VOL 1 E254-E256, 306
VOL 10 387,E391, 394,E402, 406, 467, 471, 473, 474, E476, 481, 485-487, 496, 497
CAPELL,R.G.
VOL 4 144
VOL 7 447, 448
VOL 10 325, 326
CAPUTI,R.
VOL 2 E1-E3,E9, 11,E134-E136, 148
CAREY,F.G.
VOL 7 E41-E43
CARGILL,R.W.
VOL 1 35-38,E82, 84
VOL 4 E1-E7,E80-E84, 90, 91, 94, 95,E176, 179
VOL 5/6 97-99, 106-108,E187, 198
VOL 7 E190-E192, 200-202, 208-210,
CARIUS,L.
VOL 5/6 E187, 194
VOL 7 E261-E267,E474-E480
VOL 8 E160-E162, 188,E336, 343
CARPENTER,J.H.
VOL 4 E27-E29
VOL 7 E1-E5, 25, 26,E41-E43, 47, E48
CARRITT,D.E.
VOL 7 E1-E5
CARSON,T.
VOL 12 48-50

CASALE,C.
VOL 5/6 E610,E611
CASSUTO,L.
VOL 7 E1-E5
VOL 10 E1-E4
CATCHPOOL,J.F.
VOL 2 E199-E201, 219
CAUQUIL,G.
VOL 5/6 E189, 210
VOL 7 E261-E267, 289
VOL 8 202, 345
CAVE,G.C.B.
VOL 3 E102-E113, 150, 151
CERVINKA,M.
VOL 12 E354-E366, 420
CHAENKO,N.V.
VOL 7 183, 184, 187
CHAIKOVSKII,V.F.
VOL 10 460, 461
CHAMBERLAIN,R.E.
VOL 16/17 E175, 178-198
CHANDRA,D.
VOL 11 E259, 260,E262, 263,E266, 267, E271. 272,E275. 276, E279, 280,E285, 286,E289, 290,E298, 301, E302, 303,E305. 306,E307, 308
CHANG,E.T.
VOL 1 100-103, 111-113
VOL 4 232-238
VOL 7 432, 433
VOL 10 270-274, 317, 318
CHANG,S.D.
VOL 10 E361,E362, 369, 378, 379,E391, 393
CHANTOONI,M.K.
VOL 18 E31,E32, 33,E64,E140, 144, 145,E149, 152,E154, 157, E158, 161, 162,E166, 168, 169, 180, 181,E198, 202, E203, 207
CHAO,E.E.
VOL 3 E45-E56,E102-E113
CHAO,K.-C.
VOL 5/6 E298,E355, 369, 370, 373, 374, 394,E415, 417,E419, 422, 423, 426, 428-433, 475, 476, 510, 517, 518, 556-561
CHAPPELEAR,P.S.
VOL 10 E361,362,372, 373,E391, 395,
CHAPPELOW,C.C.
VOL 5/6 E123, 147, 268
VOL 9 E77,E78, 103, 108,E138-E141, 160,E195-E199, 227
CHEN,C.T.
VOL 4 E27-E29
VOL 7 E41-E43
VOL 8 E23, E24
VOL 10 E31-E33
CHEN,H.Y.
VOL 5/6 E346, 352-353
CHENG,K.L.
VOL 3 E45-E56,E102-E113
CHENG,S.C.
VOL 9 E77,E78, 83, 85, 91, 96, 97, 109
CHEN-TUNG A. CHEN
VOL 7 E41-E43
CHEREPENNIKOV,A.A.
VOL 1 E20-E22
CHERNAYA,V.I.
VOL 11 E309,E310, 311
CHEUNG,H.
VOL 4 E261, 262
VOL 10 E361,E362, 368,E402, 403
CHEVELEV,A.V.
VOL 13 E461
CHLOUPEK J.B.
VOL 14 E59-E68,E70,E73,E74, 145, 146, 148, 152

CHODAKOV, K.V.
VOL 12 73-75, 377
CHRETIEN,A.
VOL 7 489
CHRISTENSEN,J.J.
VOL 3 E45-E56
CHRISTIANSEN,L.J.
VOL 4 E253-E255,E261, 263, 318
CHRISTOFF,A.
VOL 5/6 E1-E3,E25-E28,E35, 38,E213-E215, 216
VOL 7 E1-E5,E59,E60,E75, 79, 298
VOL 10 E1-E4,E47,E48,E56, 57, 227, 521
CHRISTOFORIDES,C.
VOL 7 373, 375, 376
VOL 8 E1,E2, 15
VOL 10 286
CHUEH,P.L.
VOL 12 194, 250, 258, 259,E261,E262, 272, 276, 283
CINES,M.R.
VOL 10 E361,E362, 365-367
CLAES,P.
VOL 9 E77,E78, 82, 105
CLANCY,P.
VOL 5/6 E321, 329-331
CLARK,F.M.
VOL 10 538
CLARK,L.C.
VOL 7 309, 315, 318-320, 322, 325, 326, 331, 332, 336, 338, 340-344, 356
VOL 10 523-530, 532
CLAUDEC,B.
VOL 13 E156-E162
CLAUSSEN,W.F.
VOL 9 E1,E2, 12
CLEMENTS,H.E.
VOL 5/6 E119, 127,E169, 174,E239, 248,E354
VOL 9 E138-E141, 154
CLERMONT M.J.
VOL 5/6 263
CLEVE,P.T.
VOL 13 E15-E19
CLEVER,H.L.
VOL 1 E20-E22, 34, 41, 42,E43, 44, 46-49,E51, 52, 55, 57, E59, 61, 62,E68,E69, 71, 91,E124-E126, 187,E188, 189, 191-195,E196, 197, 200, 202,E204, 206, 208,E214, 216, 235
VOL 2 E26, 27, 30-34,E35, 36, 39, 40, 44, 48,E52, 53, 55, 62, 66, 86, 92, 94, 96, 156, 158, 160,E162,163, 165, 166, 172
VOL 4 E1-E7, 14,E33-E36, 45, 49, 54, 55, 57, 79,E108, 109, E111, 112,E114,E115, 116, 120-122,E123, 124,E125, 126, E128, 129,E133, 134,E137, 138,E145, 147,E149, 151,E158, 160, 167,E171, 174,E207, 209, 218, 220, 222
VOL 7 E1-E5,E56
VOL 8 E160-E162,E265-E267
VOL 10 E1-E4, 19
VOL 12 E354-E366
CLIFFORD,A.F.
VOL 18 E3
CLYNNE,M.A.
VOL 4 E1-E7,E256, 258
COCKBURN.A.
VOL 7 E41-E43
COFFIN,R.L.
VOL 9 E1,E2, 13,E27-E33, 41,E64, 73
COHEN-ADAD,R.
VOL 3 E39, 42
COHEN,Y.
VOL 5/6 E1-E3, 14,E17,E18, 24
COHEN.A.E.
VOL 5/6 E332, 335, 534-536

COLE,J.J.
VOL 18 E182, 186
COLLIE,J.N.
VOL 1 E1-E4,E68,E69
COLMAN-PORTER C.A.
VOL 14 E191-E197, 202
COLTON,E.
VOL 11 E7, 8,E9, 13,E14,E15, 19,E21, 25,E28,E29, 33,E39, 44,E46, 47,E48, 51,E76, 77
CONANT,J.B.
VOL 10 287, 288
CONNOLLY,J.F.
VOL 5/6 E355, 366, 367,E406, 412
CONN,H.L.
VOL 2 E199-E201, 207, 208, 221
CONRAD,F.H.
VOL 12 E3-E5, 18,E37-E39, 71, 72, 77
CONWAY,B.E.
VOL 4 E1-E7,E33-E36, 39, 40
VOL 7 E1-E5, 34
COOK,E.V.
VOL 8 231
COOK.M.W.
VOL 5/6 E120-E122, 131, 132, 134, 136,E159,E160, 164,E169, 171,E237,E238, 243, 246,E255, 258,E280-E282, 283-291
COOPER,D.F.
VOL 12 326, 327
CORBET,A.S.
VOL 3 E45-E55, 69
CORCORAN,W.H.
VOL 7 E62,E76, 85
CORINALDI,G.
VOL 4 143
VOL 5/6 271
VOL 10 322
CORNIDES,I.
VOL 4 E314
COSGROVE,B.A.
VOL 10 29, 30
COSTE,J.H.
VOL 10 E31-E33
COSWAY,H.F.
VOL 5/6 530-533, 552
VOL 10 376, 377
COTE,P.A.
VOL 3 E45-E56, 79-80
COUNIOUX, J.-J.
VOL 13 E1,E15-E19,E38-E47,E156-E162,E241-E246,E301-E308, E354-E359,E381-E385,E414-E417
COURTOT-COUPEZ,J.
VOL 3 E1-E8, 18,E102-E113,E185, 186, 228
COVINGTON,A.K.
VOL 3 E185,E223
COX,A.J.
VOL 3 E45-E56,E102-E113, 116
COX,B.G.
VOL 18 E182, 185,E187, 191
CRAIG,H.
VOL 4 E27-E29
VOL 10 E31-E33
CRAIN,R.W.
VOL 1 E310, 314, 315
CRAMER,S.D.
VOL 7 E68,E69,E76, 144-146
CRANE,W.E.
VOL 1 E254-E256, 306
CREIGHTON,H.J.M.
VOL 8 E1,E2,6, 146, 148-150, 152, 156, 158, 229, 248, 251, 254, 255

CREIGHTON,H.J.M.
VOL 10 289
CREW,M.C.
VOL 13 E15-E19, 20
CRISS,C.M.
VOL 11 E148, 149, 150,E159, 160, 161,E162, 164, 165, E186, 189, 190,E212, 213,E216, 217
CROZIER,T.E.
VOL 5/6 E1-E3, 13,E17,E18, 19-23,E32,E35, 70
CSEKO,GY.
VOL 4 E314
CUBINA,J.M.
VOL 8 E226,E227, 241, 245, 246
CUKOR,P.M.
VOL 5/6 E123, 145, 158, 184
VOL 9 E77,E78, 100,E138-E141, 150,E195-E198, 203
CULBERSON,O.L.
VOL 9 E16, 17-19,E23
CULLEN,M.C.
VOL 13 E109, 143
CULLEN,S.C.
VOL 8 231
CUPR,V.
VOL 12 43, 199
CURDA,M.
VOL 12 E354-E366,E395-E397, 406, 413
CZAPLINSKI,A.
VOL 9 E1,E2, 15,E27-E33, 43, 49
CZERSKI,L.
VOL 9 E1,E2, 15,E27-E33, 43, 49

D

DALTON,B.J.
VOL 1 E328, 343, 344
DALY,L.K.
VOL 13 E38-E47,E109, 128
D´AMBROSI
VOL 7 E474-E480
DANA,L.I.
VOL 7 434, 435, 468
VOL 10 E488, 490
DANCKWERTS,P.V.
VOL 8 E1,E2, 16,E27-E39, 112, 120, 122, 123, 126, 127,
DANESOVA,B.A.
VOL 14 E59-E68,E70,E73,E74, 145, 146, 148, 152
DANES,VL.Z.
VOL 14 E59-E68,E70,E73,E74, 145, 146, 148, 152
DANFORTH,W.F.
VOL 1 E43, 45,E51, 53, 85-87, 119,E188, 190,E196, 198, 230-232, 251
VOL 2 E35, 38, 42, 77, 79, 80
VOL 4 E114,E115, 119,E128, 131,E158, 161,E169,E170, 191, 197, 198,E241, 244, 248-250
VOL 7 E214,E215, 224, 231,E250,E251, 255,E261-E267, 281, 285, 286
VOL 10 E119-E122, 134, 140,E162, 167,E174-E178, 201, 212, 214,E279, 282, 300-302
DANIEL S.G.
VOL 5/6 270
VOL 7 E214,E215, 226
VOL 10 E119-E122, 136, 319, 327, 510, 518, 537, 541
DANIELS,F.
VOL 3 233
DANIL DE NAMOR,A.F.
VOL 18 6, 50, 67, 68, 99, 100, 103-105, 109-114, 192, 193, 230, 231
DANNEIL,A.
VOL 9 E16, 21, 22

DASH,U.N.
VOL 3 E102-E113, 175, 176
DAS,A.K.
VOL 18 24, 30
DAS,A.R.
VOL 14 E59-E68,E75,E76,E78-E81, 107, 108, 117, 118, 126, 127
DAS,R.C.
VOL 3 E90,E91, 96.97
DATE,R.V.
VOL 12 419
DAVIDSON,D.
VOL 7 368
VOL 10 E279, 280
DAVIES,C.W.
VOL 3 E1-E8,E223
VOL 14 E59-E70,E72,E73-E81, 96, 104-106, 109-111, 113-116, 119-125, 128-131, 134, 147,153, 154, 156,E242-E251, E253,E254, 262-264, 267, 268, 272, 277
DAVIES,J.A.
VOL 9 E232, 236, 237
DAVIS,J.A.
VOL 1 E328, 339, 340
DAVIS,R.E.
VOL 7 167
DAVIS,T.W.
VOL 3 E177,E178,E193,E200
VOL 14 E242-E246,E248-E251, 269, 274, 276, 280, 307
DAVYDOV,I.A.
VOL 1 E328
DAVYDOV,YU,P.
VOL 13 E1
DAWE,R.A.
VOL 4 E1-E7, 15
DAWSON,R.L.
VOL 11 E255, 256, 257,E259, 261,E262, 264, 265,E266, 268, E269, 270,E271, 273,E275, 277,E279, 281,E282, 283, E285, 287,E289, 291,E292, 293,E294, 295,E296, 297, E298, 299,E302, 304
DEAN,M.R.
VOL 5/6 E356, 371, 375, 376
VOL 10 475
DEBAILLEUL,M.
VOL 7 E190-E192, 196, 197
DEBON,F.L.
VOL 2 E114,E115, 122,E199-E201, 210
VOL 8 E226,E227
DECANNIERE, L.
VOL 9 E77,E78, 82, 105
DECKWER,W.-D.
VOL 5/6 E30,E35, 52
VOL 12 379, 381-385, 427-429
DELANEY,D.E.
VOL 1 105
VOL 4 229
VOL 9 E195-E199, 215
VOL 10 261
DELIVORIA-PAPADOPOULOS,M.
VOL 5/6 15,E33,E35, 73, 276
DEL´PINO,KH.
VOL 13 E449,E454-E456
DEMANGE-GUERIN,G.
VOL 3 209, 210, 230-232
DEMARCAY,E.
VOL 13 E15-E19,E354-E359
DEMINA,G.A.
VOL 13 E15-E19, 25, 32, 33,E241-E246, 254,E301-E308, 315
DERER,J.L.
VOL 13 E38-E47, 52,E241-E246, 251,E301-E308, 312,E354-E359, 362,E381-E385, 387,E409, 410,E414-E417, 419,E429, 430,E434-E436,E449, 450,E454-E456, 458,E461, 462

DERRY,L.D.
VOL 10 542, 543
DERR,P.F.
VOL 14 E59-E69, 88,E242-E247,E253,E254, 293
DEWET,W.J.
VOL 5/6 E1-E3, 9,E159,E160, 166,E169, 173, 176,E240, 249
DE GRAAFF,W.
VOL 5/6 E435, 439
DE SAUSSURE,T.
VOL 12 E181, 188
DE SWAAN ARONS,J.
VOL 1 E254-E256, 353-355
DE VANEY,W.E.
VOL 1 E263, 265, 266, 271, 272,E328, 335, 336, 343, 344
VOL 5/6 562-566
DE WET,W.J.
VOL 1 E1-E4, 10,E68,E69, 72,E73, 75,E77, 78, 94,E124,E126, 129,E214, 217,E218, 220,E223, 239
DIAZ,J.M.
VOL 8 120, 122, 123, 126, 127
DICKINSON,T.
VOL 3 E185,E223
DIEPEN,G.A.M.
VOL 1 E254-E256, 353-355
VOL 12 E1,E2
DILLON,R.T.
VOL 10 E1-E4, 15,E50,E51,E53,E56, 71, 290, 291, 295
DILL,A.J.
VOL 18 25, 26, 85, 87
DIMENOV,A.T.
VOL 7 427
DIMICCO,J.A.
VOL 7 385
DITTMAR,W.
VOL 10 E31-E33
DI CAPUA,C.
VOL 13 E38-E47, 62, 79
VOL 14 E208-E212, 215,-217
DODD,E.N.
VOL 3 E45-E56
DODE,M.
VOL 10 E333
DODGE,B.F.
VOL 10 443,E476, 479, 480
DOKOUPIL,Z.
VOL 5/6 E581, 588,E589,E590,E601
DOMANSKY,R.
VOL 12 E37-E39, 60, 61
DOMENICONI,M.
VOL 12 330
DORNTE,R.W.
VOL 7 431
VOL 10 315
DOTY,V.
VOL 8 E1,E2,11,E226,E227, 234
DOUABUL,A.
VOL 12 E3-E5, 32,E34, 35, 36
DOUGHERTY,J.
VOL 2 E114,E115
DOUGLAS E.
VOL 4 E1-E7, 12,E27,E28, 29, 30
VOL 7 E1-E5, 24
VOL 8 E226,E227
VOL 10 E1-E4, 23,E31-E33, 39, 40
DOVE,M.F.A.
VOL 3 87
DOWNING.A.L.
VOL 7 E41-E43
DRABBLE,F.
VOL 3 E102-E113, 143, 144

DRAKIN,S.I.
VOL 1 E1-E4,E68,E69
VOL 2 E1-E3
VOL 4 E1-E7,E111,E158,E171
DRETCHEN,K.L.
VOL 7 385
DRISCOLL,J.R.
VOL 12 331
DRUCKER,K.
VOL 5/6 E1-E3, 105
VOL 10 E1-E4,E83-E85, 100, 103, 222
DRUSHLYAK,O.G.
VOL 10 59
DRUZHINA-ARTEMOVICH,S.I.
VOL 5/6 E303,E304,E435
DRUZHININA,E.N.
VOL 16/17 E1, 2-6
DUCRUE,H.
VOL 13 E286-E288, 290, 293, 295, 297, 298, 300,E339-E341, 343, 348, 350, 353
DUMOULIN,E.
VOL 2 105
VOL 4 E314, 315
DUNCAN,A.G.
VOL 4 E261, 266, 267
DUNCAN,D.W.
VOL 7 418
DUPERRAY M.H.
VOL 13 E15-E19, 30
DUPONT DE NEMOURS
VOL 12 261, 262, 269, 390
DURAND,G.
VOL 14 E208-E212, 224
DVORNIKOVA,L.M.
VOL 13 E381-E385,E414-E417
DYKHNO,N.M.
VOL 1 E283, 289,E328, 338,E348, 352,E359, 360,E373, 375, 381
DYMOND,J.H.
VOL 1 E59, 99,E204, 207, 244
VOL 2 E52, 54, 101,E162, 164, 184
VOL 4 E145, 148, 226
VOL 5/6 148, 151,E259, 260,E280-E282, 292
VOL 7 346
VOL 9 E138-E140, 149,E195-E199, 221
VOL 10 E148, 150, 151,E253, 258
DYMOVA,R.P.
VOL 10 357, 358, 449, 450
DZHAGATSPANYAN,R.V.
VOL 12 E116,E117, 123, 125, 131, 137, 139,E298, 304, 316, 317, E354-E366, 369-373, 378, 391, 417, 418 421, 423, 430, 442

E

EAKIN.B.E.
VOL 5/6 562-566
VOL 10 E391, 392
EBERLEIN,W.
VOL 3 E45-E56,E102-E113
ECKERT,C.A.
VOL 4 E253-E255, 270
VOL 10 459
ECKSTROM,H.C.
VOL 11 E255, 256, 257,E259, 261,E262, 264, 265,E266, 268, E269, 270,E271, 273,E275, 277,E279, 281,E282, 283, E285, 287,E289, 291,E292, 293,E294, 295,E296, 297, E298, 299,E302, 304

EDMISTER,W.C.
VOL 5/6 E380, 387,E406, 413
EDMONDS,R.G.
VOL 12 E333,E334
EDWARDS,A.W.T.
VOL 10 E1-E4, 20, 284, 285, 299
EDWARDS,E.
VOL 10 E361
EDWARDS,H.
VOL 9 E232,E233, 249-251
VOL 10 505, 506, 546, 547
EFREMOVA,G.D.
VOL 5/6 E406, 410,E435, 438,E622
VOL 10 444,E476
EGEL-THAL,M.
VOL 4 E33-E36, 38, 44, 52, 56, 61, 62, 64
EGER,E.I.
VOL 8 14,E226,E227, 236, 237
EGGLETON,P.
VOL 7 368
VOL 10 E279, 280
EGOROVA,V.I.
VOL 4 E80-E84, 101
EGOROV.V.S.
VOL 12 E354-E366, 380,E395-E397, 409, 424-426
VOL 14 E44-E47, 49, 50
EHRET,W.F.
VOL 14 E44-E47, 55, 56
EICHELBERGER,W.C.
VOL 10 341, 512-514
EICKE,H.F.
VOL 10 E425, 428, 441
ELEY.D.D.
VOL 4 E80-E84
ELIZAROVSKAYA,L.M.
VOL 16/17 E1, 7-9,E657, 659-661,E664, 666
ELLINGTON,R.T.
VOL 10 E391, 392
ELMORE,H.L.
VOL 7 E1-E5, 18
ELSEY,H.M.
VOL 1 E1-E4, 5
ELSHAYAL,I.M.
VOL 4 E253-E255, 269
VOL 10 E391
EL´NATANOV,A.I.
VOL 10 445
EMMEL,V.M.
VOL 4 E27-E29
VOL 10 E31-E33, 36
ENDERS,C.
VOL 7 E190-E192
ENGLISH,W.D.
VOL 10 387,E391, 394,E402, 406, 467, 471, 473, 474,E476, 481, 485-487, 496, 497
ENNS,T.
VOL 1 E1-E7,E27-E28,E257
VOL 7 E41,E43, 44
VOL 10 10, 24,E31-E33,E83-E85, 352
ERICKSON A.L.
VOL 1 E263, 267-270,E283, 284, 285,E328, 331, 332
ESAKA,N.
VOL 8 E1,E2,17,E27-E39, 93
ESTREICHER,S.
VOL 1 E1-E4
ETO,Y.
VOL 5/6 E435, 440, 567, 568, 620
EUBANKS,.L.S
VOL 5/6 E589,E590, 595,E601, 605, 631-633

EUCKEN,A.
VOL 2 E134-E136,E149, 152
VOL 4 E1-E7,E33-E36, 50
VOL 7 E58,E65-E69,E72,E73,E75, 120, 130, 142, 173
VOL 9 E1,E2, 10,E27-E33, 46
EULER,R.
VOL 7 E1-E5, 38
EVANS,D.E.
VOL 8 240
EVANS,F.D.
VOL 1 85, 92, 119, 230, 236, 251
VOL 2 77, 90
VOL 4 E169,E170, 191, 216,E241, 244
VOL 7 E261,E267, 281
VOL 10 E174-E178, 201, 242,E279, 282
EVANS,D.F.
VOL 10 542, 543
EVANS,W.H.
VOL 3 E39,E45-E56
EWING,G.J.
VOL 2 E134-E136, 141,E199-E201, 220
VOL 10 296
EYRING,L.
VOL 13 E241-E246
EZHELEVA,A.E.
VOL 9 E166,E167,E195-E198, 205, 209, 210, 219
EZHOV,A.I.
VOL 13 E409,E414-E417

F

FAETH,G.M.
VOL 10 437, 507, 508, 519, 548, 549
FAL´KOVICH,YU.E.
VOL 7 427,E428, 429
FARHI,L.E.
VOL 10 E1-E4, 20, 284, 285, 293, 299
FARKAS,E.J.
VOL 5/6 E178
FASTOWSKI,V.G.
VOL 1 E263, 275,E328, 345
VOL 5/6 E321, 323,E589,E590, 592, 593
VOL 10 E361,E362
FAULCONER,A.
VOL 8 233
FAULFNER,B.A.
VOL 10 542, 543
FEAKINS,D.
VOL 11 E88,E93,E96,E101,E106,E111,E113,E118,E123,E128,
E130,E134,E138,E140,E142,E144,E146
FEATHERSTONE,R.M.
VOL 2 E114,E115, 122,E199-E201, 209, 210
VOL 8 E226,E227
FEDORINA,V.F.
VOL 12 E349
FEDOROV,A.A.
VOL 11 E309,E310,E313,E315, 316,E317,E321,E323, 324,
E325,E327,E329,E331, 332,E333,E337,E341, 342,
E343, 344
VOL 14 E60, 93,E191-E197, 199,E243, 261
FEDOSEEVA,E.I.
VOL 13 E15-E19, 24
FEILLOLAY,A.
VOL 1 E1-E4,E20-E22, 26,E141-E143
FENTON,J.
VOL 12 42
FERGUSON,C.V.
VOL 7 431
VOL 10 315

FERGUSON,R.E.
VOL 13 E241-E246
FERRELL,E.
VOL 3 E45-E56
FERRELL,J.K.
VOL 10 495
FICKE,H.F.
VOL 10 434
FIELD,L.R.
VOL 7 244, 257
VOL 10 155,E162, 170
FIGUIERE,P.
VOL 10 E425, 430
FILATOVA,N.I.
VOL 11 E309,E310,E313,E315, 316,E317,E321,E323, 324, E325,E327,E329,E331, 332,E333,E337,E339,E341, 342, E343, 344
FINDLAY,A.
VOL 5/6 E1-E3, 114, 115, 117
VOL 8 E1,E2, 6, 7, 146-153, 156-159, 229, 248, 251-255
VOL 10 289
FINDL,E.
VOL 9 E232,E233, 249-251
VOL 10 505, 506, 546, 547
FINHOLT,P.
VOL 16/17 E10, 14, 15,E39, 41
FINYAGINA,R.A.
VOL 5/6 E571, 577, 578
FIORANI,M.
VOL 3 E102-E113
FISCHER,F.
VOL 7 E474-E480
FISHER,G.T.
VOL 7 259
FITZGERALD,T.B.
VOL 12 251,E261,E262, 290, 291, 294, 296
FLASCHKA,H.
VOL 18 E3
FLEISHER,J.
VOL 12 E189,E190, 203, 295
FLOOK.V.
VOL 8 240
FOCK,A.
VOL 13 E226,E227
FOGGIE,P
VOL 7 368
VOL 10 E279,E280
FOLEY,R.T.
VOL 7 185, 186
FONTAINE,R.
VOL 9 E77,E78, 82, 105
FOOTE,H.W.
VOL 3 E102-E113, 119, 120
VOL 14 E208-E212, 223
FORSTER,R.E.
VOL 5/6 15,E33,E35, 73, 276
FOX,C.J.J.
VOL 7 E1-E5, 8, 9,E41-E43
VOL 10 E1-E4, 7, 8,E31-E33, 34, 35
VOL 12 E3-E5, 9,E37-E39, 54-56, 62, 63, 78-81, 88-90, 95, 97, 100-104
FRANCK,E.U.
VOL 2 144, 145
VOL 4 E256
VOL 9 E16, 21, 22
FREDENSLUND,A.
VOL 4 E253-E255,E261, 263, 319
VOL 7 470-472
VOL 10 388,E391, 397, 401,E402, 408

FREEDMAN,A.J.
VOL 14 E11-E18, 21, 22, 41,E208-E212, 218-222,E225-E227, 236, 237, 240, 241,E242-E246,E250,E254, 291, 292, 296
FREETH,F.A.
VOL 5/6 E321, 322
FRENCH,G.M.
VOL 11 E88,E93,E96,E101,E106,E111,E113,E118,E123,E128,E130 E134,E138,E140,E142,E144,E146
FREYHOF,J.N.
VOL 16/17 E394, 398,E442,E443, 446
FRIEDMAN,E.
VOL 2 E114,E115
FRIEDMAN,H.L.
VOL 1 E1-E4, 39, 104, 245
VOL 2 103, 187
VOL 4 E1-E7,E80-E84, 106, 227
VOL 10 260
FRIEDMAN,R.M.
VOL 18 E7-E10, 14, 38,E72,E73, 84, 88, 98, 101,E106
FRIEND,J.N.
VOL 13 E38-E47, 48,E148, 150,E226,E227, 228,E241-E246, 247,E286-E288, 291,E301-E308,309,E339-E341, 344-346
FROLICH,P.K.
VOL 5/6 E119,E298,E299,E303,E304,E339, 340,E346, 347,E354,E355, 358, 359, 365,E380, 382,E406,E435, 436,E441-E414, 447, 521, 522
VOL 10 12,E31-E33,E333,E409, 410, 447, 458, 503, 504
FROLOVA,S.I.
VOL 13 E156-E162, 180, 181
FRONMULLER,C.
VOL 3 E45-E55, 58
FUJIO,J.
VOL 10 E104,E105, 107-112
FUNK,R.
VOL 14 E1-E3,E11-E18, 19,E44-E47, 48,E179-E181, 182
FUOSS,R.M.
VOL 14 E246
VOL 18 E7-E10
FURMER,I.E.
VOL 8 166, 179, 182
VOL 10 E119-E121, 127,E148, 153,E162, 168
VOL 12 E116,E117, 121,141,E145-E147, 152
FUWA,T.
VOL 12 332

G

GADDY,V.L.
VOL 1 E1-E4,E257, 258
VOL 5/6 E303,E304, 305, 306, 312,E622, 624
VOL 10 E31-E33,E333, 335, 336,E476, 478
GAINER,J.L.
VOL 7 380, 381
GAINER,J.V.
VOL 7 380, 381
GALAL,H.A.
VOL 7 E1-E5, 20, 21
GALL,J.F.
VOL 10 259
GAMESON,A.L.H.
VOL 7 E41-E43
GAMI.D.C.
VOL 10 392
GANZ,S.N.
VOL 8 E265-E267, 299-301

GARANIN,V.F.
VOL 10 472
GARDINER,G.E.
VOL 1 E1-E4,E257, 259, 261
GARDNER,G.C.
VOL 10 380-382,E391, 398, 399
GARELLI, F
VOL 8 E336
GARRICK.F.J.
VOL 3 E102-E113, 132
GARST,A.W.
VOL 9 E74, 75, 76
GASEM,K.A.M.
VOL 10 389, 390
GAUGUIN,R.
VOL 3 E45-E56, 72, 73
GAVRILOVA,E.M.
VOL 2 E260,E261
GAY,J.
VOL 8 E265-E267, 268-273
GEBAUHR,W.
VOL 18 E7-E10, 12,E43, 44,E52,E53, 59
GEFFCKEN,G.
VOL 5/6 E1-E3, 6,E25-E28,E31,E33, 36, 37, 41, 64, 77, 91, 94
VOL 7 E1-E5, 7,E58-E62,E68,E69,E71,E72-E75, 77, 78, 84
VOL 8 E1,E2, 4,E27-E38, 41-43, 46, 77, 94, 97, 102, 103, 113, 114, 145,
GEHLAWAT,J.K.
VOL 12 E354-E366, 387
GEILMANN,W.
VOL 18 E7-E10, 12,E43, 44,E52,E53, 59
GELLER,E.B.
VOL 1 64-67, 210-213
VOL 2 57-60
VOL 4 153-156
VOL 7 245-248
VOL 10 157-160
GEL´PERIN,I.I.
VOL 5/6 E622, 629
VOL 12 E145-E147, 164-168
GEMANT,A
VOL 10 539
GEORGE,J.H.B.
VOL 14 E59-E70,E77-E81, 132, 144
GEORGES,J.
VOL 3 220
GERECKE,J.
VOL 5/6 E1-E3, 12,E32,E34,E35, 43, 44, 63, 69, 85
GERLACH,J.
VOL 5/6 E25-E30,E35, 39, 42, 45, 48-50, 65, 79
VOL 7 E57,E58,E60,E62-E64,E67,E68,E76, 81, 97, 98
GERRARD,W.
VOL 2 E260,E261
VOL 8 E265-E267,E351
VOL 12 E37-E39,E116,E117, 129,E145-E147, 155, 158, 163, 169,E174, 180,E181, 186,E189,E190, 195, 197, 206, 219, 222, 228, 240, 243, 247, 249, 253,E261,E262, 273, 279, 285, 288,E298, 305, 310-312, 314, 315, E354-E366
GERVER,J.
VOL 1 E1-E4
GERWIG,W.
VOL 8 E336,E350
GESTRICH,W.
VOL 5/6 E148, 152, 156, 157,E159,E160, 167,E169, 175, 185,E214, E215, 220
VOL 7 E70,E72, 152

GIBOFSKY,A.
VOL 18 E31,E32, 34,E43, 46, 49, 51,E52,E53,58,E64, 66, 69, E228, 229
GIBSON,Q.H.
VOL 7 E41-E43
GIESE,L.
VOL 1 E310, 311
GILLIES,A.J.
VOL.8. 13
GILL,D.S.
VOL 11 E1, 2,E4, 5,E9, 10,E14,E15, 16,E21, 22,E26, 27,E28,E29 30,E35, 36,E39, 41,E48, 49,E61, 62,E64, 65,E67, 68,E70, 71,E88, 89,E93, 94,E96, 97,E101, 102, E106, 107,E111, 112,E113, 114,E118, 119,E123, 124, E128, 129,E130, 131,E134, 135,E138, 139,E140, 141, E151, 153,E156, 157,E168, 169, 173,E181, 183, E186, 188,E194, 195,E198, 200,E208, 210,E222, 224,E226, 228,E229, 230,E231, 233
GILOT,B.
VOL 12 E1,E2
GIZHAVINA,E.I.
VOL 13 E38-E47, 64, 84,E156-E162, 182, 198
GJALDBAEK,J.C.
VOL 4 E169,E170,E180,E181, 182,E185,E186, 187, 192, 199, 206,E223, 225
VOL 5/6 E122, 137,E237, 244,E255, 257
VOL 7 E214,E215, 220, 222, 228
VOL 9 E77,E78, 87, 90, 92,E138-E140, 144, 145, 162,E166,E167, 170, 174, 178, 181, 184, 189, 190,E232,E233
VOL 10 E119-E122, 124, 129, 132, 137,E148, 149,E162, 165,E174-E178, 184, 191, 203, 207, 244-246,E253, 255
VOL 12 E354-E366, 416
GLASS,H.I.
VOL 2 E114,E115
GLASSTONE,S.
VOL 3 E45-E56
GLAZACHEVA,G.I.
VOL 13 E1
GLOVER,K.H.
VOL 11 E88,E93,E96,E101,E106,E111,E113,E118,E123, E128,E130,E134,E138,E140,E142,E144,E146
GLUECKAUF,E.
VOL 2 E9
GLUKAUF,E.
VOL 1 E1-E4,E138
GNIEWOSZ,S.
VOL 5/6 269
VOL 7 444
VOL 8 257
VOL 10 329
GODWIN,W.L.
VOL 12 E344
GOKCEN,N.A.
VOL 1 100-103, 111-113
VOL 4 232-238
VOL 7 432, 433
VOL 10 270-274, 317, 318
GOLDBERG,S.S.
VOL 18 27-29
GOLUB,A.M.
VOL 3 E90,E91, 98-101,E102-E113, 165, 166, E177,E178, E193, 207, 219
GOLYAND,S.M.
VOL 4 E33-E36

GONIKBERG,M.G.
VOL 1 E263, 275,E328, 345
VOL 5/6 E321, 323,E589,E590, 592, 593
GONZALEZ,A.
VOL 4 E27-E29
VOL 7 E41-E43
VOL 10 E31-E33
GOODMAN,J.B.
VOL 10 13,E31-E33,E333, 334
GOPAL,R.
VOL 11 E9,12,E14,E15, 18,E21, 24,E28,E29, 32,E35, 38, E39,E40, 43,E61, 63,E64, 66,E67, 69,E72,E74, 76, E78,E80, 81,E82, 83,E84, 85,E86, 87,E259, 260, E262, 263,E266, 267,E271, 272,E275, 276,E279, 280,E285,286,E289, 290,E300, 301,E302, 303,E305, 306,E307, 308
GORBACHEV,V,M,
VOL 10 252
GORBOVITSKALYA,T.I.
VOL 8 E39,78, 100
GORDON,L.I.
VOL 5/6 E1-E3, 14,E17,E18, 24
GORDON,V.
VOL 8 E27-E39, 66, 70, 74, 76, 79, 80, 86, 96, 105
GORODETSYII,I.G.
VOL 4 E253-E255
GORSHUNOVA,V.P.
VOL 13 E38-E47, 86-94, 96-107,E148
GORYUNOVA,N.P.
VOL 1 E254-E256, 280, 281
VOL 10 455, 457
GOTTLIEB-BILLROTH,H.
VOL 8 256
GOULD,L.P
VOL 12 E37-E39
GRAHAM.E.B.
VOL 4 E114,E115, 117,E123,E169,E170,E207, 214,E253-E255, 274, 275, 307, 309
VOL 10 E419,431, 433
GRANKINA,Z.A.
VOL 13 E376,E377, 378, 379
GRASSHOFF,K.
VOL 7 E41-E43
GRAUSO.L.
VOL 10 388,E391, 397, 401,E402, 408
GRAVELLE,D.
VOL 4 E261, 264, 265
GRECO,G.
VOL 5/6 E610,E611
GREEN,E.J.
VOL 7 E41-E43, 45,E46
GREZIN,A.K.
VOL 10 460, 461
GRIFFITH,W.P.
VOL 8 E265-E267,312, 313, 326, 327
GROLLMAN,A.
VOL 10 E1-E4, 303
GROSS,F.
VOL 14 E59-E68,E70-E72, 89, 91, 94, 133
GROSS,P.M.
VOL 1 44, 46-50,E51, 52, 55, 57,E59, 61, 62,E68, E69, 71, 91,E124-E126, 187,E188, 189,191-195, E196, 197, 200, 202,E204, 206, 208,E214, 216, 235
VOL 2 E26, 27, 30-34,E35, 36, 39, 40, 44, 48,E52, 53, 55, 62, 86, 165, 172

GROSS,P.M.
VOL 4 E1-E7,E108, 109,E111, 112,E114,E115, 116, 120-122, E123, 124,E125, 126,,E128, 129,E133, 134,E145, 147, E149, 151,E158, 160,E207, 209
VOL 7 E1-E5
VOL 10 E1-E4, 19
GROVE,N.H.
VOL 1 356
VOL 5/6 526
VOL 10 502
GRUTTNER,B.
VOL 3 E86
GUARINI,J.R.
VOL 16/17 E175, 178-198
GUBANOVA,T.I.
VOL 12 E354-E366, 380,E395-E397, 409, 424-426
GUBBINS,K.E.
VOL 1 E1-E4,E20-E22, 28
VOL 4 E1-E7,E33-E36, 60
VOL 5/6 E1-E3, 11,E33,E35, 81, 82
VOL 7 E1-E5,E71,E72,E76, 168, 169
GUBELI,A.O.
VOL 3 E45-E56, 79, 80
GUBKINA,G.F.
VOL 1 E254-E256, 276, 277
GUDRAIS,P.
VOL 12 331
GUERECA,R.A.
VOL 1 E298, 299
GUERRY,D.JR.
VOL 5/6 E119, 126, 158, 149,E214,E215, 218, 219, 225, E240, 250, 265
VOL 7 E214,E215, 219, 241, 297, 300, 301
VOL 10 E119-E122, 128,E148, 154, 220, 229, 230, 241, 263
GUIGLION,C.
VOL 12 E1,E2
GUINASSO,N.L.
VOL 5/6 E17,E18
GUITIERREZ,L.
VOL 13 E38-E47,E354-E359, 363,E381-E385, 387,E409, 411
GULN,J.A.
VOL 5/6 523-525
GUPTA,M.K.
VOL 10 380-382,E391, 398, 399
GURCHIN,I.I.
VOL 12 E349
GURIKOV,YU.V.
VOL 2 235
GURWITSCH,J.G.
VOL 5/6 E589,E590, 592, 593
GUSEVA,M.S.
VOL 7 E71,E76, 162
GUTH,E.D.
VOL 13 E241-E246
GUZHVINA,O.V.
VOL 13 E37-E47,E354-E359
GYUNNER,E.A.
VOL 3 E102-E113, 169, 170

H

HABENSCHUSS,A.
VOL 13 E38-E47,E354-E359, 363,E381-E385, 387,E409, 411

HADWIGER,H.
VOL 12 143,E145-E147, 170, 171,E189-E191, 238, 257
HALA, E.
VOL 12 E174, 177,E189,E190, 209, 215, 216
HALFORD,J.O.
VOL 3 E45-E55, 70, 71,E102-E113
HALL,L.C.
VOL 7 E474-E480
HALL,N.F.
VOL 13 E109
HALLER,H.S.
VOL 5/6 274
VOL 7 367
VOL 10 277, 534
VOL 12 E454, 455
HALLET,J.G.
VOL 12 E454,455
HALOW,I.
VOL 3 E39,E45-E56
HALSEY,G.D.JR
VOL 1 110, 249
VOL 4 240
VOL 5/6 E571, 573-575
HAMANO,S.
VOL 16/17 E359, 360,E394, 402, 403, 407, 408,E442,E443, 448, 449, 453, 454,E479, 480-483,
HAMBERG,A.
VOL 10 E31-E33
HAMMEL,A.VON.
VOL 10 102, 237
HAN,C.H.
VOL 10 E1-E4
HANDA,Y.P.
VOL 9 E1,E2, 3, 4
HANSEN,C.J.
VOL 12 53, 64, 76, 96, 105
HANSEN,C.M.
VOL 9 E195-E199
HANSON,D.N.
VOL 5/6 E120-E122, 131, 132, 134, 136,E159,E160, 164,E169, 171,E237,E238, 243, 246,E255, 258,E280-E282, 283-291
HANTKE G.
VOL 12 E395-E397, 402
HARDEWIG,A.
VOL 2 E114,E115, 121
HARDY Z.M.
VOL 13 E278, 279, 280, 282, 284,E335
HARGIS,B.
VOL 8 E1,E2,11,E226,E227, 234
HARKINS,W.D.
VOL 14 E225-E227, 231-234,E242-E246,E248-E251, 270, 271, 279
HARMAN,R.H.
VOL 3 E45-E56
HARMEL,M.H.
VOL 8 232, 242
HARPER,A.M.
VOL 2 E114,E115
HARRIS,H.G.
VOL 9 101, 102, 104
HARRIS,J.F.
VOL 12 E3-E5, 25, 26
HARTLEY, H.
VOL 3 E193, 217, 218,E223, 224, 225
HASBROUCK,J.D.
VOL 8 12, 235

HATFIELD,N.S.
VOL 1 262
VOL 2 143, 146, 150
VOL 5/6 E30,E35, 56, 57,E303,E304, 311, 313
VOL 7 E1-E5,E64,E76, 107, 108, 112-116
HATTOX,J.S.
VOL 8 233
HAUFFE,K.
VOL 3 E102-E113
HAUNSCHILD,H.
VOL 8 293, 297, 298
HAVIR,J.
VOL 18 E7-E10,20,E140, 141
HAWKINS,J.A.
VOL 1 E1-E4
VOL 10 E1-E4, 16, 464, 465
HAYDUK,W.
VOL 9 E77,E78, 83-86, 91, 96, 97, 109,E111,E112,E166,E167, E195-E199
HAYES,T.W.
VOL 7 E1-E5, 18, 19
HEALY,T.V.
VOL 13 E221, 225, 240
HECK,C.K.
VOL 1 E263, 273, 274,E298, 300, 301,E324, 325, 326
VOL 5/6 E332, 336, 337,E395, 398,E581, 584, 585,E610, 614, 615
HEDLEY-WHITE.J.
VOL 7 373, 375, 376
VOL 8 E1,E2,15
VOL 10 286
HEGARTY,M.J.
VOL 10 E391,398, 399
HEINS,C.
VOL 5/6 E303,E304, 305
VOL 10 14,E31-E33,E333, 335
HEISER,D.J.
VOL 13 E354-E359, 363,E381-E385, 388,E409, 411,E449, 450
HEISE,F.
VOL 1 E254-E256, 282
VOL 4 E314, 318
VOL 10 E476, 483
HELLWIG,K.
VOL 3 E45-E55, 60,E102-E113, 114, 115
HENLEY,J.P.
VOL 5/6 523-525
HENRICK,F.
VOL 9 E1,E2
HERD,J.A.
VOL 2 E114,E115
HERLIHY,J.C.
VOL 9 E111,E112, 137
HERMANS,L.J.F.
VOL 5/6 E581
HERRING,R.N.
VOL 1 E348
HERSHKOWITZ,M.
VOL 5/6 E178, 181, 314-320
HERTZBERG,G,
VOL 2 E134 -E136,E149, 152
VOL 4 E1-E7,E33-E36, 50
VOL 7 120, 130, 142, 173
VOL 9 E1,E2, 10,E27-E33, 46
HESS,L.G.
VOL 9 E195-E199, 214
VOL 10 454

HEVESY,G.
VOL 2 E238-E241,E339, 340-342
HICKEY,F.C.
VOL 14 E208-E212, 223
HIEMENZ,W.
VOL 7 439
VOL 10 324
HIESE,F.
VOL 5/6 E622, 626
HIKITA,H.
VOL 7 416
VOL 8 E1,E2, 17,E27-E39, 93
HILDEBRAND,J.H.
VOL 1 40,E59, 238
VOL 2 174
VOL 4 E123,E149, 150,E158,E162, 164,E207, 208, 210, 211, 224, 230
VOL 5/6 E241, 252, 253
VOL 7 E41-E43,E214,E215, 227
VOL 9 E77,E78, 106,E138-E140, 143, 146, 163,E138-E140,
VOL 10 E119-E121, 124,E148, 149, 151,E162, 165, 239, 244-246,E253, 255, 256
VOL 12 E354-E366, 367,E395-E397, 401, 412, 446
HILLEBRAND,W.A.
VOL 7 293-296
HILL A.E.
VOL 3 E45-E56,E102-E113, 126, 127,
VOL 12 251,E261,E262, 290, 291, 294, 296
VOL 14 E11-E18, 20, 31, 32, 37-39,E59-E69, 84, 164, E242,E243,E247,E251, 294, 299
HILL,E.P.
VOL 5/6 15,E33,E35, 73, 276
HILL,J.L.E.
VOL 1 E254-E256,E263, 267-270,E283, 292-294,E328, 329, 330, 357, 358,E359, 361
HILL,M.E.
VOL 12 448
HILL,S.A.
VOL 16/17 E394, 399, 406,E442,E443, 447, 452
HIMMELBLAU,D.M.
VOL 10 E1-E4
HINE,F.
VOL 12 E344, 347, 348
HIPKIN,H.G.
VOL 5/6 E332, 335, 534-536
HIRAOKA,H.
VOL 4 210
VOL 10 239
HIRAYAMA,C.
VOL 10 E1-E4,E54,E56, 77
HIZA,M.J.
VOL 1 E254-E256,E263, 273, 274, 302-305,E307, 309, E310, 320, 321, 323,E369, 372
VOL 4 E261, 266-268, 312, 313
VOL 5/6 E332, 336, 337,E395, 398
VOL 10 E361,E362, 370, 371, 374, 375, 389, 390
HOATHER,J.
VOL 7 E474-E480
HOFBAUER,G.
VOL 2 E227-E229,E260,E261, 272,E273, 277-283,E291, 292,E294, 295, 298, 299
HOFMANN,R.
VOL 2 E227-E229, 230,E265, 266,E322, 324

HOGAN,J.J.
VOL 5/6 E119,E298,E299,E303,E304,E339, 340,E346, 347, E354,E355, 358, 359, 365,E380, 382,E406, 426, E435,E441-E444, 447, 521, 522
VOL 10 12,E31-E33,E333,E361,E362, 365-367,E409, 410, 447, 458, 503, 504
HOGLUND,O.
VOL 13 E15-E19
HOLAS,J.
VOL 12 E354-E366,E395-E397, 406, 413
HOLLAND,C.J.
VOL 4 E1-E7, 14,E33-E36, 45, 49, 54, 57
HOLLECK,G.L.
VOL 12 331
HOLLENBERG,M.
VOL 2 E114,E115
HOLMBERG,O.
VOL 13 E148, 155
HOMMA,T.
VOL 10 E1-E4, 20, 284, 285, 299
HOM,J.F.
VOL 10 E425, 430
HOOVER,J.R.E.
VOL 16/17 E657,E664
HOPKINS,B.S.
VOL 13 E15-E19, 20,E109, 133, 135,E221, 223, 224, E278, 283, 285,E335, 337, 338
HORIUTI,J.
VOL 4 E5
VOL 5/6 E159,E190, 162,E213-E215, 217,E221,E222, 224, E230,E231, 232, 245, 251
VOL 7 E250,E251, 253, 291, 299, 302
VOL 8 E160-E162, 180, 195, 209, 218, 225
VOL 9 E138-E141, 152, 156, 158, 159,E195-E198, 204, 206, E232
VOL 10 E162, 164,E216, 218, 223, 228, 236,
VOL 12 E145-E147, 149,E189,E190, 208, 214,E298, 303, 308
HORN,A.B.
VOL 9 E16, 17,E23
HORVATH,G.L.
VOL 7 167
HOSKINS,J.C.
VOL 9 E27-E33, 55, 56,E61, 62, 63
HOU,J.P.
VOL 16/17 E394, 400, 401,E410, 412, 414, 416, 417,427, 432, 440, 441,E442,E443,E479,E487, 694,E698
HOUGEN,O.A.
VOL 5/6 E178
HOUGHTON,G.
VOL 7 E1-E5, 377,E378
HOUGHTON,L.
VOL 7 32
HOWARD,W.B.
VOL 9 E195-E199, 220
HOWELL,O.R.
VOL.8. E1,E2, 7, 147, 151, 153, 157, 159, 252, 253
HOWICK,L.C.
VOL 18 E7-E10, 13,E43, 45,E52.E53, 57,E72,E73, 74, 75, 83, 89, 90, 94, 96, 97,E106, 108, 115,E177, 178
HOYLE,B.E.
VOL 14 E58-E69, 154
HSU,C.C.
VOL 9 208
HSU,H.
VOL 8 E160-E162, 163, 185, 196, 208, 220

HSU,N.T.
VOL 5/6 E339, 341
HUDSON,J.C.
VOL 12 E3-E5, 12, 13,E37-E39, 91, 98, 99
HUFNER,G.
VOL 5/6 E1-E3, 7, 90, 95, 96, 103, 110
VOL 8 E265-E267, 281, 282, 328-330
VOL 10 E1-E4, 6,E83-E85, 93, 306-312
HUME,D.N.
VOL 3 E102-E113, 150, 151
HUNG,J.H.
VOL 9 E1,E2, 5,E27-E33, 36-40, 42,E64, 72
HUNZIKER,H.H.
VOL 18 E21,22,E40,E41
HURSH,J.B.
VOL 2 E260,E261, 290,E291, 293,E294, 296, 297, 300-311, 314-317,E328, 330, 335, 336
HUSAIN,M.M.
VOL 11 E9, 12,E14,E15, 18,E21, 24,E28,E29, 32,E35, 38, E39, 43,E61, 63,E64, 66,E67, 69,E72,E74, 76,E78, E80, 81,E82, 83,E84, 85,E86, 87
HUTTNER,F.
VOL 8 E265-E267,288-291, 310
HUTTNER,K.
VOL 3 34-37,E39, 40, 41,E90,E91, 92, 93
HU,J.-H.
VOL 5/6 630
HYDE,B.G.
VOL 13 E241-E246

I

IJAMS,C.C.
VOL 5/6 E120-E120, 130,E189, 206,E242, 254
VOL 7 E214,E215, 238
VOL 10 E119-E122, 131,E174-E178, 209, 243, 517, 520, 522
IKELS,K.G.
VOL 1 E124-E126, 250, 252
VOL 10 E1-E4,E279, 281
ILTIS,R
VOL 7 309, 315, 318-320, 322, 325, 326, 331, 332, 336, 338, 340-344, 356
VOL 10 523-530, 532
INGHAM,J.D.
VOL 7 310, 316, 323, 329, 330, 334, 335, 339
INGLIS,J.K.H.
VOL 7 E474-E480
INGVAR,D.H.
VOL 2 E114,E115
INOZEMTSEVA,I.I.
VOL 16/17 E127,E128, 130, 131,E135, 136, 138, 139
INUTA,S.
VOL 12 E344,E347, 348
IONESCU,L.G.
VOL 2 E134-E136, 141,E199-E201
IONESCU,L.V.
VOL 10 296
IPATIEV,V.V.
VOL 1 E254-E256
VOL 5/6 E303,E304,E380, 383, 391,E406, 408, 411,E415, 416,E419, 420, 424, 435,E622
VOL 12 E145-E147, 150
ISAKOVA,S.
VOL 13 E434-E436, 446, 447

ISBISTER,W.H.
VOL 2 E149,E199-E201, 213, 226
ISHIKAMA.H.
VOL 8 E1,E2, 17,E27-E39, 93
ISHMURATOV,G.YU.
VOL 13 E301-E308, 327, 330
ISMAILOV,M.
VOL 13 E1, 4, 5,E156-E162, 216, 217, 219, 220,E434-E436, 444, 445, 447
ITAMI,J.
VOL 12 E181, 187
ITO,Y.
VOL 8 E1,E2, 18, 19,E116, 117-119, 121, 128, 131, 132, 134,E160-E162, 186
IVANOVSKII,M.D.
VOL 7 490,E491
IVANOV-EMIN,B.N.
VOL 13 E409,E414-E417,E449,E454-E456
IVIN,K.J.
VOL 12 E1,E2
IZARD,E.F.
VOL 12 E344, 346, 350-353
IZATT,R.M.
VOL 3 E45-E56
IZMELEUOVA,M.B.
VOL 13 E38-E47, 95

J

JACOBI,J.
VOL 8 E350
JADOT,R.
VOL 9 E77,E78, 79, 89, 93,E138-E141, 153, 157,E232
JAENICKE,W.
VOL 3 E102-E113
JAGER,L.(JAEGER)
VOL 12 82, 83
JAHRSTORFER,M.
VOL 8 E1,E2, 9,E27-E38, 47-50, 52-63, 65, 68, 71, 73, 75, 83, 85, 87, 91, 92, 98, 101, 106, 108, 111, 137
JAKOWKIN,A.A.
VOL 12 E344,E395-E397, 399
JAMES,C.
VOL 13 E15-E19, 26, 27,E38-E47, 67, 68,E301-E308, 316
JANDER,G.
VOL 3 E86, 88, 234
JANSHOLT,A.L.
VOL 16/17 E10, 14, 15,E39, 41
JANTSCH,G.
VOL 13 E145,E148, 149, 151-154,E226,E227, 229, 233,E286-E288, 289, 292, 294, 296, 299,E339-E341, 342, 347, 349, 351, 352, 371-375, 405-408,E414-E417
JAY,B.E.
VOL 8 E1,E2, 11,E226,E227, 234
JAYLES,C.
VOL 12 E1,E2
JELFS,E.C.G.
VOL 10 542, 543
JENKINS,W.I.
VOL 4 E111,E123,E145,E158
JHAVERI,A.S.
VOL 7 E70,E76, 161
JOANNIS,A.
VOL 3 E102-E113

JOHNSON,C.A.
VOL 10 E425, 426
JOHNSON,D.L.
VOL 5/6 E178
JOHNSON,L.F.
VOL 4 E241
JOHNSTONE,H.F.
VOL 12 E3-E5, 19, 45
JOHNSTONE,N.B.B.
VOL 1 E1-E4, 9,E20-E22, 23, 25, 27, 30, 32,E124-E126,128, E141-E143, 173
VOL 2 E1-E3, 4,E12,E13, 14-16, 19, 21, 22, 24,E134-E136, 137,E149, 151
VOL 4 E1-E7, 9,E33-E36, 48
VOL 7 E63,E76
JOHNS I.B.
VOL 5/6 E415
JOLIN,S.
VOL 13 E156-E162
JOLLEY,J.E.
VOL 4 E123,E149,E162,E207,E223
JONES,A.L.
VOL 14 E242-E246,E253,E254, 259, 260
JONES,C.H.
VOL 1 E380, 382, 383
VOL 5/6 E581, 585, 586
JONES,K.H.
VOL 16/17 E394, 399, 406,E442,E443, 447, 452
JONES,L.H.
VOL 3 E45-E56
JONES,W.J.
VOL 12 E354-E366, 386, 388,E395-E397, 400
JOOSTEN,G.E.H.
VOL 8 E1,E2, 16,E27-E39, 112
JORDAN,C.F.
VOL 11 E259
JOSLYN,M.A.
VOL 7 E190-E192
JUNCHER,H.
VOL 16/17 E10, 13,E39, 42,E66, 67,E72, 77
JURIO,R.
VOL 3 E45-E56, 81
JUST,G.
VOL 5/6 E1-E3,E159,E160, 161,E169, 170, 177,E186,E187, E189, 191, 195, 204,E221,E222, 223,E227, 228, E230,E231, 233, 235, 236,E255, 256, 264, 266
VOL 10 E1-E4,E83-E85, 88,E162, 163, 169, 171,E174-E178, 179, 185, 202,E216, 217,221, 224-226, 232,E253, 254, 264, 266

K

KADLER,B.
VOL 12 E3-E5, 27, 28
KAHRE,L.C.
VOL 9 113
KAISER,V.
VOL 12 E1,E2
KALININA,S.UE.
VOL 5/6 E622, 629
KALO,A.M.
VOL 13 E38-E47, 49, 70-72, 75, 76,E156-E162, 164,E241-E246, 248, 255,E301-E308, 309, 317,E354-E359, 360,E376,E377
KALRA,H.
VOL 9 E111,E232, 244
VOL 10 416, 417,E419, 420, 421, 470

KALZ,G.
VOL 12 E354-E366, 414
KAMINISHI,G.
VOL 4 E253-E255,E314, 316, 320
VOL 5/6 E610, 613
VOL 7 469
VOL 10 E488, 491, 493
KAPLAN,L.K.
VOL 5/6 478, 480, 511, 512, 519
VOL 10 448, 452, 453, 462, 463
KARAPYSH,V.V.
VOL 12 E349
KARASZ,F.E.
VOL 1 110, 249
KARATAEVA,I.M.
VOL 14 E11-E18, 33-36
KARMAZINA,V.D.
VOL 12 84-87
KARTSEVA,V.D.
VOL 16/17 E740, 765, 766
KASARNOVSKII,J.S.
VOL 1 E1-E4
KASARNOWSKY,I.
VOL 4 E1-E7,E145,E149,E162,E169,E170,E180,E181,E185, E186,E253-E256, 257, 271, 277-304
KASHINA,N,I.
VOL 14 E59-E69, 163
KASHTANOV,S.A.
VOL 7 481
KASTNER,N.
VOL 12 449, 450
KATAMONOV,V.L.
VOL 13 E381-E385, 394, 396,E414-E417, 421-427
KATAYAMA,T.
VOL 5/6 E119, 128,E186-E190, 192, 197, 200, 202,E354,E435
VOL 9 E111,E112, 124,E232, 234, 238, 245-247
VOL 10 E104,E105, 107-112, 115-118,E119-E122, 125,E174-E178, 182, 189, 193, 196, 199
VOL 12 126,E181, 187,E189-E191, 201, 223, 229, 232
KATZ,D.L.
VOL 5/6 E321, 324,E332, 333, 334,E339, 342, 343,E346, 349, 350,E395, 396, 397,E400, 401, 402, 530-533, 540-542, 552, 553
VOL 10 376, 377
KAWAMURA,F.
VOL 7 E474-E480
KAWASHIRO,T.
VOL 1 121
VOL 5/6 277
VOL 8 247
KAY,W.B.
VOL 9 E232, 239-243
KEDRINSKII,I.A.
VOL 7 183, 184, 187
KEELER,R.A.
VOL 2 E1-E3, 7, 8,E12,E13, 17, 18
VOL 7 E65,E76, 117, 118
KEEVIL,T.A.
VOL 9 E195-E198, 202
KEFFER,R.M.
VOL 14 E59-E68,E70,E72,E75, 135, 137, 177, 178,E225-E227, 230, 235,E242,E243,E246-E248,E252, 266, 281, 283, 285
KELEMEN S.
VOL 10 E48-E53,E56,E333, 342, 345-348

KELLY,R.
VOL 7 380, 381
KENDREW,J.C.
VOL 2 E199-E201
KENTON,F.H.
VOL 5/6 E33,E35, 72
VOL 7 E70,E75, 149
VOL 8 E27-E39, 88, 89
VOL 9 E27-E33, 53
VOL 10 E53,E56, 74, 76, 515, 516
VOL 12 92
KEPINSKI,J.
VOL 12 E454, 456
KERKELA R.
VOL 3 30,229
VOL 18 5, 39, 80, 102
KERTESZ,D.J.
VOL 3 233
KESTER,D.
VOL 4 E27-E29
VOL 5/6 E17,E18
VOL 10 E31-E33
KETY,S.S.
VOL 8 232, 242
KE,P.J.
VOL 7 361-363, 371, 443
KEYS,M.
VOL 2 E199-E201
KHAIME DEL´PINO.
VOL 13 E409,E414-E417
KHALFINA,L.R.
VOL 13 E434-E436, 441, 442
KHARAKHORIN,F.F.
VOL 1 E328, 333, 334
KHARCHENKO,A.A.
VOL 5/6 E434,E441-E444, 452, 453, 456, 457, 569, 570
KHAZANOVA,N.E.
VOL 1 E254-E256
VOL 5/6 E622
VOL 10 E488, 489
KHISAEVA,D.A.
VOL 13 E156-E162
KHODEEVA,S.M.
VOL 1 E254-E256
VOL 10 353, 354, 357-360,E419, 439, 449, 450
KHOMUTOV,N.E.
VOL 7 E59,E66-E76, 131-139, 164
KHRIPIN,L.A.
VOL 13 E38-E47, 54, 55
KHUCHUA,K.S.
VOL 10 414, 415
KHUDAIBERGENOVA,N.
VOL 13 E15-E19, 35, 36,E241-E246, 276,E301-E308, 334,E376, E377,E380,E381-E385, 403,E414-E417, 428,E429, 433, E449, 453,E454-E456, 460,E461, 464
KIDNAY,A.J.
VOL 1 E254-E256, 323,E369, 372
VOL 4 E261, 268, 312, 313
VOL 5/6 E332, 336, 337,E395, 398
VOL 10 E361,E362, 370, 374, 375, 389, 390,E391, 398, 399, E488, 494, 500, 501
KIDO,K.A.
VOL 12 126,E189-E191, 201, 223, 229, 232
KIDYAROV,B.I.
VOL 14 E11-E18, 29, 30, 40,E59-E69, 162
KILDE,G.
VOL 14 E59-E70,E72,E73,E75,E76,E78-E81, 95, 98, 99, 102, 112, 140, 149, 155, 159, 160

KILGORE,C.H.
VOL 10 E425, 427
KING,A.D.
VOL 4 E1-E7, 259, 260
VOL 7 212
VOL 9 E27-E33, 54-56,E61, 62, 63
KING,E.J.
VOL 3 E90,E91
KINSMAN,D.J.J.
VOL 7 E41-E43, 52,E53
KIRGINTSEV,A.N.
VOL 13 E38-E47,E301-E308,E354-E359,E409
VOL 18 1, 2,E3, 4,E21, 23,E40,E41, 48, 63, 79
KIRK.B.S.
VOL 5/6 E321, 326
KIRK,W.P.
VOL 2 E12,E13,E114,E115, 125, 129, 130, 133
KIRMSE,E.M.
VOL 13 E38-E47, 56
KIRSCHNER,A.
VOL 3 E102-E113, 128, 129
KISE,H.
VOL 7 451
KITAEV,G.A.
VOL 3 E31, 33
KITANI,K.
VOL 2 E1-E3,E12,E13,E109, 113,E114,E115, 120, 124, 131, E134-E136, 140,E193,E194, 198,E199-E201, 206, 218, 223
KITO,S.
VOL 8 E1,E2, 18, 19,E27-E39, 51, 64, 72, 104, 110,E116, 117-119, 121, 128, 131, 132, 134, 138,E160-E162, 186
KITT,G.A.
VOL 2 E9
KLACH,S.J.
VOL 2 E1-E3, 7, 8,E12,E13, 17, 18
VOL 7 E65,E76, 117, 118
KLAUS,A.
VOL 2 E343, 344, 345
KLEBER,W.
VOL 7 E190-E192
KLEIN,F.S.
VOL 3 E102-E113, 167, 168
VOL 5/6 296
KLEMENC,A.
VOL 8 E336, 339, 347, 350
KLINEDINST,K.
VOL 12 330
KLINGHOFFER,ST.S.
VOL 14 E59-E68,E70-E72, 89, 91, 94, 133
KLINK,A.E.
VOL 5/6 E346, 352, 353
KLIPFEL,K.
VOL 13 E38-E47, 83,E241-E246, 256,E301-E308, 318,E354-E359, 368, 369
KLOTS,C.E.
VOL 4 E1-E7, 10
VOL 7 E1-E5, 22, 23
VOL 10 E1-E4, 21, 22
KLYUEVA,A.V.
VOL 7 E60,E76, 83
KNAFFLE-LENZ,E.
VOL 2 337
KNAPP,H.F.P.
VOL 5/6 E581

KNASTER,M.B.
VOL 5/6 E33,E35, 78
VOL 7 E71,E72,E76, 166
KNOPP,W.
VOL 5/6 E1-E3,E28,E33-E35, 46, 75, 84, 88, 92
VOL 8 E1,E2, 5,E27-E39, 90, 107,E116, 133, 144
KNORN,M.
VOL 1 E324, 327
KNOTT,T.H.
VOL 16/17 E548, 549
KO,E.C.F.
VOL 3 20-22,E23, 24, 26-28, 208, 211, 212,E213, 214, 221, 222,E223, 226
VOL 11 E52,E54, 55,E56, 57,E58, 59,E171,E186, 192,E194, 197, E198, 202,E212, 215,E216, 219,E220, 221
VOL 18 E140, 143,E149, 150, 153,E154, 155,E158, 159, 164, 165,E166, 167
KOBATAKE,Y.
VOL 1 90
VOL 4 230
VOL 7 E214,E215, 227
VOL 9 E77,E78, 106,E138-E140, 143
VOL 10 E253, 256, 267
KOBAYASHI.R.
VOL 10 E361,E362, 372, 373,E391, 395, 396
KOBE,K.A.
VOL 5/6 E33,E35, 72
VOL 7 E70,E75, 149
VOL 8 E1,E2,10,E27-E39, 67, 69, 84, 88, 89, 99, 109, E116,E226,E227
VOL 9 E27-E33, 53
VOL 10 E53,E56, 74, 76, 515, 516
VOL 12 92
KOBZOVA,S.N.
VOL 16/17 E740, 765, 766
KOFLER,M.
VOL 2 E227-E229, 233, 237,E238-E241, 242-258
KOGAN,L.N.
VOL 12 E354-E366,E395-E397, 405, 415
KOHLRAUSCH,F.
VOL 14 E6,E7
KOHLSCHUTTER,V.
VOL 8 R265-E267, 283-287, 314-316, 319-324, 332
KOHN,D.M.
VOL 10 E425, 427
KOHN,J.P.
VOL 9 E111,E112, 127, 128, 134-136,E232, 248
KOLESNIKOV,A.A.
VOL 13 E37-E47, 73, 74, 77, 78,E301-E308,E354-E359
KOLESNIKOV,I.M.
VOL 12 E354-E366, 391, 392, 395-397, 408, 410, 411, 435, 436, 439-441
KOLOSOV,L.V.
VOL 14 E191-E197, 203
KOLTHOFF,I.M.
VOL 3 E45-E56, 75, 76,E102-E113
VOL 14 E59-E69, 87
VOL 18 E32,E32, 33,E64,E140, 144, 145,E149, 152,E154, 157,E158, 161, 162,E166, 168, 169, 180, 181,E182, E198, 202,E203, 207
KOL´TSOV,N.S.
VOL 12 E354-E366,E395-E397, 405, 415

KOLTSOVA,E.V.
VOL 16/17 E657, 658,E664, 665
KOLYGINA,T.S.
VOL 16/17 E1, 2-6
KOMARENKO,V.G.
VOL 2 E71,E73, 74-76, 170
VOL 4 E180,E181, 184,E185,E186, 189
KOMISSAROVA,L.N.
VOL 13 E1, 2, 3, 6-14
KOMISSAROVA,V.D.
VOL 12 E1,E2,E116
KONDO,K.
VOL 2 175, 176
KONDRATENKO,V.T.
VOL 12 393, 394, E395-E397, 407
KONIG,J.
VOL 1 E1-E4.E124-E126,E138
VOL 2 E1-E3,E9,E134-E136, 147
VOL 4 E1-E7,E27-E29
VOL 8 E350, 352
KONISHI,E.
VOL 16/17 E733, 734-739
KONNIK,E.I.
VOL 7 E59,E66-E76, 131-139, 162, 164
KONOBEEV,B.I.
VOL 12 E354-E366, 380,E395-E397, 409, 424-426
KONTOS,H.A.
VOL 7 386
KOPOLD,H.H.
VOL 2 E114,E115
KOPOV,V.I.
VOL 4 E33-E36
KOPPANY,C.R.
VOL 5/6 E332, 335, 534-536
KORL,A.N.
VOL 12 389
KORONDAN,I.
VOL 3 E102-E113,E177,E178, 183, 184,E185, 187,E188, 192,E193, 198
KOROSY,F.
VOL 2 E12,E13, 61,E73, 81-85, 88, 106
VOL 4 212
VOL 10 233
KOROTKEVICH,I.B.
VOL 13 E38-E47, 73, 74, 77, 78,E301-E308,E354-E359
KOSHANOVA,T.
VOL 13 E434-E436, 446
KOSOROTOV,V.I.
VOL 12 E116,E117, 123, 125, 131, 137, 139,E298, 304, 316,E354-E366, 369-373, 378, 392,E395-E397, 408, 410, 411, 417, 418, 421, 423, 430-434, 437, 438, 442
KOUSAKA,K.
VOL 7 451
KOZAM,R.L.
VOL 8 E226,E227, 241, 245, 246
KOZITSKII,V.P.
VOL 18 1, 2,E3, 4,E7-E10, 15, 16,E21, 23,E40,E41, 42, 48, 63, 79
KOZYREVA,O.V.
VOL 10 E361
KRASE,N.W.
VOL 10 13,E333, 334, 337
KRATOHVIL,J.
VOL 3 E102-E113, 152, 153,E177,E178, 179, 180,E193, 194, 195, 199,E200, 201, 202, 205, 206

KRAUSE,D.
VOL 1 E1-E4, 13,E124-E126, 134
VOL 2 E1-E3, 5,E134-E136, 142
VOL 7 E1-E5, 36, 37, 39, 40,E41-E43
KRAUSS,W.
VOL 5/6 E148, 152, 156, 157,E159,E160, 167,E169, 175, 185,E214,E215, 220
KRAUS,C.A.
VOL 14 E246
KREITUS,I.
VOL 8 E39, 78, 95, 100, 115
KRENTSEL,L.B.
VOL 12 E354-E366, 422
KRESTOV,G.A.
VOL 1 E1-E4, 14,E124-E126, 130, 137, 179-182,E227, 229
VOL 2 E73,E260,E261
VOL 4 E1-E7, 16,E20,E21, 24, 26,E33-E36, 46, 53, 58, 63, 65, 66-76,E80-E84, 88, 89, 92, 93, 97-101, 103-105, E169,E170,E185,E186, 190, 204, 205
KRETSCHMER,C.B.
VOL 7 E190-E192, 213,E214,E215,E261-E267, 225, 268, 272, 278, 279, 292, 303
VOL 10 E83-E85, 89,E104,E105, 106, 113, 114,E119-E122, 135, E174-E178, 180, 187, 195, 197,E216, 219
KRICHEVSKII,I.R.
VOL 1 E1-E4
VOL 5/6 E380, 385, 386,E406, 410,E435, 437, 438,E622
VOL 10 444, 446,E476,E488, 489
KRIEVE,W.F.
VOL 7 465
VOL 12 447
KRISHNAN,T.
VOL 9 E232, 244
VOL 10 416, 470
KRIVONOS,F.F.
VOL 12 E354-E366, 376
KRUYER,S.
VOL 5/6 E148, 150, 154, 155,E159,E160, 165,E189, 211, 226
KUBELKA,V.
VOL 12 94
KUBIE,L.S.
VOL 7 436
VOL 10 321
KUDBYASHOV,S.F.
VOL 13 E156-162, 173-181, 187, 188
KUKARIN,V.A.
VOL 12 E116,E117, 123, 125, 131, 139, 173,E298, 304, 316, 317
KULIKOV,N.E.
VOL 10 E488
KUMAR,S.
VOL 12 E354-E366, 387
KUMUZAWA,H.
VOL 8 E1,E2, 20,E116, 139-143
KUNDU,K.K.
VOL 18 24, 30
KUNERTH,W.
VOL 8 E1,E2,8,E160-E162, 183, 184, 189, 190, 193, 194, 198, 199, 206, 207, 210-217, 221-224
KURAMSHIN,E.M.
VOL 12 E1,E2
KURATA,F.
VOL 1 E254-E256,E263, 264, 278, 279,E283, 287, 297,E328, 339, 340,E348, 351
VOL 5/6 E339, 344, 345, 528, 529
VOL 10 E402, 404, 405

KUSANO,K.
VOL 7 364
VOL 10 278
KUSTER,F.W.
VOL 3 E102-E113, 121
KUTSCHEROFF,M.
VOL 8 E265-E267, 283-287, 314-316, 319-324, 332
KUVANOVA,N.I.
VOL 4 E33-E36
KUWAYAMA,N.
VOL 16/17 E733, 734-739
KUZNETSOVA,G.P.
VOL 13 E15-E19, 22, 31,E301-E308, 319
KUZNETSOVA,L.S.
VOL 13 E241-E246, 257-275, 277
KUZ´MIN,A.A.
VOL 7 E71,E76, 162
KYSER,T.K.
VOL 2 E1-E3, 6,E9, 10

L

LABROCCA,P.J.
VOL 18 27-29
LACEY,W.N.
VOL 9 E111, 119, 120,E232, 235-237
LACHOWICZ,S.K.
VOL 5/6 E299,E354,E355
LACOMBE,H.
VOL 7 E41-E43
LADDHA,S.S.
VOL 8 120, 122, 123, 126, 127
LADEFOGED,J.
VOL 2 E134-E136, 139,E149,E193,E196, 197,E199-E201, 205, 212, 215, 217
LALIBERTE,L.
VOL 4 E33-E36, 39, 40
LAL,S.C.
VOL 3 E102-E113, 171
LAMANNA,U.
VOL 3 232
LAMBA,M.S.
VOL 11 E1, 2,E4, 5,E9, 10,E14,E15, 16,E21, 22,E26, 27,E28,E29, 30,E35, 36,E39, 41,E48, 49,E61, 62,E64, 65,E67, 68, E70, 71,E88, 89,E93, 94,E96, 97,E101, 102,E106, 107, E111,112,E113, 114,E118, 119,E123, 124,E128, 129,E130, 131,E134, 135,E138, 139, 141,E151, 153,E156, 157,E168, 169, 173,E181, 183,E186, 188,E194, 195,E198, 200,E208, 210,E222, 224,E226, 228,E229, 230,E231, 233
LANDAU,S.M.
VOL 8 E226,E227, 241, 245, 246
LANGE,A.
VOL 16/17 E548, 549
LANGE,P.
VOL 1 122, 123
LANGE,T.
VOL 13 E156-E162
LANNUNG,A.
VOL 1 E1-E4, 6,E59, 60,E68,E69, 70, 81,E82, 83, 88, 89, E124-E126, 127,E204, 205,E214, 215, 226,E227, 228, 233, 234
VOL 4 E1-E7, 8,E80-E84, 86,E145, 146,E158, 159,E171, 172, E176, 177, 201, 203

LARSON,A.T.
VOL 5/6 E622, 634
LASSEN,N.A.
VOL 2 E114,E115
LAUDIE,H.
VOL 9 E166,E167
LAUGIER,S.
VOL 5/6 E356, 378, 379, 418
VOL 10 E425, 430
LAVRENT´EVA,V.G.
VOL 13 E409
LAVROVA,E.M.
VOL 12 E3-E5, 31, 40, 41
LAWRENCE,J.H.
VOL 2 E109, 110,E193,E194, 195,E328
VOL 4 E241, 243
LAWRENCE,K.G.
VOL 11 E88,E93,E96,E106,E111,E113,E118,E123,E128,E130, E134,E138,E140,E142,E144,E146
LAWSON,C.C.
VOL 5/6 426
LAWSON,D.D.
VOL 7 310, 316, 323, 329, 330, 334, 335, 339
LAWSON,J.D.
VOL 9 E74, 75, 76
LEBEDEVA.E.S.
VOL 10 446
LECERF,J.
VOL 7 E190-E192, 196, 197
LEDEN,I.
VOL 3 E1-E8, 15, 16,E102-E113, 154-163
LEE,K.H.
VOL 9 E111,E112, 134, 135
LEE,T.P.
VOL 1 E20-E22
LEGIN,E.K.
VOL 2 235
LEGRET,D.
VOL 9 E111,E112, 133
LEGTYAREVA,L.V.
VOL 5/6 569, 570
LEIBOWITZ,R.
VOL 5/6 314-320
LEHIGH,W.R.
VOL 10 409
LEITES,I.L.
VOL 10 234
LEMIRE,R.J.
VOL 11 E259
LENOIR,J.-Y.
VOL 9 E77,E78, 98,E138-E141, 151,E166,E167, 191-194, E195-E199, 211, 212, 217, 218, 222, 224, 230
VOL 12 E116,E117, 138, 144,E145-E147,E189, 190, 205, 224, 235, 244,E261,E262, 284,E318, 321, 325
LENTZ,H.
VOL 2 144, 145
VOL 4 E256
LEPESHKOV,I.N.
VOL 14 E11-E18, 25, 26
LEPPLA,P.W.
VOL 12 E3-E5, 19, 45
LESNEVSKAYA,L.S.
VOL 10 E488, 489
LESTEVA,T.M.
VOL 11 E309,E310, 311

LEUHDDEMANN,R.
VOL 9 E27-E33, 60
LEVCHENKO,V.M.
VOL 12 E349
LEVINA,M.I.
VOL 5/6 E380, 383, 391,E406, 408,E415, 416,E419, 420, 424
VOL 7 155,E156
VOL 10 E49,E50,E56, 65, 66
LEVINA,N.D.
VOL 18 E7-E10
LEVITOV,M.M.
VOL 16/17 E657, 658,E664, 665
LEVITSKAYA,E.P.
VOL 5/6 E332
LEVIUS,L.A.
VOL 10 E361
LEVY,J.B.
VOL 7 351
VOL 10 262, 531
LEWIS,J.
VOL 8 E265-E267, 312, 313, 326, 327
LEY,H.
VOL 3 E102-E113
LE DEMEZET,M.
VOL 3 E23, 25, 84,E213, 216
LE,V.T.
VOL 12 E3-E5, 33, 110
L´HER,M.
VOL 3 E23, 25, 84,E213, 216, 228
LHOTAK,V.
VOL 9 E111, 116
LI,K.Y.
VOL 7 E474-E480
LIBINSON,G.S.
VOL 16/17 E1, 2-6
LIKHTER,A.I.
VOL 5/6 E395
LINCKH,E.
VOL 8 294-296
LINDNER,J.
VOL 12 E3-E5, 10,E298, 299
LINDROOS,A.E.
VOL 10 E476, 479, 480
LINEK,V.
VOL 7 E70,E76, 150
LINFORD,R.G.
VOL 1 238
VOL 2 174
VOL 4 E158, 211
VOL 5/6 E241, 252, 253
VOL 9 E138-E140, 146, 147, 155, 163-165,E232
LINGANE,J.J.
VOL 3 E102-E113
LINKE.W.F.
VOL 2 E328
VOL 3 E45-E56,E86
VOL 12 E37-E39
VOL 14 E6,E7, 8-10,E179-E181, 183-185,E186-E188, 189, 190,E191-E197, 198
LINNIK,N.V.
VOL 13 E449, 451, 452
LIN,H.M.
VOL 5/6 E298,E355, 369, 370, 373, 374, 394,E415, 417, E419, 422, 423, 426, 428-433, 475, 476, 510, 517, 518, 556-561

LISS,P.S.
VOL 7 E1-E5
LITMANOVICH,A.D.
VOL 12 E354-E366, 422
LITVINOV,N.D.
VOL 12 E354-E366,E395-E397, 405, 415
LIU,K.D.
VOL 5/6 431
LIU,M.S.
VOL 7 418
LIVINGSTON,J.
VOL 7 E1-E5, 13-15
LLOYD,S.J.
VOL 12 E145-E147, 148, 154,E189-E191, 220,E261, 262, 282, 289
LOBONOVA,N.N.
VOL 1 E283, 290, 291,E348, 349, 350,E359, 365, 366,E373, 377,E380, 384, 385
LOBRY DE BRUYN,C.A.
VOL 12 E174, 175,E181, 183
LODZINSKA,A.
VOL 3 E39, 43, 44
LOGVINYUK,V.P.
VOL 1 114
VOL 7 442
VOL 10 328, 544
LOHSE,M.
VOL 12 379, 381-385, 427-429
LOKTEV,S.M.
VOL 5/6 E442, 471-474
LONG,F.A.
VOL 7 E57,E76
VOL 9 E27-E33
VOL 10 E46
LONGI,A.
VOL 3 E45-E55, 59
LONGO,L.D.
VOL 5/6 15,E33,E35, 73, 276
LOOMIS,W.F.
VOL 2 E109, 110,E193,E194, 195,E328
VOL 4 E241, 243
LOPATTO,E.K.
VOL 12 44
LORIMER,J.W.
VOL 12 E145-E147, 156, 157, 159-162,E189,E190, 211, 212,E298, 301, 302
LOSEVA,G.K.
VOL 12 E349
LOVCHIKOV,V.S.
VOL 7 490,E491
LOWENSTEIN,E.
VOL 13 E156-E162
LUCAS,M.
VOL 1 E1-E4,E20-E22, 26,E141-E143
LUCAS,R.
VOL 3 E45-E55, 63-67,E102-E113, 122, 123
LUEHRS,D.C.
VOL 3 E213, 215
LUKASHINA,V.F.
VOL 16/17 E127,E128, 133, 134
LUKAS,D.S.
VOL 8 E226,E227, 241, 245, 246
LUKSHA,E.
VOL 11 E148, 149 , 150,E159, 160, 161,E162, 164, 165, E186, 189, 190,E212, 213,E216, 217

LUMRY,R.
VOL 2 E199-E201
LUNGE,G.
VOL 8 350
LURIE,A.
VOL 2 E260,E261,E265, 267, 271, 284, 287, 313,E322, 323, 327,E328, 329, 332-334
LUTHER,H.
VOL 7 439, 440
VOL 10 323, 324
LUTHER-LEIPZIG,R.
VOL 7 E474-E480
LUTSYK,A.I.
VOL 9 E1,E2, 14,E27-E33, 34
LUZNY,Z.
VOL.7. E1-E5, 188, 189,E190-E192
LU,B.C.Y.
VOL 4 E253-E255,E261, 264, 265, 269
VOL 10 E361,E362, 369, 378, 379, 382, 383,E391, 393, E402, 407
LYASHCHENKO,A.K.
VOL 1 E124-E126, 132,E141-E143, 144-160, 162, 166-168, 172, 174

M

MA,Y.H.
VOL 5/6 E178, 180, 182, 183
VOL 9 E232, 248
MAASS,C.E.
VOL 12 E3-E5, 14, 15
MAASS,O.
VOL 12 E3-E5, 14-17
MACARTHUR,C.G.
VOL 7 E58,E65-E75, 119, 125, 127, 129, 141, 147, 148, 170, 176, 177, 179-182
MACDOUGALL,G.
VOL 14 E242-E250,E253,E254, 264, 267, 268, 272, 277
MACFARLANE,A.
VOL 3 E193, 217, 218
MACHIGIN,A.A.
VOL 7 88,E89,E90
MACKENDRICK,R.F.
VOL 1 E298, 300, 301
MACWOOD,G.R.
VOL 5/6 630
MAC,Y.C.
VOL 3 20-22,E23, 24, 26-28, 208, 211, 212,E213, 214, 221, 222,E223, 226
VOL 11 E52,E54, 55,E56, 57,E58, 59,E171,E186, 192,E194, 197,E198, 202,E212, 215,E216, 219,E220, 221
VOL 18 E140, 143,E149, 150, 153,E154, 155,E158, 159, 164, 165,E166, 167
MACHIGIN,A.A.
VOL 7 E61,E62,E76
MADEC,C.
VOL 3 E1-E8, 18,E23, 25, 84,E102-E113,E185,E213, 216
MADIGAN,G.A.
VOL 14 E242-E246,E253,E254, 259, 260
MAEDA,S.
VOL 1 320
MAESTAS,S.
VOL 2 E199-E201, 220
MAGNO,F.
VOL 3 E102-E113

MAHARAJH,D.M.
VOL 7 E1-E5
MAHIEUX,F.
VOL 7 485, 489
MAILFERT,M.
VOL 7 E474-E480
MAIZLISH,V.E.
VOL 4 E33-E36
MAIMONI,A.
VOL 5/6 E589,E590, 596, 597
MAKARENKOV,V.V.
VOL 1 114
VOL 7 442
MAKAREWIC,Z.Z.
VOL 3 E102-E113, 172-174
MAKRANCZY,J.
VOL 1 40,E43,E51, 54, 56, 58, 185,E288,E196, 199, 201,203
VOL 2 25,E26, 29,E35, 37, 41, 43, 46, 47, 49, 155, 157, 159, 161
VOL 4 107,E108, 110,E111, 113,E114,E115, 118,E125, 127, E128, 130, 132,E133, 135, 136,E137, 139-141
VOL 5/6 E119-E124, 125, 129, 133, 135, 138-144, 146,E187-E190, 193, 199, 201, 203, 205, 207-209,E441-E444
VOL 7 E214,E215, 216, 218, 221, 223, 229, 230, 232-237, E261-E267, 269, 276-280, 282, 284, 287, 288
VOL 8 E160-E162, 164-166, 168, 169, 171-179, 182, 187, 191, 192, 197, 200, 201, 203-205
VOL 10 E119-E122, 123, 126, 127, 130, 133, 138, 139, 141-146, E148, 153,E162, 168,E174-E178, 183, 190, 194, 200, 205, 210, 213, 215
VOL 12 E116,E117, 118, 120-122, 124, 127, 128, 130, 132-135, 137, 141, 145-147, 152,E189,E190
MALIK,S.K.
VOL 9 E1,E2, 13,E27-E33, 41,E64, 73
MALIK,V.K.
VOL 9 E77,E78, 84, 86
MAL´KEVICH,N.V.
VOL 13 E1
MALLET,B.L.
VOL 2 E199-E201, 214, 225
MALUIGIN,P.V.
VOL 12 172,E181, 184,E189-E191, 231
MALYSHEV,V.V.
VOL 1 114
VOL 7 442
VOL 10 328, 544
MAMON,L.I.
VOL 8 E265-E267, 299-301
MANCHOT,W.
VOL 8 E1,E2, 9,E27-E38, 47-50, 52-63, 65, 68, 71, 73, 75, 83, 85, 87, 91, 92, 98, 101, 106, 108, 111, 137,E265-E267, 276-280, 288-298, 309-311, 317, 318, 325,E350, 352
MANK,V.V.
VOL 1 E298,E307, 308
MANOWITZ,B.
VOL 1 115
VOL 2 E26, 28, 45, 69, 70,E71, 87, 102,E134-E136, 168, 169, 171, 173, 186, 188, 191, 192
VOL 4 142
VOL 7 449
VOL 10 235, 316, 330
MANUELE, R.J.
VOL 3 E45-E56, 81

MANZHELII,V.G.
VOL 2 E71,E73, 74-76, 170
VOL 4 E180,E181, 184,E185, 186, 189
MAPLESON,W.W.
VOL 8 240
MARGARIA,R.
VOL 10 E1-E4, 15,E50,E51,E53,E56, 71, 290, 291, 295
MARGREITER,H.
VOL 16/17 E68, 69-71,E72, 74-76,E167, 168-170,E171, 172-174,E301, 302-304,E330, 331-33
MARINCIC,N.
VOL 12 330
MARKELEVA,L.S.
VOL 16/17 E127,E128, 133, 134
MARKHAM.A.E.
VOL 8 E1,E2, 10,E27-E39, 67, 69, 84, 88, 99, 109,E116, E226,E227
MARRINER,D.E.
VOL 12 58,59
MARSHALL,A.C.
VOL 16/17 E394, 396,E442,E443, 444
MARSH,J.K.
VOL 13 E15-E19,E454-E456,E461
VOL 16/17 E249, 250-275,E367, 368-393,E394, 397, 404, 405, E410, 411, 413, 415, 418-426,
MARSH,J.R.
VOL 16/17 428-431, 433-439,E442,E443, 445, 450, 451,E457, 458-478, 491, 492-, ??517,E518, 519-544,E548, 550-575,E576, 577-602,E603, 604-629,E630, 631-656,E667, 668-693, E706, 707-732,E740, 741-764
MARTYNOV,YU.M.
VOL 12 E354-E366, 369-373
MAR´YANOVSKAYA,K.SH.
VOL 12 389
MASAKI,K.
VOL 3 E45-E56,E102-E113, 131
MASLENNIKOVA,V.YA.
VOL 1 E254-E256, 280, 281
VOL 10 E333, 455, 457
MASON,D.M.
VOL 7 E62,E76, 85
VOL 12 447
MASSON,M.B.R.
VOL 2 E109,112,E114,E115, 117
MASTERTON,W.L.
VOL 1 E20-E22
VOL 4 E1-E7, 19,E33-E36, 41,E108,E114,E115,E171
MASUOKA,H.
VOL 5/6 E321, 327,E435, 440, 567-568,E589,E590, 598,E601, 606,E610,E611, 616, 620
VOL 10 E488, 492
MATANGE,J.N.
VOL 10 495
MATHESON,I.B.C.
VOL 4 E1-E7, 259, 260
VOL 7 212
VOL 9 E27-E33, 54,E61
MATHURIN,D.
VOL 14 E59-E68,E77-E81, 85, 86,E191-E197, 200, 201,E242, E243,E253,E254, 258
MATOUS,J.
VOL 4 E314, 317
VOL 10 E476,482
MATROZOV,V.E.
VOL 7 481

MATSUDA,M.
VOL 16/17 E545, 546, 547
MATTES,K.
VOL 7 452,E453,E454
MATVEEV,I.G.
VOL 12 E145-E147, 164-168
MATYASH,I.V.
VOL 1 E307, 308
MATYASH,YU.I.
VOL 10 460, 461
MAUDE,A.H.
VOL 10 320
MAXTED,E.B.
VOL 5/6 E159,E160, 163,E187, 196,E227, 229,E230,E231, 234,E238,E239, 247
MAYFORTH,F.R.
VOL 9 E195-E199, 220
MAYRHOFEROVA,J.
VOL 7 E70,E76, 150
MAZEPA,A.V.
VOL 3 E45-E56
MAZZETTI,C.
VOL 14 E44-E47, 51-54, 57, 58
MCAULIFFE,C.A.
VOL 9 E1,E2
MCCAFFREY,D.S.
VOL 9 E111,E112, 127, 128
MCCLURE,J.E.
VOL 18 E7-E10,19,E43, 47,E52,E53, 60,E72,E73, 78,E140, 146,E177, 179
MCCRAE,J.
VOL 12 106,107
MCCROSKY,C.R.
VOL 3 E90,E91, 94, 95
MCDANIEL,A.S.
VOL 9 E77,E78,81, 88,E166,E167, 171, 175,E195-E198, 200, 201
MCDEVIT, W.F.
VOL 7 E57,E76
VOL 9 E27-E33
VOL 10 E46
MCDOWELL,L.S.
VOL 13 E38-E47,E354-E359, 363,E381-E385, 387,E409, 411
MCELHILL,E.A.
VOL 5/6 E415
MCHALE,J.L.
VOL 1 109, 248
VOL 2 107
VOL 4 239
VOL 10 275
MCKAY,H.A.C.
VOL 13 E221, 225, 240
MCKAY,R.A.
VOL 9 E232, 235
MCKEE,O.L.JR.
VOL 10 E333, 340, 511
MCKERROW,A.K.
VOL 3 E102-E113, 143, 144
MCKETTA,J.J.JR.
VOL 8 E160-E162, 167, 170, 181, 219, 259
MCKETTA,J.J.
VOL 9 E16, 17-20,E23, 24-26
VOL 10 166, 385, 386,E409, 412,E419, 422, 423, 435, 436
MCTAGGART,H.A.
VOL 10 E361

MEADOWS,R.W.
VOL 5/6 E18
MEADOWS-TAYLOR,
VOL 7 E474-E480
MEEKS,J.C.JR.
VOL 1 E328, 343, 344
MEGYERY-BALOG,K.
VOL 1 40,E43,E51, 54, 56, 58, 185,E188,E196, 199, 201, 203
VOL 2 25,E26, 29,E35, 41, 43, 46, 47, 49, 155, 157, 159, 161
VOL 4 107,E108, 110,E111, 113,E114,E115, 118,E125, 127, E128, 130, 132,E133, 135, 136,E137, 139-141
VOL 5/6 E119-E124, 125, 129, 133, 135, 138-1448146,E187-E190, 193, 199, 201, 203, 205, 207-209,E441-E444
VOL 7 E214,E215, 216, 218, 221, 223, 229, 230, 232-237
VOL 8 E160-E162, 164, 165, 168, 169, 171-178, 187, 191, 192, 197, 200, 201, 203-205
VOL 10 E119-E122,123, 126, 130, 133, 138, 139, 141-146, E174-E178, 183, 190, 194, 200, 205, 210, 213,
VOL 12 E116,E117, 118, 120, 122, 124, 127, 128, 130, 132-135, 137,E189,E190
MEHRA,V.S.
VOL 9 E111,E112, 125, 126
MEJANE,J.V.
VOL 7 E190-E192, 196, 197
MELAMUD,S.F.
VOL 11 E245, 246,E251,E252, 254
MELLOR,J.W.
VOL 13 E409
MELLOR,W.
VOL 12 E333,E334, 339,E345
MENSHUTKIN,B.N.
VOL 11 E245,E251,E252
MERISLIN,R.V.
VOL 13 E38-E47, 69
MERKULOVA,V.P.
VOL 7 490,E491
MESERVEY,A.B.
VOL 2 98, 99
MESKEL-LESAVRE,M
VOL 9 E111,E112, 132
MESMER,R.E.
VOL 13 E1
METSCHL,J.
VOL 7 E261-E267, 271, 290
VOL 10 E174-E178, 186, 265
MEUSSER,A.
VOL 14 E1-E3, 5,
MEYER,K.H.
VOL 8 256
MEZENTSEV,A.S.
VOL 16/17 E664, 665
MICHELS,A.
VOL 1 E1-E4
VOL 2 105
VOL 4 E314, 315
VOL 5/6 E345, 439
MIHARA,S.
VOL 5/6 545-551
MIKSOVSKY,I.
VOL 9 E111, 114, 115
MILAD,N.E.
VOL 14 E243, 301
MILANOVA,E.
VOL 12 142,E145-E147, 152,E189-E191, 225, 227, 230,E261, E262, 265, 268, 275, 280, 286, 307,E318, 322, 323

MILBAUER,J.
VOL 12 46
MILES,F.D.
VOL 12 43, 48-50
MILLER,A.A.
VOL 5/6 520
VOL 7 441
MILLER,F.J.L.
VOL 5/6 E610,E611
MILLER,H.C.
VOL 10 259
MILLER K.W.
VOL 4 E1-E7, 15
MILLER.P.
VOL 10 443
MILLER,R.C.
VOL 1 E254-E256, 323,E369, 372
VOL 4 E261, 268, 312, 313,
VOL 10 E361,E362, 370, 374, 375
MILLERO,F.J.
VOL 4 E27-E29
VOL 7 E42-E43
VOL 10 E31-E33
MILLIGAN,L.H.
VOL 5/6 E1-E3
MILLS,J.F.
VOL 5/6 E610,E611
MININKOV,N.E.
VOL 13 E156-E162, 201, 202, 205-215, 218
MINNICH,B.H.
VOL 1 E257, 260
VOL 5/6 E303,E304, 310
VOL 7 458
VOL 10 E333, 339
MIRANDA,R.D.
VOL 9 E232, 244
VOL 10 416, 417
MIRANOV,N,N
VOL 13 E1
MIRONOV,K.E.
VOL 13 E38-E47, 50, 51, 54, 55,E156-E162, 165, 166, 170,E241-E246, 249, 252, 253,E301-E308, 311, 314,E354-E359, 361, 365-367,E376-E377, 378, 379
MIRONOV,V.E.
VOL 14 E60, 93,E191-E197, 199,E243, 261
MISHNINA,T.A.
VOL 1 E20-E22, 31
VOL 4 E33-E36, 51
VOL 7 E68,E69,E76, 143
VOL 9 E27-E33, 48
VOL 10 E50,E51,E56, 70
MISRA,S.N.
VOL 3 E90,E91, 96, 97
MITINA,N.K.
VOL 13 E156-E162, 183-186, 195, 203, 204,E226,E227
MIYAMOTO,E.
VOL 16/17 E545, 546, 547
MIYAMOTO,H.
VOL 14 E59-E68,E74, 173, 176,E191-E197,206,207, E242-E246,E251,E252, 306, 310
MOACANIN,J.
VOL 7 310, 316, 323, 329, 330, 334, 335, 339
MOCHALOV,K.I.
VOL 13 E156-E162, 193, 194, 199, 200

MODELL,J.H.
VOL 7 324, 333, 337
VOL 10 248-250
MODITIS,I.Z.
VOL 7 426, 427,E428, 429
MOELLER,T.
VOL 13 E190, 115, 143, 144
MOHANTY,J.
VOL 3 E102-E113, 175, 176
MOHS,M.A.
VOL 13 E38-E47, 52,E241-E246, 251,E301-E308, 312,E354-E359, 362,E381-E385, 387,E409, 410,E414-E417, 419,E429, 430,E434-E436, 438,E449, 450,E454-E456, 458,E461, 462
MOLES,E.
VOL 5/6 E1-E3, 105
VOL 10 E1-E4,E83-E85, 100, 103, 222
MOLL,F.
VOL 16/17 E10, 12,E359, 362,E442,E443, 455,E487, 489,E767, 768
MOLLERUP,J.
VOL 4 E253-E255,E261, 263, 319
VOL 7 472
VOL 10 388,E391, 397, 401,E402, 408
MOLODKIN,A.K.
VOL 13 E38-E47, 60, 61, 65,E148,E461
MONATH,E.
VOL 8 E336
MONFORT,J.P.
VOL 9 E77,E78, 94, 95
MONTGOMERY,M.A.C.
VOL 7 E41-E43
MONK,C.B.
VOL 14 E59-E68,E74,E75,E77-E81, 141-143, 158, 165-169, 171, 172, 174,E191-E197, 202,E242-E246,E251-E254, 282, 284, 286-288, 298, 300, 302-305, 308, 309
MONROE,G.S.
VOL 12 E145-E147, 150
MOON,C.H.
VOL 5/6 E159,E160, 163,E187, 196,E227, 229,E230,E231, 234, 247
MOORE,R.G.
VOL 5/6 E481-E483, 484-487, 489, 491, 492, 495, 496,E622, 627, 628
MORAN,G.
VOL 4 E80-E84, 102, 202
MORDUKHOVICH,I.L.
VOL 7 E71,E76, 162
MORET,R.
VOL 13 E15-E19, 21,E381-E385, 386,E414-E417, 418,E434-E436, 437,E454-E456, 457
MORGAN,J.L.R.
VOL 3 E45-E56
VOL 7 252
MORGAN,R.
VOL 7 E1-E5, 13-15
MORGAN, O.M.
VOL 12 E3-E5, 16, 17
MORITA,S.
VOL 5/6 E178, 181
MORKEN,D.A.
VOL 2 E12,E13,E114,E115, 125, 129, 130, 133
MORRIS C.J.
VOL 7 E1-E5, 20,E21
MORRISON,T.J.
VOL 1 E1-E4, 9,E20,E22, 23, 25, 27, 30, 32,E124-E126, 128,E141-E143, 173
VOL 2 E1-E3, 4,E12,E13, 14-16, 19, 21, 22, 24,E134-E136, 137,E149, 151

MORRISON,T.J.
VOL 4 E1-E7, 9,E33-E36, 48,E80-E84, 90, 91, 94, 95,E176, 179
VOL 5/6 E1-E3, 8,E30,E31,E32,E35, 55, 62, 68, 86
VOL 7 E1-E5, 16,E63,E76
VOL 9 E1,E2, 11,E27-E33, 44, 47, 58
VOL 10 E1-E4, 17,E49-E52,E55,E56, 64, 69, 80
MORSI,S.E.
VOL 14 E243, 301
MORTIMER,G.A.
VOL 5/6 E119, 127, 174,E239, 248,E354
VOL 9 E77,E78, 80,E138-E141, 154
MOSKOWITZ,S.
VOL 14 E11-E18, 20, 37-39
MOTSAREV,G.V.
VOL 12 E354-E366, 423, 430, 431
MUCCITELLI,J.
VOL 5/6 279
MUEHLBAECHER,C.
VOL 2 E114,E115, 122,E199,E201, 210
VOL 8 E226,E227
MUIRBROOK,N.K.
VOL 10 466
MUKHIN,N.N.
VOL.7. E250,E251, 256, 258, 260
MULLER,C.
VOL 5/6 E1-E3, 93, 104, 109, 112
VOL 7 E1-E5, 11, 207, 211
VOL 10 E1-E4, 10,E83-E85, 92, 101, 314
MULLER, F.
VOL 3 E1-E8, 9
MULLER,H.R.
VOL 12 E333,E334
MULLINS,J.C.
VOL 1 E283, 286
VOL 5/6 E571, 576,
MUNCK,O.
VOL 2 E114,E115
MUNSON,E.S.
VOL 8 14,E226,E227, 236, 237
MURRAY,C.N.
VOL 4 E1-E7, 17,E27-E29, 31
VOL 7 E1-E5, 29-31,E41-E43, 49, 50, 51,E64,E69,E76
VOL 10 E1-E4, 25,E31-E33, 41-44
MYASOEDOVA,V.V.
VOL 4 E33-E36,E80-E84
MYLIUS,F.
VOL 14 E1-E3, 4,E11-E18, 19,E44-E47, 48,E59-E68, 83, E179-E181, 182
NAGY,B.
VOL 10 E48-E53,E56,E333, 342, 345-348
NAIDICH,S.
VOL 14 E242-E250, 265, 275, 295
NAIR,V.S.K.
VOL 14 E59-E68,E75,E76,E78-E81, 107, 108, 117-118, 126, 127
NAKAMOTO,Y.
VOL 12 E3-E5, 29, 47,E174, 176, 192, 193, 196, 200, 204, 213, 217, 233, 236, 246,E261,E262, 263, 277, 278, 287, 293, 313
NAKASHIMA,E.
VOL 16/17 E359, 360,E394, 402, 403, 407, 408,E442,E443, 448, 449, 453, 454,E479, 480-483, E487, 488, 490,E694, 695-697,E698, 699-701,E702, 703-705

NAMIOT.A.YU.
VOL 1 E1-E4
VOL 4 E1-E7,E256
VOL 10 E419, 438
NARULA,S.F.
VOL 11 E1, 2,E4, 5,E9,10,E14,E15, 16,E21, 22,E26, 27,E28, E29, 30,E35, 36,E39, 41,E48, 49,E61, 62,E64, 65,E67, 68,E70, 71,E88, 89,E93, 94,E96, 97,E101, 102,E106, 107,E111, 112,E113, 114,E118, 119,E123, 124,E128, 129,E130, 131,E134, 135,E138, 139,E140, 141,E151, 153, E156, 157,E168, 169, 173,E181, 183,E186, 188,E194, 195,E198, 200,E208, 210,E222, 224,E226, 228,E229, 230,E231, 233
NASINI,A.G.
VOL 4 143
VOL 5/6 271
VOL 10 322
NAUMANN,A.
VOL 12 E354-E366, 414
NAUMANN,E.
VOL 7 452,E453,E454
NAUMENKO,N.K.
VOL 7 183, 184, 187,E214,E215,E250,E251,E261-E267
NAUMOVA,A.A.
VOL 5/6 E343
NAVARI,R.M.
VOL 7 386
NEDELKO,B.E.
VOL 2 E73,E260,E261
VOL 4 E1-E7, 16,E33-E36,E80,E84, 88, 89, 92, 93, 98, 103, 104,E169-E170, 204
NEDERBRAGT,G.W.
VOL 5/6 E356
NEGRI,G.
VOL 5/6 E610,E611
NELSON,E.E.
VOL 5/6 E346,E348
NEVENS,T.D.
VOL 9 E232, 239-244
NEWITT,D.M.
VOL 5/6 E299,E354,E355
NEZZAL,G.
VOL 14 E59-E68,E77-E81, 100, 101
NG,S.
VOL 9 E77,E78, 101, 102, 104
NG,W.Y.
VOL 1 E43
VOL 4 E108
NICHOLS,W.B.
VOL 5/6 E355, 361-363
NIEBELSCHUTZ,H.
VOL 7 399, 410, 411, 419-425
NIEMANN,H.
VOL 4 E169,E170,E180,E181, 182,E185,E186, 187, 192, 199, 206,E223, 225
VOL 9 E138-E140, 145, 162,E166,E167, 170, 174, 178, 181, 184, 189, 190,E232,E233
VOL 10 E174-E178, 184, 191, 203, 207
NIKIFOROVA,M.B.
VOL 1 E254-E256
NIKITINA.I.E.
VOL 1 E254-E256, 276, 277
VOL 10 E361,E362,413

NIKURASHINA,N.I.
VOL 13 E38-E47, 66, 69
NILSSON,R.
VOL 3 E102-E113, 154--163
NIMS,L.F.
VOL 3 E1-E8, 10, 11
NISHIMRA,K.
VOL 16/17 E545, 546, 547
NITTA,T.
VOL 5/6 E119, 128,E186-E190, 192, 197, 200, 202,E354,E435
VOL 10 E104,E105,107-112, 115-118,E119-E122, 125,E174-E178, 182, 189, 193, 196, 199
NOBBS,C.L.
VOL 2 E199-E201
NOBEL,A.P.P.
VOL 5/6 E148, 150, 154, 155,E159,E160, 165,E189, 211, 226
NODDES,G.
VOL 9 E27-E33, 60
NOEL,R.H.
VOL 16/17 E175, 176, 177
NOHKA.J.
VOL 4 E253-E255, 308
VOL 10 456
NOONAN,E.C.
VOL 3 E45-E56, 74, 89
NORTHGRAVES, W.W.
VOL 12 E454, 455
NORMAN,W.K.
VOL 10 E31-E33
NOTLEY,J.M.
VOL 11 E1,E4,E7,E9,E14,E15,E21,E26,E28,E29,E36,E39,E46, E48,E52,E54,E56,E58,E61,E64,E67,E70,E72,E74,E78,E80, E82,E84,E86
NOTZ,K.J.
VOL 2 98, 99, 181, 182
NOVAK,D.M.
VOL 4 E1-E7,E33-E36, 39, 40,E314,E317
VOL 7 E1-E5, 34
NOVAK,J.P.
VOL 10 E476, 482
NOWAKOWASKA,J.
VOL 7 E190-E192,213,E214,E215,E261-E267, 225, 268, 272, 278, 279, 292, 303
VOL 10 E83-E85,89,E104,E105,106, 113, 114,E119-E122, 135, E174-E178, 180, 187, 195, 197,E216, 219
NUSSBAUM,E.
VOL 2 E260-E261, 290,E291, 293,E294, 296, 297, 300-311, 314-317,E328, 330, 335, 336
NYS,P.S.
VOL 16/17 E1, 7-9,E657, 658-661,E662, 663,E664, 665, 666
NYSTROM,O.
VOL 1 122, 123
OCCLESHAW,V.J.
VOL 3 E102-E113, 133-140, 143, 144
ODENT,G.
VOL 13 E15-E19, 30,E38-E47, 80
ODINETS,Z.K.
VOL 13 E38-E47, 60, 61, 65,E148,E409,E414-E417,E449, E454-E456,E461
O´DONNELL,J.H.
VOL 12 E1,E2
OHGAKI,K.
VOL 9 E111,E112, 124,E232, 234, 238, 245-247
OHTA,Y.
VOL 10 293
OLIPHANT,J.L.
VOL 5/6 558-561

OLSEN,J.D.
VOL 9 E195-E199, 213
VOL 10 231
OMAR,M.H.
VOL 5/6 E589,E590
O´MORCHOE,C.C.C.
VOL 2 E114,E115
ONISHCHENKO,M.K.
VOL 13 E156-E162, 192
ONUMA,N.
VOL 2 E260,E261
OPYKHTINA,M.A.
VOL 12 E354-E366, 375
ORCUTT,F.S.
VOL 7 E1-E5
VOL 8 E1,E2, 230
VOL 10 E1-E4
OROBINSKII,N.A.
VOL 2 50, 51
VOL 4 E253-E255, 276
VOL 10 400
OSTIGUY,G.L.
VOL 8 238, 239
OSTRONOV,M.G.
VOL 5/6 E571, 577, 578
O´SULLIVAN.T.D.
VOL 10 E50-E52,E56,E333,343, 344, 349-351
OTTO,F.D.
VOL 5/6 E481-E483, 484-487, 489, 491, 492, 495, 496, E622, 627, 628
OTUKA,Y.
VOL 12 E3-E5, 20, 57, 69, 70, 93
OURISSON,M.M.J.
VOL 12 449, 450
OWEN,B.B.
VOL 3 E90,E91,E102-E113
OWEN,G.
VOL 16/17 E394, 398,E442,E443, 446

P

PACK,R.T.
VOL 3 E45-E56
PAGE,J.E.
VOL 16/17 E127,E128, 129,E135,E136, 137,E165, 166
PAKHOMOV,V.I.
VOL 14 E11-E18, 27, 28,E191-E197, 204, 205
PALEPU,N.R.
VOL 16/17 E484, 485, 486
PANCHENKOV,G.M.
VOL 1 114
VOL 7 442
VOL 10 328, 544
PANDA,K.N.
VOL 3 E102-E113, 175, 176
PANTELEEVA,N.I.
VOL 18 E7-E10
PARENT,J.D.
VOL 10 E361,E362,363, 364,E391, 392
PARISH,P.W.
VOL 2 E12,E13,E114,E115, 125, 129, 130, 133
PARKER,A.J.
VOL 3 20-22,E23, 24, 26-28, 208, 211, 212,E213, 214, 221, 222,E223, 226

PARKER,A.J.
VOL 11 E52,E54, 55,E56, 57,E58, 59,E171,E186, 192,E194, 197,E198, 202,E212, 215,E216, 219,E220, 221
VOL 18 E31,E32, 35-37,E52,E53, 54,E64, 65, 70, 71,E140, 143,E149, 150, 151, 153,E154, 155, 156,E158, 159, 160, 163-165,E166, 167, 170-173,E174, 175, 176, E182, 183-185,E187, 188-191,E194, 195-197,E198, 199-201,E203, 204-206, 208,E209, 210, 211, 212,E213, 214-216,E217, 218-224,E225, 226, 227
PARKER,P.D.M.
VOL 10 E31-E33, 37, 38
PARKER,V.B.
VOL 3 E39,E45-E56
PARKHURST,R.B.
VOL 12 E3-E5, 11,E37-E39
PARRISH.W.R.
VOL 1 E254-E256, 295, 296, 346, 347
VOL 10 E361,E362, 371, 374, 375
PATSATSIYA,K.M.
VOL 1 E124-E126, 130, 179-182,E227, 229
VOL 4 E1-E7,E80-E84
PATTERSON,J.L.
VOL 7 386
PATYI,L.
VOL 1 40,E43,E51, 54, 56, 58, 185,E188,E196, 199, 201, 203
VOL 4 107,E108, 110,E111, 113,E114,E115, 118,E125, 127, E128, 130, 132,E133, 135, 136,E137, 139-141
VOL 5/6 E119-E124, 125, 129, 133, 135, 138-144, 146
VOL 7 E214,E215, 216, 218, 221, 223, 229, 230, 232-237
VOL 8 E160-E162, 164-166, 168, 169, 171-179, 182, 201
VOL 10 E119-E122, 123, 126, 127, 130, 133, 138, 139, 141-146 E148, 153,E162, 168
PAUKNER,E.
VOL 7 E190-E192
PAUL,R.C.
VOL 11 E1,2,E4, 5,E9, 10,E14,E15, 16,E21, 22,E26, 27,E28, E29, 30,E35, 36,E39, 41,E48, 49,E61, 62,E64, 65,E67, 68, E70, 71,E88, 89,E93, 94,E96, 97,E101, 102,E106, 107, E111, 112,E113, 114,E118,119,E123, 124,E128, 129,E130, 131, E134, 135,E138, 139,E140, 141,E151, 153, 155,E156, 157,E162, 167,E168, 169,E171, 173, 175,E176, 178,E179, E181, 183, 185,E186, 188, 193,E194, 195,E198, 200, 203, E204,E206, 210, E222, 224, 225,E226, 228,E229, 230,E231, 233, 234,E235, 237,E240, 242
PAVLOPOLOUS,T.
VOL 11 E14,E15, 20,E28,E29, 34,E39,E40, 45,E52, 53,E54,E56,E58, 60
PAWLEK,F.
VOL 5/6 E25-E30,E35, 39, 42, 45, 48-50, 65, 79
VOL 7 E57,E58,E60,E62-E64,E67,E68,E76, 81, 97, 98
PEARCE,J.N.
VOL 3 E102-E113, 141, 142
PECCI,G.
VOL 11 E151, 154,E156, 158,E162, 166,E168, 170,E171, 174,E181, 184,E186, 191,E194, 196,E198, 201,E235, 236,E238, 239, 241,E243, 244
PEDERSEN,C.J.
VOL 3 83
PEDERSEN,K.J.
VOL 14 E59-E68,E74, 151, 170, 175,E242-E246,E251,E252, 311
PEER,A.A.
VOL 5/6 E119,E298,E299,E303,E304,E339, 340,E346, 347,E354, E356, 358, 359, 365,E380, 382,E406,E435, 436, E441-E444, 447, 521, 522

PEER,A.A.
VOL 10 12,E31-E33,E333,E409, 410, 447, 458, 503, 504
PENNEMAN,R.A.
VOL 3 E45-E56,
PELEG,M.
VOL 7 E474-E480
PEOPLES,R.S.
VOL 1 262
VOL 2 143, 146, 150
VOL 5/6 E30,E35, 56, 57,E303,E304.311, 313,
VOL 7 E1-E5,E64,E76,107, 108, 112-116
PEREIRA PAVEZE,V.
VOL 13 E38-E47, 61
PEREL´MAN,F.M.
VOL 13 E15-E19, 24, 25, 28, 29, 32-34,E38-E47,E241-E246, 254,E301-E308, 315,E381-E385, 390
PEREPLETCHIKOV,M.L.
VOL 7 492,E493
PEREVOZSKAYA,N.A.
VOL 16/17 E740, 765, 766
PERROTTET,E.
VOL 7 E474-E480, 486
PERSSON,O.
VOL 7 472
PETELINA,V.S.
VOL 13 E38-E47, 66, 69,E354-E359
PETERSON,M.A.
VOL 7 E1-E5, 39, 40
PETERSON,R.E.
VOL 2 E1-E3,E12,E13, 20, 23,E109,111,E114,E115, 116, 118, 119, 123, 126, 128, 132,E134-E136, 138,E149, 153, 154,E193,E194, 196,E199-E201, 202-204, 211, 216, 222, 224
PETER,S.
VOL 5/6 E355, 364, 377, 392
VOL 10 355, 356,E425, 428, 429, 434, 441, 442
PETRI,E.
VOL 3 E45-E56, 78
PETROVSKY,YH,V.
VOL 10 E361,E362
PETTERSON,O.
VOL 10 E1-E4
PFLAUM,R.T.
VOL 18 E7-E10, 13,E43, 45,E52,E53, 57,E72,E73, 74, 75, 83, 89, 90, 94, 96, 97,E106, 108, 115, 178,E182, 186
PHILLIPS R.
VOL 18 27-29
PIERCEY.G.A.
VOL 10 E425, 426
PIERRE,L.E.
VOL 5/6 520
PIGFORD,R.L.
VOL 7 E474-E480
PIIPER,J.
VOL 1 121
VOL 5/6 277
VOL 8 247
PINKUS,A.
VOL 8 E350
PINENOV,A.T.
VOL 7 E428
PINGREE,H.
VOL 8 E1,E2, 11,E226,E227, 234

PISTOIA,G.
VOL 11 E1, 3,E4, 6,E9, 11,E14,E15, 17,E21, 23,E28,E29, 31, E35, 37,E39,E40, 42,E72, 73,E74, 75,E78, 79,E88, 90, E93, 95,E96, 98,E101, 103,E106, 108,E113, 115,E118, 120,E123, 125,E142, 143,E144, 145,E146, 147,E151, 154, E156, 158,E162, 166,E168, 170,E171, 174,E181, 184,E186, 191,E194, 196,E198, 201,E235, 236,E238, 239,E240, 241, E243, 244,E309,E310, 312,E313, 314,E317, 318,E319, 320, E321, 322,E325, 326,E327, 328,E329,E333, 334,E335, 336, E337, 338,E339, 340
PODZOLKO,L.G.
VOL 10 472
POGREBNAYA,V.L.
VOL 7 E1-E5,E61,E62,E76, 88-90, 93-95
POLCARO,A.M.
VOL 11 E1,3,E4, 6,E9, 11,E14,E15, 17,E21, 23,E28,E29, 31,E35, 37,E39,E40, 42,E71, 73,E74, 75,E78, 79,E88, 90,E93, 95, E96, 98,E101, 103,E106, 108,E113, 115,E118, 120,E123, 125,E142, 143,E144, 145,E146, 147
POLEJES,J.D.
VOL 5/6 E178, 179
POLESSITSKIJ,A.
VOL 14 E242-E246,E248-E250, 257, 273, 278
POLEVA,G.V.
VOL 14 E11-E18, 40,E59-E69
POLGLASE,M.F.
VOL 9 E1,E1, 12
POLISHCHUK,A.P.
VOL 4 E33-E36,E80-E84
POLIZZOTTI,D.
VOL 4 E1-E7, 19,E33-E36, 41
POLOVCHENKO,V.I.
VOL 8 305
POLYAKOV,A.A.
VOL 5/6 E356, 368, 372, 373,E400, 403-405, 425, 427
POMERANTS,G.B.
VOL 3 E90,E91
PONOMAREV,YU.L.
VOL 8 E265-E267, 302-304
PONS,E.
VOL 10 E361
PONTOW,B.
VOL 7 E70,E72, 152
POOLE,J.W.
VOL 16/17 E359, 361,E363, 364,E394,E395, 398, 400, 401, 409, E410, 412, 414, 416, 417, 427, 432, 440, 441,E442,E443, 446, 456,E479,E487,E694,E698
POON,D.P.L.
VOL 10 383, 384,E402, 407
POPIEL,W.J.
VOL 14 E59-E68,E77-E81, 100, 101,E225-E227, 229
POPOVA,M.V.
VOL 16/17 E127,E128, 132,E135,E136, 140
POPOVIC,M.
VOL 7 399, 410, 411, 419-425
POPOVYCH,O.
VOL 18 E7-E10, 14, 25-29,E31,E32, 34, 38,E43, 46, 49, 51, E52,E53, 58,E64, 66, 69,E72,E73, 84-88, 98, 101,E106, E140, 142, 147, 148,E228, 229
POPOV,A.P.
VOL 13 E38-E47, 50, 51, 54, 55,E156-E162, 170,E241-E246, 249, 252, 253,E301-E308, 311,E354-E359, 361, 365-367, E376-E377, 378, 379
POPOV,G.A.
VOL 1 E1-E4,E68,E69
VOL 2 E1-E3
VOL 4 E1-E7,E111,E158,E171

POSLAVSKAYA,R.D.
VOL 13 E226,E227, 235, 237-239
POSPISIL,J.
VOL 7 E1-E5, 188, 189,E190-E192
POSTMA,H.
VOL 7 E41-E43
POSTON.R.S.
VOL 10 E419, 422, 423
POSTON.T.M.
VOL 1 113
VOL 4 232-235
VOL 10 270-272, 274
POTTER,O.E.
VOL 7 243
VOL 10 E148, 440
POTTER R.W.
VOL 4 E1-E7,E256, 258
POWELL,M.E.
VOL 5/6 263
POWELL,R.J.
VOL 1 98, 107, 108, 243, 247
VOL 2 97, 100, 153, 185
VOL 4 231
VOL 5/6 267,E280-E282, 294
VOL 9 E138-E140, 142, 161
VOL 10 E253, 257, 269
POWER,G.G.
VOL 5/6 15, 16,E33,E35, 73, 118, 275, 276
VOL 7 E1-E5, 27, 33
VOL 10 E1-E4,26,E279, 283, 298
POZIN,M.E.
VOL 8 E265-E267, 302-304
PRANDTL,W.
VOL 13 E286-E288, 290, 293, 295, 297, 298, 300, E339-E341, 343, 348, 350, 353
PRASAD,B.
VOL 3 E102-E113, 148, 149, 171
PRATHER, J.W.
VOL 5/6 523-525
PRATT,L.A.
VOL 13 E15-E19, 26, 27
PRAUSNITZ,J.M.
VOL 4 E253-E255, 270, 272
VOL 5/6 E123, 145, 147, 158, 184, 268
VOL 9 E77, E78, 100-104, 108, E136-E141, 150, 160, E195-E199, 203, 216, 223, 225-228
VOL 10 E1-E4, E119-E122, 147, 161, 172, 173, E361, E362, 459, 466
PRAY,H.A.H.
VOL 1 E257, 260, 262
VOL 2 143, 146, 150
VOL 5/6 E30, E35, 56, 57, E303, E304, 310, 311, 313
VOL 7 E1-E5, E64, E76, 107-116
VOL 10 E333, 339
PREECE,F.H.
VOL 1 116, 117
PRICE,B.A.
VOL 8 E1, E2, 21, 22, E23, E24, 25, 26, E160-E162
PROKHOROV,V.M.
VOL 4 E256
PROROKOV,G.A.
VOL 4 E33-E36
PRUZANSKY,J.
VOL 2 87, 102, 173, 186
VOL 10 235, 316

PRYANIKOVA, R.D.
VOL 5/6 E434
PRYANNOKOV,K.
VOL 5/6 E332
PURI,P.S.
VOL 5/6 E148, 153, E159, E160, 168, E188, 212, E221, E222, EEE441-E444
VOL 9 136
PUSHKINA,G.YA.
VOL 13 E1, 2, 3, 6-14
PYNE,H.R.
VOL 7 E250, E251, 252

Q

QUILL,L.L.
VOL 13 E38-E47, 53, E109, 133, 135, E156-E162, 163, E221, 223, 224, E278, 283, 285, E301-E308, 313, E335, 337, 338, E354-E359, 364, E381-E385
QUINCHE, J.P.
VOL 13 E38-E47, 49, 70-72, 75, 76, E156-E162, 164, E241-E246, 248, 255, E301-E308, 310, 317, E354-E359, 360, E376, E377

R

RAASCHOU,F.
VOL 16/17 E10, 13, E39, 42, E66, 67, E72, 77
RAFF,P.
VOL 18 E7-E10
RAKESTRAW,N.W.
VOL 4 E27-E29
VOL 10 E31-E33, 36
RAMSAY,W.
VOL 1 E1-E4, E68, E69
RAMSTEDT,E.
VOL 2 E227-E229, 231, E260, E261, 262-264, E265, 269, 270, E273, 276, 286, 288, 289, 312, 318-320, E322, 326
RANDALL,M.
VOL 3 E45-E55, 70, 71, E102-E113, 130
RAPSON,H.D.C.
VOL 16/17 E1
RARD,J.A.
VOL 13 E15-E19, E37-E47, 52, E245, E246, 250, 251, E301-E308, 312, E354-E359, 362, E376, E377, E381-E385, 387, E409, 410, 411, E414-E417, 419, E4 29, 430, 432, E434-E436, 438, E449, 450, E454-E456, 458, E461
RAWSON,A.E.
VOL 7 E474-E480
REAMER,H.H.
VOL 5/6 E339, 341, E354, 361-363, E380, 388, 389, E622
VOL 9 E111, E112, 119, 120, 129-131, E232, 235
RECHNITZ,G.A.
VOL 18 E7-E10, 19, E43, 47, E52, E53, 60, E72, E73, 78, E140, 146, E177, 179
REDDY,G.S.
VOL 1 E20-E22, 34
VOL 4 E1-E7, E33-E36, 55 , 79, E171, 174
REED,M.G.
VOL 18 E21, 22, E40, E41
REEVES,L.W.
VOL 4 E149, 150, E162, 164, E207, 208, 213, E223, 224
REIBER,H.G.
VOL 14 E59-E68, E70, E72, E75, 135, 137, 177, 178, E225-E227, 230, 235, E242, E243, E246-E248, E252, 266, 281, 283, 285
REILLEY,C.N.
VOL 18 E3

REIMLINGER,S.
VOL 8 E350, 352
REINHARTZ,K.
VOL 5/6 E335, 364, 377, 392
REMY-GENNETTE,P.
VOL 14 E208-E212, 224
RENAULT,P.
VOL 9 E77, E78, 98, E138-E141, 151, 191-194, E195-E199, 211, 212, 217, 218, 222, 224, 230
RENON,H,
VOL 5/6 E356, 378, 379, 418
VOL 9 E77, E78, 98, 99, 107, E111, E112, 132, 133, E138-E141, 151, E166, E167, 191-194, E195-E199, 211, 212, 217, 218, 222, 224, 230
VOL 10 E425, 430
RENS,G.
VOL 14 E59-E68,E70-E73,E77,E78, 90, 92, 97, 103, 136, 138, 139, 150, 157
RETTICH,T.R.
VOL 9 E1, E2, 3, 4
REUSS,M.
VOL 7 399, 410, 411, 419-425
REWICK,R.T.
VOL 7 430
RHODE,C.B.
VOL 8 232, 242
RHODES,H.L.
VOL 1 E263, 265, 266, 271, 272, E328, 335, 336
RICCI,J.E.
VOL 3 E45-E56, E177, E178, E193, E200
VOL 14 E11-E18, 21, 22, 31, 32, 41,E208-E212, 218-222,E225-E227, 236-241,E242-E251,265, 269, 274-276, 280, 289-292, 295, 296, 307
RICCOBINI,L.
VOL 8 E336, 342
RICHARDSON,A.H.
VOL 7 E1-E5, 13-15
RICHON,D.
VOL 5/6 E356, 378, 379, 418
VOL 9 E77, E78, 99, 107, E111, E112 132, 133
VOL 10 E425, 430
RIDENOUR,W.P.
VOL 4 144
VOL 7 447, 448
VOL 10 325, 326
RIDGION,J.M.
VOL 3 E45-E56
RIESENFELD,E.H.
VOL 3 E1-E8, 9
RILEY,H.L.
VOL E E45-E56
RILEY,J.P.
VOL 4 E1-E7, 17, E31-E33, 41-44
VOL 7 E1-E5, 29-31, E41-E43, 49, 50, E51, E64, E69, E76
VOL 10 E1-E4, 25, E31-E33, 41-44
RIVAS,O.R.
VOL 9 E195-E199, 216, 223, 225, 226, 228
ROACH.J.T.
VOL 10 E361, E362, 365-367
ROBERTSON,G.D.
VOL 7 E62, E76, 85
ROBERTS.L.R.
VOL 10 385, 386, E409, 412
ROBEY,R.F.
VOL 13 E38-E47, 53, E156-E162, 163, E301-E308, 313, E354-E359, 364, E381-E385, 388

ROBINS,D.A.
VOL 4 E108, E114, E115, E171
ROBINSON,D.B.
VOL 9 E111, 117, 118, E232, 244
VOL 10 416, 417, E419, 420, 421, 468-470
ROBINSON,J.A.
VOL 10 E409, 411
ROBINSON,J.E.
VOL 13 E301-E308, 316
ROBINSON,M.J.
VOL 16/17 E175, 178-198
ROBINSON,R.A.
VOL 3 E45-E56, E102-E113
VOL 4 E27-E29
VOL 5/6 E17,E18
VOL 10 E31-E33
ROBOV,A.M.
VOL 14 E60, 93,E191-E197, 199,E243, 261
ROBSON,J.H.
VOL 10 387, E391, 394, E402, 406, 467, 471, 473, 474, E476, 481, 485-487, 496, 497
ROCHESTER,D.F.
VOL 2 E114, E115, 121
ROCHOW,E.G.
VOL 11 E148, E151, E156, E159, E162, E168, E171, E176, E181, E186, E194, E198, E208, E212, E216, E220, E222, E226, E229, E231, E235, E238, E240, E243
ROCKER,A.W.
VOL 8 258
ROCKERT,H.
VOL 1 122, 123
RODEWALD,N.C.
VOL 1 E328, 339, 340
RODMAN,C.J.
VOL 10 320
RODNIGHT,R.
VOL 7 438
RODRIGUES,A.B.J.
VOL 9 E111, E112, 127, 128
ROELLIG,L.O.
VOL 1 E310, 311
ROGERS,B.L.
VOL 4 E253-E254, 272
ROLAND,C.H.
VOL 10 E361, E362, 365-367
ROSENBLUM,W.I.
VOL 7 386
ROSENMAN,S.B.
VOL 16/17 E394, 398,E442,E443, 446
ROSEN,E.
VOL 16/17 E175, 178-198
ROOF,J.G.
VOL 10 E402
ROSS,M.
VOL 8 244
ROTHSCHILD,B.F.
VOL 13 E104
ROTH,J.A.
VOL 7 E474-E480
ROTH,W.
VOL 8 E1, E2, 3, E27-E39, 44, 45, 81, 82, E116, 124, 125, 129, 130, 135, 136
ROTHMUND,V.
VOL 7 E474-E480

ROTTGER,H.
VOL 7 440
VOL 10 323
ROUGHTON,F.J.W.
VOL 7 374
ROUSSEAU,R.W.
VOL 10 495
ROWLINSON,J.R.
VOL 1 E254-E256
RUDAKOV,E.S.
VOL 9 E1, E2, 14, E27-E33, 34
RUDORFF,W.
VOL 18 E7-E10, 11, E43, E52, E53
RUETHER,J.A.
VOL 5/6 E148, 153, 168,E188, 212,E441-E444
RUETSCHI,P.
VOL 5/6 E1-E3, 10,E25-E27,E33,E35, 40, 80
RUHEMANN,M.
VOL 5/6 E589,E590,E601
RUNOV,N.N.
VOL 13 E38-E47, 108
RUPPEL,D.
VOL 7 446
RUPPRECHT,S.D.
VOL 10 437, 507, 508, 519, 548, 549
RUSTON,M.S.
VOL 14 E225-E227, 229
RUSZ,L.
VOL 1 40, E43, E51, 54, 56, 58, 185, E188, E196, 199, 201, 203
VOL 2 25, E26, 29, E35, 37, 41, 43, 46, 47, 49, 155, 157, 159 , 161
VOL 4 107, E108, 110, E111, 113, E114, E115, 118, E125, 127, E128, 130, 132, E133, 135, 136, E137, 139-141
VOL 5/6 E119-E124, 125, 129, 133, 135, 138-144, 146, E187-E190, 193, 199, 201, 203, 205, 207-209, E441-E444
VOL 7 E214, E215, 216, 218, 221, 223, 229, 230, 232-237, 269, 276-280, 282, 284, 287, 288
VOL 8 E160-E162, 164, 165, 168, 169, 171-178, 187, 191, 192, 197, 200, 201, 203-205
VOL 10 E119-E122, 123, 126, 130, 133, 138, 139, 141-146, E174-E178, 183, 190, 194, 200, 205, 210, 213, 215
RYABOVA,N.M.
VOL 16/17 E664, 665
RYABSTEV,N.I.
VOL 10 414, 415
RYSAKOV,M.W.
VOL 10 E361

S

SAARI,J.M.
VOL 8 233
SADAMOTO,S.
VOL 5/6 E435, 440, 567, 568, 620
SADA,E.
VOL 8 E1, E2, 18-20, E27-E39, 51, 64, 72, 104, 110, E116, 117-119, 121, 128, 131, 132, 134, 138-143, E160-E162, 186
SADDINGTON.A.W.
VOL 10 E333, 337
SADILENKO,A.S.
VOL 8 166, 179, 182
VOL 10 E119-E121, 127, E148, 153, E162, 168

SAFRANOVA,T.P.
VOL 10 331
SAGARA,H.
VOL 5/6 E321, 328, E332, 338, E395, 399, 537-539, 543-551
SAGE,B.H.
VOL 5/6 E339, 341, E354, 361-363, E380, 388, 389 E622
VOL 9 E111, E112, 119, 120, 129-131, E232, 235-237
SAHU, G.
VOL 3 E90, E91, 96, 97
SAIDMAN,L.J.
VOL 8 14, E226, E227, 236, 237
SAIFI,R.N.
VOL 7 E71, E76, 162
SAITO,H.
VOL 4 E253-E255, 320, 321
VOL 5/6 E178
SAITO,S.
VOL 5/6 E321, 328, E332, 338, E395, 399, 537-539, 543-551
VOL 10 E488, 456, 493, 498, 499
SAKHAROVA,N.F.
VOL 11 E245, 246, E251, E252, 254
SAL´NIK,L.V.
VOL 13 E15-E19, 23, E381-E385, 389, E461, 463
SAMOILOV,O.YA.
VOL 1 133, E141-E143, 161, 163-165, 169-171, 175-178
SANDALOVA.L.
VOL 10 E488, 489
SANO,F.
VOL 9 E111, E112, 124, E232, 234, 238, 245-247
SARASHINA.E.
VOL 4 E253-255, 308, 321
VOL 10 456, 498, 499
SARGENT,J.W.
VOL 7 328, 357, 450
VOL 10 247, 251, 268
SARJEANT,E.P.
VOL 3 E1-E8
SARKISOV,A.G.
VOL 11 E245, 246, E251, E252, 254
SATHER,G.A.
VOL 7 470, 471
SATO,M.
VOL 3 E31, 32, E45-E56
SATTERFIELD,C.N.
VOL 5/6 E178, 180, 182, 183
SATTLER,H.
VOL 5/6 E354, 360, E380, 384, E406, 409, E419, 421
SATYANARAYANA,D.
VOL 3 E90, E91, 96, 97
SAUTER,C.G.
VOL 14 E242-E246, E248-E251, 269, 274, 276, 280, 307
SAVINOV,V.M.
VOL 11 E309, E310, E313, E315, 316, E317, E321, E323, 324, E325, E327, E329, E331, 332, E333, E337, E339, E341, 342, E343, 344
SAVITSKAYA,E.M.
VOL 16/17 E1, 2-9,E657, 658-661,E662, 663,E664, 665, 666
SAYLOR,J.H.
VOL 1 41, 42, E43, 44, 46-50, E51, 52, 55, 57, E59, 61, 62, E68, E69, 71, E73, 74, 91, 93, 95-97, 106, E124-E126, 187, E188, 189, 191-195, E196, 197, 200, 202, E204, 206, 208, E214, 216, E218, 219, 235, 237, 240-242, 246
VOL 2 E26, 27, 30-34, E35, 36, 39, 40, 44, 48, E52, 53, 55, 62, 63, 86, 89, 91, 93, 95, 104, 165, 167, 172, 177-180, 190

SAYLOR,J.H.
VOL 4 E1-E7, E108, 109, E111, 112, E114, E115, 116, 120-122, E123, 124, E125, 126, E128, 129, E133, 134, E137, 138, E145, 147, E149, 151, 152, E158, 160, E162, 163, E207, 209, 215, 216, 219, 221, 228
VOL 5/6 E169, 172
VOL 7 E1-E5
VOL 10 E1-E4, 19
SCHAFER,K.
VOL 3 E102-E113
SCHAFFER,P.S.
VOL 5/6 274
VOL 10 277, 534
SCHEHERBINA,N.V.
VOL 3 E102-E113
SCHEICHERT,C.E.
VOL 7 458
SCHERER,K.V.
VOL 7 310, 316, 323, 329, 330, 334, 335, 339
SCHICKENDANTLZ,F.
VOL 9 E1, E2
SCHINDLER,D.L.
VOL 1 E254-E256, E263, 264, 278, 279
VOL 10 E402, 404, 405
SCHLAPFER,P.
VOL 7 E190-E192, E250, E251, 254, E261-E267, 273, 304, 305
SCHMACK,P.
VOL 5/6 477
SCHMIDT-NIELSEN,S.
VOL 10 536
SCHMIDT,U.
VOL 5/6 E18
SCHNEIDER,G.M.
VOL 1 E254-E256, E298, E369
VOL 5/6 E298,E415,E419
SCHNITTKER,J.B.
VOL 5/6 263
SCHOCH,E.P.
VOL 9 E195-E199, 220
SCHOENBORN,B.P.
VOL 2 E199-E201, 209
SCHOFIELD,P.F.
VOL 2 E149, E199-E201, 213, 226
SCHOLANDER,P.F.
VOL 1 E1-E4, E257
VOL 4 E1-E7, E27-E29
VOL 7 E41-E43, 44
VOL 10 24, E31-E33, E83-E85, 352
SCHOON,N.-H.
VOL 3 E1-E8, 15, 16
SCHORLEMMER,C.
VOL 9 E1, E2
SCHOUTEN,J.A.
VOL 1 E359, 367, 368, 370, 371
SCHRODER,W.
VOL 5/6 E441-E444
VOL 8 E1, E2
VOL 10 E55, E56, 81, 82
SCHRODT,O.
VOL 2 E322, E328, 331
SCHULZE,A.
VOL 2 E260, E261
SCHULZE,G.
VOL 10 E1-E4

SCHUMM,R.H.
VOL 3 E39, E45-E56
SCHURCHKOVA,YU,A.
VOL 16/17 E135,E136, 140
SCHWAB,G.M.
VOL 7 E1-E5
VOL 8 E1-E2
SCHWARZ,H.-G.
VOL 9 E27-E33, 60
SCHWEICHERT,C.E.
VOL 1 E257, 260
VOL 5/6 E303,E304, 310
VOL 10 E333, 339
SCHWEITZER,P.H.
VOL 10 540, 545
SCOTT,A.D.
VOL 18 E21, 22, E40,E41
SCOTT,N.D.
VOL 10 287, 288
SCOTT, R.L.
VOL 7 E41-E43
VOL 9 E138-E140, E195-E199
SCROSATI,B.
VOL 11 E151, 154, E156, 158, E162, 166, E168, 170, E171, 174, E181, 184, E186, 191, E194, 196, E198, 201, E235, 236, E238, 239, E240, 241, E243, 244, E309, E310, 312, E313, 314, E317, 318, E319, 320, E321, 322, E325, 326, E327, 328, E329, 330, E333, 334, E335, 336, E337, 338, E339, 340
SEAGER,H.
VOL 16/17 E394, 399, 406,E442,E443, 447, 452
SEARS,P.G.
VOL 11 E259
SEBASTIAN,H.M.
VOL 5/6 E298, E355, 369, 370, 373, 374, 394, E415, 417, E419, 422, 423, 428-430, 432, 433, 475, 476, 510, 517, 518, 556, 557
SEDOVA,L.K.
VOL 13 E38-E47, 69
SEEVERS,M.H.
VOL 7 E1-E5
VOL 8 E1, E2, 230
VOL 10 E1-E4
SEFFL,R.J.
VOL 7 328, 357, 450
VOL 10 247, 251, 268
SIEDELL,A.
VOL 2 E328
VOL 3 E45-E56, E86
VOL 8 E40
SEIFTER S.
VOL 13 E38-E47, E156-E162, E301-E308, E354-E359, E381-E385
SELEEM,L.M.N.
VOL 14 E243, 301
SEMENCHENKO,V.K.
VOL 2 E260, E261
SEMENISHIN,D.I.
VOL 3 E45-E56
SEMENOVA,E.B.
VOL 13 E156-E162
SEMYANNIKOVA,E.L.
VOL 2 50, 51
VOL 4 E253-E255, 276
SENO,M.
VOL 7 451

SEREBRENNIKOV,V.V.
VOL 13 E381-E385
SERGEEVA,A.N.
VOL 3 E45-E56
SERGEEVA,V.F.
VOL 13 E38-E47, 95
SERVIGNE,M.
VOL 7 48
SETCHELL,B.P.
VOL 2 E114, E115
SEVERINGHAUS,J.W.
VOL 7 374
VOL 8 14, E226, E227, 236, 237
SEVOST´YANOV,V.P.
VOL 13 E381-E385, E414-E417
SEWELL,R.G.
VOL 13 E38-E47, E156-E162, E241-E246, E301-E308, E354-E359, E376, E377, E381-E385, E414-E417, E429, E434-E436, E449, 450, E454-E456, E461
SHABIKOVA,G.KH.
VOL 13 E38-E47, 95
SHAKHALEEVA,N.N.
VOL 13 E38-E47, 77, 78, E354-E359
SHAKHANOVA,N.A.
VOL 13 E409, 412, 413,E434-E436, 439, 440
SHAKHOVA,S.F.
VOL 4 E253-E255, 305, 306, 310, 311
VOL 5/6 478-480, 511, 512, 519
VOL 10 448, 451-453, 462, 463
SHAKHPARANOV,M.I.
VOL 2 E200,E201
SHAPKA,A.V.
VOL 7 E62,E76, 86,E87, 91,E92
VOL 10 59
SHARMA,M.M.
VOL 7 E70,E76, 161
SHARP,J.H.
VOL 18 E149, 151,E154, 156,E158, 160, 163, 170-173,E174, 176,E187, 190,E194, 197,E198, 201,E203, 206, 208, E209, 212,E213, 216,E217, 220, 221-224,E225, 227
SHATSKAYA,L.V.
VOL 5/6 E571, 577, 578
SHAW,A.W.
VOL 8 E336, 338, 340, 341, 344, 346, 348, 349
SHCHENEV,A.V.
VOL 13 E38-E47, 108
SHCHUKAREV,S.A.
VOL 7 E1-E5, 28,E190-E192, 195,E261-E267
SHCHUROV,V.A.
VOL 13 E156-E162, 173-179, 196, 197, 199, 200
SHEDLOVSKY,T.
VOL 3 E102-E113
SHELLENBERG,N.N.
VOL 16/17 E1, 2-9,E657, 658-661,E664, 665, 666
SHENDEREI,L.I.
VOL 10 445
SHEN,B.
VOL 5/6 E1-E3, 114, 115, 117
SHERBORNE,J.E.
VOL 9 E232, 236, 237
SHERWOOD,T.K.
VOL 5/6 E178, 180, 182, 183
VOL 7 E474-E480
SHEVCHUK,V.G.
VOL 13 E156-E162, 183-186, 192, 195, 203, 204,E226,E227

SHEVELEVA,A.D.
VOL 13 E156-E162, 171-179, 196, 197
SHEVTSOVA,Z.N.
VOL 13 E38-E47, 63, 81, 82, 85,E148
SHIAU,J.F.
VOL 5/6 513-516
SHIERS,L.E.
VOL 13 E38-E47, 52,E301-E308, 312,E354-E359, 362, 363,E381-E385, 387, 388,E409, 410, 411,E414-E417, 419,E429, 430,E434-E436,438,E449,E454-E456, 458,E461,E462
SHILIONSKAYA,M.A.
VOL 11 E309,E310,E313,E315,316,E317,E321,E323, 324,E325,E327, E329,E331, 332,E333,E337,E339,E341, 342,E343, 344
SHILLING,C,W.
VOL 1 E1-E4
VOL 10 E1-E4, 16, 464, 465
SHKLOVSKAYA,R.M.
VOL 14 E11-E18, 29, 30, 40,E59-E69
SKHOLNIKOVA,R.I.
VOL 4 77, 78, 245-247
SHKOL´NIKOVA,R.I.
VOL 5/6 E29,E30,E35, 51, 53, 113, 116
VOL 7 E63,E64,E76, 101, 105
VOL 8 154, 155, 249, 250
VOL 10 E48,E56, 60, 61, 297, 310
SHMYD´KO,I.I.
VOL 14 E60, 93,E191-E197, 199,E243, 261
SHONE,E.
VOL 7 E474-E480
SHOOLERY,J.N.
VOL 4 E241
SHOOR,S.K.
VOL 1 E1-E4,E20-E22, 28
VOL 4 E1-E7,E33-E36, 60
VOL 5/6 E1-E3, 11,E33,E35, 81, 82
VOL 7 E1-E5,E71,E72,E76, 168, 169
SHOPENKO,V.V.
VOL 3 E90,E91, 98-101
SHUROV,V.A.
VOL 13 193, 194
SIBBITT,W.L.
VOL 5/6 E303,E304,
VOL 10 18,E333
SIEBECK,R.
VOL 8 E226,E227
SILBERBERG,I.H.
VOL 9 E111, 121-123
SILVERIO,J.
VOL 16/17 E394, 398,E442,E443, 446
SIMNICK.J.J.
VOL 5/6 E298,E355, 369, 370,E415, 417,E419, 422, 423, 426, 431-433, 475, 476, 510, 517, 518, 556, 557
SIMONSON,R.
VOL 7 E67,E68,E71,E76, 160
SIMON,M.
VOL 5/6 E581, 582
SINDA,F.
VOL 3 E39, 43, 44
SINGLA,J.P.
VOL 11 E1, 2,E4, 5,E9, 10,E14,E15, 16,E21, 22,E26, 27,E28,E29, 30,E35, 36,E39, 41,E48, 49,E61, 62,E64, 65,E67, 68,E70, 71,E88, 89,E93, 94,E96, 97,E101, 102,E106, 107,E111, 112, E113, 114,E118, 119,E123, 124,E128, 129,E130, 131,E134, 135,E138, 139,E140, 141,E151, 153,E156, 157,E168, 169, 173,E181, 183,E186, 188,E194, 195,E198, 200,E208, 210, E222, 224,E226, 228,E229, 230,E231, 233

SINGLETON,J.H.
VOL 4 240
SINITSYNA,E.D.
VOL 13 E241-E246, 252,E301-E308, 314,E354-359, 365, 366
SINN,E.
VOL 7 452,E453,E454
SINOR,J.E.
VOL 1 E254-E256,E263, 264,E283, 287, 297,E348, 351
SIROTIN,A.G.
VOL 1 E254-E256, 276, 277
VOL 10 E361,E362
SISKA,E.
VOL 18 E7-E10, 17, 18,E52,E53, 55, 56,E72,E73, 76, 77
SISSKIND,B.
VOL 4 E1-E7,E145,E149,E162,E169,E170,E180,E181,E186,E253-E256, 257, 271, 277-304
SITZUNGSBER,
VOL 2 E260,E261
SIXMA,F.L.J.
VOL 7 484
SKINNER, J.F.
VOL 18 E7-E10
SKRIPKA.V.G.
VOL 1 E254-E256, 276, 277,E283, 289-291,E328, 338,E348, 349, 350, 352,E359, 360, 365, 366,E373, 375, 377,E380, 381, 384, 385
VOL 10 E361,E362,E409,413,E419, 438
SKURAT,V.E.
VOL 4 E253-E255
SKVORTSOV,G.A.
VOL 8 305
SLATER,P.G.
VOL 7 E1-E5
SLOWINSKI,
VOL 4 E108,E114,E115,E171
SLYUSAR,V.P.
VOL 4 E33-E36,E80-E84,E169,E170,E185,E186, 190
SMILEY,S.H.
VOL 14 E208-E212,E225-E227, 238, 239
SMIRNOVA.A.M.
VOL 4 E253-E255, 273
VOL 10 E419, 424, 432
SMITH,E.B.
VOL 4 E1-E7, 15
SMITH,H.
VOL 16/17 E394, 398,E442,E443, 444
SMITH,J.M.
VOL 5/6 E178, 181
SMITH,J.O.
VOL 5/6 E415
SMITH,L.J.
VOL 7 E41-E43
SMITH, L.
VOL 3 E102-E113, 141, 142
SMITH.N.O.
VOL 1 E1-E4,E257, 259, 261
VOL 10 E48-E53,E56,E333, 342-351
SMITH,S.R.
VOL 1 E310
SMITH,T.R.
VOL 2 E114,E115
SMITH W.E.
VOL 3 E102-E113, 164
SNEED,C.M.
VOL 1 312, 313, 322

SOBR,J.
VOL 4 E314, 317
VOL 10 E476, 482
SOKOLOV,B.I.
VOL 5/6 E356, 368, 372, 393,E400, 403-405, 425, 427, E441-E444, 452, 453, 456, 457
SOKOLOV,L.B.
VOL 11 E309,E310,E313,E315, 316,E317,E321,E323, 324,E325,E327, E329,E331, 332,E333,E337,E339,E341, 342,E343, 344
SOLIMAN,S.T.
VOL 14 E59-E69, 87
SOLVANG,S.
VOL 16/17 E10, 14, 15,E39, 41
SOMAIT,F.A.
VOL 10 E488, 494, 500, 501
SONDEN,K.
VOL 10 E1-E4
SONNTAG,R.E.
VOL 1 E310, 312-317, 322
SORINA,G.A.
VOL 5/6 E380, 385, 386
SPANO,J.O.
VOL 5/6 E610, 614, 615
SPEDDING,D.J.
VOL 5/6 E18
SPEDDING.F.H.
VOL 13 E15-E19,E38-E47, 52,E245,E246, 250, 251,E301-E308, 312,E354-E359, 362, 363,E367,E377,E381-E385, 387, 388, E409, 410, 411,E414-E417, 419,E429, 430,E434-E436, 438, E449, 450,E454-E436, 458,E461, 462
SPINKS,T.J.
VOL 10 27,E31-E33
SPIRO,M.
VOL 11 E1,E4,E7,E9,E14,E15,E21,E26,E28,E29,E36,E39,E46,E48,E52, E54,E56,E58,E61,E64,E67,E70,E72,E74,E78,E80,E82,E84,E86
SPITSYN,V.I.
VOL 13 E1
SPITZER-NEUMANN,E.
VOL 8 E336, 339, 347, 350
SPRAGUE,R.W.
VOL 7 E62,E76, 96
SPROW,F.B.
VOL 10 E361,E362
SRIDHAR,T.
VOL 7 243
VOL 10 E148, 440
STANDAERT,D.G.
VOL 7 385
STANDLEY,D.R.
VOL 5/6 E1-E3, 14,E17,E18, 24
STARIKOVA,L.I.
VOL 13 E381-E385, 391-402, 404,E414-E417, 421, 422,E434-E436, 441-443
STARKOV,M.G.
VOL 1 E307, 308
STECKEL,F.A.
VOL 5/6 E589,E590
STEDMAN,R.J.
VOL 16/17 E657,E664
STEEN,H.
VOL 7 E1-E5, 17
STEEN.J.B.
VOL 4 251
VOL 10 292

STEGALL,H.
VOL 5/6 16, 118, 275
VOL 7 E1-E5, 33
VOL 10 E1-E4, 26,E279, 283, 298
STEINBACH,H.G.
VOL 10 418
STEINBERG,M.
VOL 1 115
VOL 2 E26, 28, 45, 69, 70,E71, 87, 102,E134-E136, 168, 169, 171, 173, 186, 188, 191, 192
VOL 4 142
VOL 7 449
VOL 10 235, 316, 33/???
STEINBRECHER,M.
VOL 10 418
STEINERT,H.E.
VOL 13 E15-E19, 20
STEINER,P.
VOL 5/6 E30,E33,E35, 47, 54, 58, 59, 61, 66, 71, 74, 76, 83, 87, 89, 111
STEIN.S.
VOL 5/6 314-320, 527
STEPANOVA,Z.G.
VOL 8 166, 179, 182
VOL 10 E119-E121, 127,E148, 153,E162, 168
STEPANOV,A.M.
VOL 7 481
STEPHAN,E.F.
VOL 7 E1-E5, 109-111
STEPHAN,E.L.
VOL 1 262
VOL 2 143, 146, 150
VOL 5/6 E30,E35, 56, 57,E303,E304, 311, 313
VOL 7 E1-E5,E64,E76, 107-108, 112-116
STEPHAN,H.
VOL 11 E245,E251,E252
STEPHAN,T.
VOL 11 E241,E251,E252
STEPIN,B.D.
VOL 13 E15-E19, 22, 31,E38-E47, 73, 74, 77, 78,E301-E308, 319,E354-E359
STEWART,D.F.
VOL 13 E109, 110-114, 116-127, 129-131, 134, 136-142, 145, 146
STEWART,W.G.
VOL 1 E254-E256, 295, 296, 346, 347
STICKINGS,B.J.
VOL 4 E1-E7, 15
STISIBAROVSKAYA,N.P.
VOL 7 155,E156
VOL 10 E49,E50,E56
STOCK,J.T.
VOL 3 E45-E56, 75, 76
STODDARD,J.L.
VOL 10 E1-E4, 294
STOKES,R.H.
VOL 3 E45-E56,E102-E113
STOLLER,L.
VOL 9 E1,E2, 13,E27-E33, 41,E64, 73
STOLYARENKO,G.S.
VOL 7 487,E488
STOLYARENKO,T.E.
VOL 7 487,E488
STRACK,G.A.
VOL 11 E88, 91, 92,E93,E96,99, 100,E101, 104, 105,E106, 109, 110,E113, 116, 117,E118, 121, 122, 123, 126, 127,E130, 132, 133,E134, 136, 137,E138

STRAKHOV,A.N.
VOL 1 E1-E4, 14,E124-E126, 137
VOL 4 E1-E7,E20,E21, 24, 26,E33-E36
STRANG,R.
VOL 2 E114,E115, 127
STRASBURGER, J.
VOL 2 E328
STREENATHAN.B.R.
VOL 11 E151, 155,E162, 167,E171, 175,E176, 178,E181, 185,E186, 193,E198, 203,E204,E206,E208, 211,E222, 225,E231, 234, E235, 237,E240, 242
STREETT,W.B.
VOL 1 E254-E256,E263, 267-270,E283, 284, 285, 288, 292-294, E310, 316-319,E328, 329-332, 341, 342, 357, 358,E359, 361-364,E373, 376, 378,E380, 382, 383
VOL 5/6 E321, 329-331,E571, 579, 580,E581, 585, 586,E589,E590, 599, 600,E601, 607, 608,E610, 617, 618
STREHLOW,H.
VOL 11 E14,E15, 20,E28,E29, 34,E39,E40, 45,E52, 53,E54,E56,E58, 60
STREITWIESER,A.
VOL 9 E195-E198, 202
STROUD,L.
VOL 10 E361,E362
STRUKOVA,V.E.
VOL 7 427,E428, 429
STRYJEK.R.
VOL 10 E361,E362, 372, 373,E391, 395, 396
STSIBAROVSKAYA,N.P.
VOL 10 65, 66
STUBICAN,V.
VOL 3 E102-E113, 145-147
STUMM,W.
VOL 7 E1-E5,20, 21,E474-E480
ST.PIERRE,L.E.
VOL 7 441
SUCIU,S N.
VOL 5/6 E303,E304, 307-309
VOL 10 18,E333, 338
SUKHOVA,G.I.
VOL 7 183, 184, 187
SUKHOVERKHOV,V.F.
VOL 10 472
SULAIMANKULOV,K.S.
VOL 13 E1, 4, 5,E15-E19, 35, 36,E156-E162, 216, 217, 219, 220,E241-E246, 276,E301-E308,334,E354-E359, 370,E376,E377, 380,E381-E385, 403,E414-E417, 428, E429, 433,E434-E436, 444-447,E449, 453,E454-E456, 460,E461,E464
SULLIVAN,D.E.
VOL 7 E474-E480, 482, 483
SULTANOV.R.G.
VOL 10 E419, 438
SUPPLEE, H.
VOL 7 E190-E192
SUZUKI,K.
VOL 14 E59-E68,E74, 176,E191-E197, 207,E242-E246,E251, E252, 310
SUZUKI,S.
VOL 3 E1-E8, 14
SVEEN,K.
VOL 16/17 E10, 14, 15,E39, 41
SVETLOVA,G.M.
VOL 8 E336, 337

SWANDA,K.S.
VOL 11 E88, 91, 92,E93,E96, 99, 100,E101, 104, 105,E106, 109, 110,E113, 116, 117,E118, 121, 122,E123, 126, 127,E130, 132, 133,E134, 136, 137,E138
SWENKER,M.D.P.
VOL 5/6 E589,E590,E601,
SWERED.K.
VOL 16/17 E657,E664
SWIFT,G.W.
VOL 1 E254-E256, 278, 279
VOL 10 E402, 404, 405
SWINNE,R.
VOL 2 E260,E261
SWINTOSKY,J.V.
VOL 16/17 E175, 178-198
SYMONS,E.A.
VOL 5/6 E1-E3, 100, 101,E259, 261, 262, 263, 278,E280-E282, 293, 295
SY,W.P.
VOL 8 12, 235
SZABO,M.
VOL 3 E102-E113,E177,E178, 183, 184,E185, 187,E188, 192, E193, 198
SZARVAS,P.
VOL 3 E102-E113,E177,E178, 183, 184,E185, 187,E188, 192, E193, 198
SZEBEHELY,V.G.
VOL 10 540, 545
SZEPAROWICZ,M.
VOL 2 E227-E229, 234,E260,E261
SZWARC,H.
VOL 10 E425,430

T

TAKAMATSU,T.
VOL 18 E116, 117-139
TAL´ROZE,V.L.
VOL 4 E253-E255
TANANAEV,N.
VOL 3 E102-E113
TARASOVA,G.N.
VOL 14 E11-E18, 25, 26, 42, 43
TARASOVA,N.V.
VOL 13 E414-E417, 420
TARAT,E.YA.
VOL 8 E265-E267, 302-304
TARRER,A.R.
VOL 5/6 523-525
TARUNIN,B.I.
VOL 7 492,E493
TASCA,M.
VOL 2 338
TASKIS,C.B.
VOL 16/17 E394, 399, 406,E442,E443, 447, 452
TATSUISHI,A.
VOL 10 E104,E105, 115-118
TAUCH,E.J.
VOL 5/6 E119,E303,E304,E339, 340,E346, 347,E354,E355, 358, 359, 365,E380, 382,E406,E435, 436,E441-E444, 447, 521, 522
VOL 10 12,E31-E33,E409, 410, 447, 458, 503, 504
TAYLOR, A.C.
VOL 3 E1-E8, 10, 11

TAYLOR,C.D.
VOL 7 E41-E43, 54, 55
TAYLOR,D.R.
VOL 9 E195-E198, 202
TAYLOR,K.
VOL 2 E109, 112,E114,E115, 117
TEMPLETON.C.C.
VOL 13 E38-E47,E109, 128
TENU,R.
VOL 13 E1,E15-E19,E37-E47,E156-E162,E241-E246,E301-E308, E354-E359,E381-E385,E414-E417
TERESHCHENKO,L.YA.
VOL 8 E265-E267, 302-304
TERRANOVA,T.F.
VOL 7 310, 316, 323, 329, 330, 334, 335, 339
TEZAK,B.
VOL 3 E102-E113, 145-147, 152, 153,E177,E178, 179, 180, E188, 189,E193, 194, 195, 199,E200, 201, 202, 205, 206
THAM,M.K.
VOL 7 324, 333, 337
VOL 10 248-250
THEODOROVICH,V.P.
VOL 1 E254-E256
VOL 5/6 E303,E304,E406, 411,E322
VOL 10 E361
THIEL,A.
VOL 3 E102-E113, 121
THODOS,G.
VOL 9 E111,E112, 125, 126, 137
THOM,N.S.
VOL 7 E41-E43
THOMS A.B.
VOL 11 E148,E151,E156,E159,E162,E168,E171,E176,E181,E186, E194,E198,E208,E212,E216,E220,E222,E226,E229,E231, E235,E238,E240,E243
THOMAS,V.
VOL 8 E265-E267, 274, 275, 306-308
THOMPSON,R.E.
VOL 5/6 E380, 387,E406, 413
THOMSEN,E.S.
VOL 5/6 E122, 137
VOL 7 E214,E215, 220, 222, 228
VOL 9 E77,E78, 87, 90, 92,E138-E140, 144
VOL 10 E119-E122, 129, 132, 137
THORBURN,G.D.
VOL 2 E114,E115,
THORNHILL,D.G.T.
VOL 9 E138-E140, 147, 155, 164, 165,E232
TIEPEL,E.W.
VOL 4 E1-E7
TIKHOVIROV,V.I.
VOL 5/6 E303,E304,E435
TIKONOVOCH,N.P.
VOL 5/6 E395
TILQUIN,B.
VOL 9 E77,E78, 82, 105
TILTON,V.V.
VOL 9 E195-E199, 214
VOL 10 454
TIMOFEEV,W.
VOL 7 E1-E5, 6,E261-E267,
TIMOFEJEW,W.
VOL 5/6 E1-E3, 4

TOBIAS,C.A.
VOL 2 E109, 110,E193,E194, 195,E328
VOL 4 E241, 243
TOBIAS,C.W.
VOL 7 167
TODHEIDE,K.
VOL 9 E16, 21, 22
TOKUNAGA,J.
VOL 7 E190-E192, 193, 194, 198, 199, 203-206
VOL 10 E83-E85, 86, 87, 90, 91, 96-99
TOLBERG,W.E.
VOL 7 430
TOLMACHEVA,T.A.
VOL 7 E1-E5, 28,E190-E192, 195,E261-E267
TOMOTO,N.
VOL 7 364
VOL 10 278
TOOKE,J.W.
VOL 5/6 E356, 371, 375, 376
TORIUMI,T.
VOL 5/6 E601, 613
VOL 7 469
VOL 10 E488, 491
TORNOE,H.
VOL 10 E31-E33
TOROCHESNIKOVA,N.S.
VOL 10 E361
TOROGONOVA,T.V.
VOL 16/17 E127,E128, 133, ????
TORRANCE,H.B.
VOL 2 E149,E199-E201, 213, 226
TOWER,O.F.
VOL 8 E350
TOYAMA,A.
VOL 5/6 E321, 327, 598,E601, 606
TRAKHTENBERG,D.M.
VOL 16/17 E127,E128, 130, 131,E135,E136, 138, 139
TRAMBOUZE,Y.
VOL 13 E156-E162
TRAPPENIERS,N.J.
VOL 1 E359, 367, 368,E369, 370, 371
TRAUBENBERG,H.F.R.
VOL 2 E260,E261,E273, 274, 321,E322, 325
TRAUCH,E.J.
VOL 5/6 E298,E299
VOL 10 E333
TRAUTZ,M.
VOL 8 E336,E350
VOL 14 E208,E209,E211,E212, 213, 214,E225-E227, 228, E242-E246, 256
TRAVERS,M.
VOL 1 E1-E4,E68,E69
TREGUBOV,B.A.
VOL 7 481
TREMEARNE,T.H.
VOL 5/6 E622, 623
VOL 10 E476, 477
TREMPER,K.K.
VOL 10 E119-E122, 147, 161, 172
TRET´YAKOV,G.V.
VOL 10 252
TREUSHCHENKO,N.N.
VOL 8 E265-E267
TROPSCH,H.
VOL 7 E474-E480

TROSTIN,V.N.
VOL 4 E80-E84, 101
TRUBITSYN,B.A.
VOL 5/6 E622, 629
TRUESDALE,G.A.
VOL 7 E41-E43
TRUSHNIKOVA,L.N.
VOL 13 E409
TRUST,D.B.
VOL 5/6 E339, 344, 345,E434, 528, 529
TSANG,C.Y.
VOL 5/6 E321, 329-331,E601, 607, 608,E610, 617, 618
TSEITLIN,A.N.
VOL 8 E350
TSIKLIS.D.S.
VOL 1 E254-E256, 280, 281,E298
VOL 4 E256,E314
VOL 5/6 E435, 437
VOL 8 E336, 337
VOL 10 E333,E419, 445, 455, 457,E476,E488
TSIN,N.M.
VOL 5/6 E589,E590,E601
TSUJI,A.
VOL 16/17 E359, 360,E394, 402, 403, 407, 408,E442,E443, 448, 449, 453, 454,E479, 480-483,E488, 490,E545, 546, 547,E694, 695-697,E698, 699-701,E702, 703-705
TULLY,P.C.
VOL 1 E263, 265, 266, 271, 272,E328, 335, 336
TURCHINOV,V.V.
VOL 7 E60,E76, 82
TURPIN,F.H.
VOL 2 E109, 110,E193,E194, 195,E328
VOL 4 E241, 243
TYVINA,T.N.
VOL 5/6 E434,E441-E444, 452, 453, 456, 457

U

UNGERER,B.
VOL 3 E45-E56, 81
URAZOV,G.G.
VOL 13 E38-E47, 63, 81, 82, 85,E148
URBAIN,G.
VOL 13 E414,E417
USHER,F.L.
VOL 8 E260,E261, 263,E265-E267, 331
USHKOVA,A.V.
VOL 13 E156-E162, 180, 181
USOV,A.P.
VOL 7 E1-E5,E61,E62,E76, 88,E89,E90, 93-95
USUBALIEVA,U.
VOL 13 E354-E359, 370

V

VALENTINER,S.
VOL 1 E1-E4,
VOL 2 E227-E229
VALUEV,K.I.
VOL 5/6 E441-E444, 452, 453, 456, 457

VAN BERK
VOL 3 E102-E113
VAN DER SOMME
VOL 5/6 E435, 439
VANDERZEE,C.E.
VOL 3 E102-E113, 164
VAN DIJK,
VOL 4 E314, 315
VAN LIEMPT, J.A.M.
VOL 2 E1-E3, 68,E71,E73, 78
VAN SLYKE,D.D.
VOL 10 E1-E4,15,E50,E51,E53,E56, 71, 290, 291, 295
VAN SOEST,G.
VOL 5/6 E589,E590,E601
VAN WIJK,W.
VOL 2 68,E71,E73, 78
VAN WYLEN,G.J.
VOL 1 E310,312-317, 322
VAN'T ZELFDE,P.
VOL 5/6 E581, 587, 588
VARGAR PONSE,O.
VOL 13 E38-E47, 60, 65,E148
VASIL´EV,YU.N.
VOL 4 E314
VASILEVA,I.I.
VOL 5/6 E441-E444, 452, 453, 456, 457
VAUGHN,J.W.
VOL 11 E255, 256, 257,E259, 261,E262, 264, 265,E266, 268,E269, 270,E271, 273,E275, 277,E279, 281,E282, 283,E285, 287, E289, 291,E292, 293,E294, 295,E296, 297,E298, 299,E302, 304
VAUQUELIN,L.N.
VOL 13 E221
VDOVENKO,V.M.
VOL 2 235
VEALL,N.
VOL 2 E199-E201, 214, 225
VELLINGER,E.
VOL 10 E361
VENOT,A.
VOL 13 E38-E47, 80
VERDELLI,L.S.
VOL 10 259
VERMANDE,P.
VOL 14 E59-E68,E77-E81, 85, 86, 100, 101, 200, 201,E242, E243,E253,E254, 258
VERNON,H.M.
VOL 10 535
VERON,J.
VOL 13 E156-E162
VERSCHOYLE,T.T.H.
VOL 5/6 E321, 322,E589,E590, 591,E601, 602, 603
VEYS,P.
VOL 3 E39
VIBRANS,F.C.
VOL 5/6 272, 273
VOL 7 366
VOL 10 276, 533
VINOGRADOV,E.E.
VOL 14 E11-E18, 23-28, 33-36, 42, 43,E59-E69,E78, 161, 162,E191-E197, 204, 205
VINOGRADOV,V.I.
VOL 4 E80-E84, 97, 99, 100
VIRTANEN,P.O.I.
VOL 3 30, 229
VOL 18 5, 39, 80, 102

VOGELHUT,P.O.
VOL 2 E199-E201, 209
VOLKENING,R.
VOL 16/17 E548, 549
VORONINA,T.M.
VOL 16/17 E127,E128, 133, 134
VOLKOV,A.A.
VOL 13 E156-E162, 193, 194, 199, 200
VOLK,H.
VOL 5/6 E571, 573-575
VON ANTROPOFF,A.
VOL 1 E1-E4,E124-E126
VOL 2 189
VOROB´EVA,V.YA.
VOL 13 E376,E377, 378, 379
VORONTSOVA,N.A.
VOL 14 E60, 93,E191-E197, 199,E243, 261
VOSBURGH,W.C.
VOL 14 E59-E69, 88,E242-E247,E253,E254, 293
VOSPER,A.J.
VOL 8 E336, 338, 340, 341, 344, 346, 348, 349
VOUK,V.B.
VOL 3 E102-E113

W

WAGHORNE,W.E.
VOL 18 E149, 151,E154, 156,E158, 160, 163, 170, 171-173,E174, 176,E187, 190, E194, 197,E198, 201,E203, 206, 208, E209, 212,E213, 216,E217, 220-224,E225, 227
WAGMAN,D.D.
VOL 3 E39,E45-E56
WAGNER,P.D.
VOL 9 E195-E199, 231
WAHLROOS,O.
VOL 7 365
WAIND,G.M.
VOL 14 E59-E68,E76,E77, 110, 115, 116, 120-123, 128-131
WAITES,G.M.M.
VOL 2 E114,E115
WAITKINS,G.R.
VOL 3 E90,E91, 94, 95
WALFISZ,A.
VOL 5/6 269
VOL 7 444
VOL 8 257
VOL 10 329
WALKER,R.D.JR.
VOL 4 E1-E7,E33-E36, 60
VOL 10 248-250
WALKER,R.D.
VOL 1 E1-E4,E20-E22, 28
VOL 5/6 E1-E3, 11,E33,E35, 81, 82
VOL 7 E1-E5,E71,E72,E76, 168, 169, 324, 333, 337
WALKEY,J.
VOL 1 E43
VOL 4 E197,E111,E123,E145,E158,
VOL 7 E1-E5
VOL 10 29, 30
WALLACE,R.A.
VOL 11 E247, 248, 250,E251,E252, 253
WALLER,J.G
VOL 16/17 E127,E128, 129,E135,E136, 137,E165, 166

WALLS.W.S.
VOL 10 475
WALSH,E.A.
VOL 16/17 E175, 176, 177
WANG.D.I.-J.
VOL 4 E261, 262
VOL 10 E361,E362, 368,E402, 403
WARD,G.K.
VOL 4 E27-E29
VOL 7 E41-E43
VOL 10 E31-E33
WASHBURN,E.W.
VOL 1 E141-E143
VOL 2 E238-E241,E328
WATSON,J.C.
VOL 2 E199-E201
WATERS,J.A.
VOL 5/6 E119, 127,E169, 174,E239, 248,E354
VOL 9 E77,E78, 80,E138-E141, 145
WEALE,K.E.
VOL 4 E114,E115, 117,E123,E169,E170,E207, 214,E253-E255, 274, 275, 307, 309
VOL 5/6 E17-E18, E299
VOL 10 E419, 431, 433
WEATHERFORD,W.D.
VOL 4 144
VOL 7 447, 448
VOL 10 325, 326
WEISS,P.J.
VOL 16/17 E10, 11, 16-38,E39, 40, 43-65,E72, 73,E78,79-101,E102 103-126,E135,E136, 141, 164,E175, 199,E200, 201-223, E224, 225-248,E249,E276, 277-300,E301, 305,E306, 307-329,E334, 335-358,E367, 368-393,E394, 397, 404, 405,E410, 411, 413, 415, 418-426, 428-431, 433-439, E442,E443, 445, 450, 451,E457, 458-478,E491, 492-517, E518, 519-544,E548, 551-575,E576, 577-602,E603,604-629, E630, 604-629,E630, 631-656,E667, 668-693,E706, 707-732,E740, 741-764
WEISS,R.F.
VOL 1 E1-E4, 11, 12, 14,E16, 17-19,E124-E126, 131,E138, 139, 140
VOL 2 E1-E3, 6,E9, 10
VOL 4 E1-E7, 18,E27-E29, 32
VOL 7 E41-E43
VOL 8 E1,E2, 21, 22,E23,E24, 25, 26,E160-E162
VOL 10 E1-E4,E31-E33
WELLES,H.
VOL 4 E1-E7, 19,E33-E36, 41
WELLS,R.C.
VOL 13 E38-E47,E109, 132,E221, 222,E278, 281,E335, 336, 448
WELSCH,H.
VOL 2 144, 145
WENDLANDT,W.W.
VOL 13 E1,E38-E47,E109, 110-114, 116-127, 129-131, 134, 136-142, 145-147,E156-E162,E241-E246,E301-E308,E354-E359, E376,E377,E381-E385,E414-E417,E429,E434-E436, E449,E454-E456,E461
WENZEL.H.
VOL 10 355, 356, E425, 429, 442
WEN,W.-Y.
VOL 5/6 279
VOL 9 E1, E2, 5, 36-40, 42, E64, 72
WESSELER,E.P.
VOL 7 309, 315, 318-320, 322, 325, 326, 331, 332, 336, 338, 340-344, 356
VOL 10 523-530, 532

WETLAUFER,D.B.
VOL 9 E1, E2, 13, E27-E33, 41, E64, 73
WHEAT,W.N.
VOL 13 E148, 150, E226, E227, 228, E241-E246, E286-E288, 291, E339-E341
WHEATLAND,A.B.
VOL 7 E41-E43
WHIPPLE,G.C.
VOL 7 E41-E43
WHIPPLE,M.C.
VOL 7 E41-E43
WHITBY,F.P.
VOL 1 356
WHITBY,G.S.
VOL 3 E102-E113
WHITELEY,F.J.
VOL 5/6 526
VOL 10 502
WHITTEMORE,C.F.
VOL 13 E38-E47, 67, 68
WICHTERLE,I.
VOL 9 E111, 114-116
WICKERT,K.Z.
VOL 3 88, 234
WIEBE,R.
VOL 1 E1-E4, E257, 258
VOL 5/6 E303, E304, 305, 306, 312, E622, 623, 624
VOL 7 E190-E192, E214,E215, 225, E261-E267, 268, 272, 278, 279, 292, 303
VOL 10 14, E83-E85, 89, E104-E105, 106, 113, 114, E119-E122, 135, E174-E178, 180, 187, 195, 197, E216, 219, E333, 335, 336, E476, 477, 478
WIESENBERG.D.A.
VOL 5/6 E17, E18
WIESS,R.F.
VOL 5/6 E17,E18
WILCOCK,R.J.
VOL 7 35
VOL 10 28, 134, 140, 156, 212, 214, 275
WILD,J.D.
VOL 7 243
VOL 10 E148, 440
WILF,J.
VOL 9 E1, E2, 6, E195-E198, 229
WILHELM,E.
VOL 1 E43, 45, E51, 53, 63-67, 76, E77, 79, 80, 85-87, 109, E188, 190, E196, 198, 209-213, 221, E222, 224, 225, 230-232, 248
WILHELM,E.
VOL 2 E35, 56-60, 77, 79, 80, 107
VOL 4 119, 131, 153-157, 165, 166, E169, E170, 191, 197, 198, 239
VOL 7 E1-E5, 224, 231, 242, 244-249, 257, E261-E267, 281, 285, 286
VOL 9 E1, E2, 3, 4
VOL 10 E119-E122, 134, 140, E148, 152, 155-160, E162, 170, E174-E178, 201, 212, 214, 275
WILKE,C.R.
VOL 7 E474-E480
WILKINSON,G.
VOL 8 E265-E267, 312, 313, 326, 327
WILLIAMSON,A.T.
VOL 13 E15-E19, E38-E47, E156-E162, E241-E246, E301-E308, E354-E359, E381-E385, E414, E417
WILLIAMS,G.B.
VOL 5/6 E380, 390, E406, 414, 554, 555

WILLIAMS,R.B.
VOL 5/6 E332, 333, 334, E339, 342, 343, E395, 396, 397, E400, 401, 402
WILLIAMS,V.D.
VOL 10 238
WILSON,C,L.
VOL 3 E102-E113, 132
WILSON,I.R.
VOL 14 E242-E246, E253, E254, 259, ???
WILSON,R.H.
VOL 8 E1, E2, 11, E226, E227, 234
WILSON,T.R.S.
VOL 7 E1-E5, 29, E41-E43, E64, E69, E76
VOL 10 E1-E4, 25, E31-E33, 41-44
WINKLER,K.
VOL 2 131, E199-E201, 223
WINKLER,L.W.
VOL 5/6 E1-E3, 5
VOL 7 E1-E5
VOL 8 E260, E261, 262
VOL 9 E1, E2, 9
VOL 10 E1-E4, 5
WINNINGHOFF W.J.
VOL 14 E242-E246, E248, E251, 270, 271, 279
WISE,D.L.
VOL 7 E1-E5, 32
WISE,W.C.A.
VOL 14 E59-E68, E70, E72, E73-E81, 96, 104, 125, 134, 147, 153, 156
WISHNIA,A.
VOL 2 E199-E201
WISNIAK,J.
VOL 5/6 314-320, 527
WOLFF,H.
VOL 13 E226, E227, 234, 236
WONG,S.S.
VOL 7 385
WOOD,D.
VOL 2 E1-E3, E9, 11, E134-E136, 148
WOOD,R.H.
VOL 1 105
VOL 4 229
VOL 9 E195-E199, 215
VOL 10 261
WOOLLEY,E.M.
VOL 3 E102-E113, E177, E178, 181, 182, E188, 190, 191, E193, 196, 197, E200,203,204
WOOLMER,R.M.
VOL 5/6 526
WORLEY,F.P.
VOL3 E45-E56
WRIGHT,W.W.
VOL 16/17 E10, 11, 16-38,E39,E72, 73,E78, 79-101,E102, 103-126, E135,E136, 141, 164,E175, 199,E200, 201-223,E224,225-248, E249,E276, 277-300,E301, 305,E306, 307-329, 335-358,E367, E410,E457,E491,E518,E576,E603,E630,E667,E707,E740
WYATT,P.A.H.
VOL 14 E59-E68, E70, E72, E73-E81, 96, 104, 125, 134, 147, 153, 156
WYROUBOFF,G.
VOL 13 E148

Y

YAACOBI,M.	
VOL 9	E1,E2, 6-8, E27-E33, 35, 45, 50-52, 57, 59, E64, 65-71, E138-E140, 148,E166, E167, 169, 173, 177, 180, 183, 186, E195-E199, 207, 229, E232, E233
YAGI,M.	
VOL 2	175, 176
YAKHKIND,N.D.	
VOL 3	E102-E113, 169, 170
YAKHONTOVA,L.F.	
VOL 16/17	E740, 765, 766
YAKIMOVA,Z.P.	
VOL 13	E15-E19, 22, 31, E301-E308, 319
YAKIMOV,M.A.	
VOL 13	E38-E47, 64, 84, E156-E162, 182, 198
YAMAMOTO,S.	
VOL 5/6	E1-E3, 13,E17,E18, 19-23,E32,E35, 70
YAMANA.T.	
VOL 16/17	E359, 360, E394, 402, 403, 407, 408, E442, E443, 448, 449, 453, 454, E479, 480-483, E487, 488, 490, E545, 546, 547, E694, 695-697,E698, 699-701,E702, 703-735
YANAI,K.	
VOL 14	E59-E68, E74, 176, E191-E197, 207, E242-E246, E251, E252, 310
YAO,J.	
VOL 5/6	394, 429, 430
YARBROUGH,O.D.	
VOL 1	E1-E4, 8, 118
VOL 4	E241, 242
VOL 10	E1-E4
YASTREBOVA,L.F.	
VOL 13	E15-E19, 22, 31, E301-E308, 319
YASUNISHI,A.	
VOL 7	E57, E59, E63, E65-E68, E70-E73, E76, 99, 100, 121-124, 126, 128, 151, 153, 158, 159, 163, 174, 175
VOL 10	E52, E53, E56, 72, 73, 75
YATLOVA,L.E.	
VOL 3	E31, 33
YEH, S.Y.	
VOL 2	E1-E3, E12, E13, 20, 23, E109, 111, E114, E115, 116, 118, 119, 123, 126, 128, 132, E134-E136, 138, E149, 153, 154, E193, E194, 196, E199-E201, 202-204, 211, 216, 222, 224
YEN,L.C.	
VOL 8	E160-E162, 167, 170, 181, 219, 259
VOL 10	166
YORIZANE,M.	
VOL 5/6	E321, 327, E435, 440, 567, 568, E589, E590, 598, E601, 606, E610, E611, 616, 620
VOL 10	E488, 492
YOSHIMURA,S.	
VOL 5/6	E321, 327, E589, E590, 598, E601, 606, E610, E611, 616
VOL 10	E488, 492
YOUNG,I.H.	
VOL 9	E195-E199, 231
YOUNG.C.L.	
VOL 5/6	E356
YU,P.	
VOL 10	E391
YUDINA,L.V.	
VOL 10	59

Z

ZAIS,E.J.
VOL 9 E111, 121-123
ZAITSEV,B.E.
VOL 13 E409, E414-E417, E449
ZAKHAROV,N.D.
VOL 10 460, 461
ZAKHAROVA,A.V.
VOL 1 E254-E256
ZANDER,R.
VOL 7 E1-E5, 38
ZANNIER,H.
VOL 18 E7-E10, 11, E43, E52, E53
ZECHENTMAYER,K.
VOL 8 E265-E267, 276-280, 309
ZEIDLER,L.
VOL 18 81, 82, 91-93, 95, E106, 107
ZEININGER H.
VOL 4 314
VOL 5/6 E622
VOL 7 466, 467
VOL 10 484
ZENNER,G.H.
VOL 7 434, 435, 468
VOL 10 E488, 490, ??
ZEPTER,H.
VOL 8 E1, E2, 9, E27-E38, 47-50, 52-63, 65, 68, 71, 73, 75, 83, 85, 87, 91, 92, 98, 101, 106, 108, 111, 137
ZHADANOVICH,L.A.
VOL 10 E361, E362
ZHAVORONOKOV,N.M.
VOL 5/6 E435, 437
ZHELEZNYAK,N.I.
VOL 4 E80-E84, 103-105, 205
ZHESKO,T.E.
VOL 5/6 569, 570
ZHIRONOVA,N.A.
VOL 5/6 E571, 577, 578
ZHOLALIEVA,Z.
VOL 13 E1, 4, 5, E156-E162, 216, 217, 219, 220
ZHUKOVSKAUA,S.A.
VOL 16/17 E127,E128, 133, 134
ZHURAVLEVA,A.N.
VOL 10 59
ZHURAVLEV,E.F.
VOL 13 E38-E47, 86-94, 96-107, E148, E156-E162, 173-179, 201, 202, 205-215, 218, E241-E246, 257-275, 277, E301-E308, 320-326, 328, 329, 331-333, E381-E385, 391-402, 404, E414-E417, 421-427, E434-E436, 441-443
ZHUZE,T.P.
VOL 10 331
ZIEGLER,W.T.
VOL 1 E283, 286
VOL 5/6 E321, 326, 513-516, E571, 576
ZIMMERMAN,P.A.
VOL 13 E109, 115, 144
ZINATULLINA,E.S.
VOL 16/17 E127,E128, 130, 131,E135,E136, 138, 139
ZINCHINKO,E.YA.
VOL 16/17 E664, 665
ZINK.W.A.H.
VOL 14 E242, E243, E247, E251, 294, 299

ZORIN,A.D.
VOL 9 E166, E167, E195-E198, 205, 209, 210, 219
ZOSS,L.M.
VOL 5/6 E303, E304, 307-309
VOL 7 456, 457
VOL 10 18, E333
ZSAKO,J.
VOL 3 E45-E56, 78
ZUBCHENKO,YU.P.
VOL 4 E253-E255, 305, 306, 310, 311
VOL 5/6 478-480, 511, 512, 519
VOL 10 448, 451-453, 462, 463
ZUBOV,V.V.
VOL 8 E265-E267, 302-304
ZUEV,A.A.
VOL 5/6 471-474
ZVORYKIN,A.YA.
VOL 13 E15-E19, 32, 33, E241-E246, 254, E301-E308, 315
ZWIETASCH,K.J.
VOL 13 E38-E47, 56